# Modern Interferometry for Length Metrology

Exploring limits and novel techniques

# Modern Interferometry for Length Metrology

## Exploring limits and novel techniques

**René Schödel (Editor)**
*Physikalisch-Technische Bundesanstalt (PTB), Braunschweig, Germany*

**IOP** Publishing, Bristol, UK

ISBN 978-0-7503-1578-4 (ebook)
ISBN 978-0-7503-1576-0 (print)
ISBN 978-0-7503-1577-7 (mobi)

DOI 10.1088/2053-2563/aadddc

Version: 20181201

IOP Expanding Physics
ISSN 2053-2563 (online)
ISSN 2054-7315 (print)

British Library Cataloguing-in-Publication Data: A catalogue record for this book is available from the British Library.

Published by IOP Publishing, wholly owned by The Institute of Physics, London

IOP Publishing, Temple Circus, Temple Way, Bristol, BS1 6HG, UK

US Office: IOP Publishing, Inc., 190 North Independence Mall West, Suite 601, Philadelphia, PA 19106, USA

# Contents

# Preface

The famous Michelson–Morley experiment [1] has initiated many developments in modern physics. Yet, interferometry plays an important role in metrology today, due to its direct link to the SI definition of the metre, as well as the accuracy that can be achieved.

The representation of a length according to the definition of the metre in the International System of Units requires a measurement principle which establishes a relation between the travelling time of light in vacuum and the length to be measured. Two procedures are available for this purpose: (a) measuring a travelling time directly and (b) using interferometry with light waves, which is essentially an indirect measurement means for the travelling time of the light.

Apart from optical shop testing, this book focusses on fundamental principles of the most important interferometric procedures for length measurements by interferometry. All of them are based on the wave properties of light. Sophisticated solutions have been developed for various applications. Many of them are discussed in depth in the different chapters of this book.

The basic principle of interferometric length measurement can be demonstrated easily, the achievable accuracy limits are reached, however, very quickly. For the interferometric realisation of lengths that are relevant in practice, a relative measurement uncertainty on the order of $10^{-7}$, $10^{-8}$ or even $10^{-9}$ (1 nm in 1 m), is required. Given the availability of modern laser light sources whose frequencies can be determined with a relative measurement uncertainty of better than $10^{-10}$, it becomes clear that the frequency of the light used today is the smallest challenge when it comes to reducing the measurement uncertainty of length measurements. To make the realization of a length more precise, other limitations come to the fore, namely those that have a direct influence on the result of the interferometric measurement, e.g. the accuracy of the interference phase measurement, the effect of the refractive index of the air, but also limits set by the quality and the adjustment of optical components that cannot be manufactured with arbitrary high precision. Evaluations are often based on assumptions which, relating to new requirements, can no longer be adhered to. This makes the struggle for improvements a tough one which demands a great deal of frustration and resilience from the persons involved, and openness regarding their own errors. The ancestral fathers of interferometric procedures surely never imagined that the measurement uncertainties that are achievable today would ever become possible.

René Schödel

## Reference


[1] Michelson A A and Morley E 1887 On the relative motion of the Earth and the luminiferous ether *Am. J. Sci.* **203** 333–45

# Contributor List

## Dr René Schödel (Editor)

Physikalisch-Technische Bundesanstalt (PTB), Braunschweig, Germany

## Dr Florian Pollinger

Physikalisch-Technische Bundesanstalt (PTB), Braunschweig, Germany

Florian Pollinger was born in 1979 in Erlenbach, Germany. He studied physics at the University of Würzburg in Germany and the University of Texas at Austin in the USA. He received his Master's degree with a thesis in the field of femtosecond time-resolved pump-and-probe reflectance difference spectroscopy in 2003. He returned to the University of Würzburg and received his doctorate in the field of surface physics for his thesis 'Surface stress and large-scale self-organization at organic–metal interfaces' in 2009. Subsequently, he joined the Physikalisch-Technische Bundesanstalt (PTB) in Braunschweig, the German national metrology institute, and started working in length metrology. His major research interests are absolute distance interferometry and refraction compensation for long-distance measurements, and their application to precision engineering and surveying. Since 2010, he has been the head of the PTB working group 5.42 'Multi-wavelength interferometry in geodetic length'. Between 2013 and 2016 he coordinated the joint research project SIB60 'Metrology for long distance surveying' within the European metrology research programme (EMRP). He is also participating in national and international standardization of verification procedures of surveying instrumentation.

## Dr Arnold Nicolaus

Physikalisch-Technische Bundesanstalt (PTB), Braunschweig, Germany

Arnold Nicolaus was born in 1960 in Bad Harzburg, Germany. He studied physics in Göttingen at the Georg-August-Universität. After obtaining his diploma in photoacoustic spectroscopy in 1984 and two years of interdisciplinary studies at the physical and medical faculties in Göttingen, he became a medical physicist, moved to

Braunschweig in 1986 to study engineering sciences, and finished his PhD (Dr.-Ing.) in 1993 in the field of Fizeau interferometry. Arnold Nicolaus works at PTB in Braunschweig, where he set up the spherical interferometer. Since 2003, he headed different working groups. At present he is head of the 'Interferometry on Spheres' working group, tasked with the high precision determination of the volume of silicon single-crystal spheres. In this context, he is an associate and representative in the 'Avogadro constant' working group, in which work for a redefinition of the kilogram, the SI-unit of mass, on the basis of fundamental constants is coordinated. Dr Nicolaus is a member of the German Society of Applied Optics, DGaO.

## Dr Guido Bartl

Physikalisch-Technische Bundesanstalt (PTB), Braunschweig, Germany

Guido Bartl was born in Osnabrück, Germany, in 1980. He received his Dipl-Phys degree from the Carl von Ossietzky University, Oldenburg, Germany, in 2006 and his PhD from the Technical University of Braunschweig, Germany, in 2010, with work on the interferometric determination of topographies of absolute sphere radii. Currently, he is with the Physikalisch-Technische Bundesanstalt (PTB), Braunschweig, the National Metrology Institute of Germany, as a member of the 'Interferometry on Material Measures' department and head of the 'Interferometry on Gauge Blocks' working group.

## Dr Thilo Schuldt

Deutsches Zentrum für Luft- und Raumfahrt (DLR), Institut für Raumfahrtsysteme, System Enabling Technologies, Germany

Thilo Schuldt was born in Singen, Germany, in 1975. He studied physics at the Universities of Konstanz and Hamburg and performed his diploma thesis on a compact optical frequency reference developed in the context of the Darwin formation flying space mission. He received his PhD by the Humboldt-University Berlin for his work on a picometer and nanoradian sensitive laser interferometer as a demonstrator for the space based gravitational wave detector LISA. Both theses were carried out in collaboration with the space company Airbus in Friedrichshafen, Germany. After a postdoc position at the University of Applied Sciences in Konstanz, Thilo Schuldt joined the German Aerospace Center (DLR), Institute of Space Systems, Bremen, in 2013 (department of Prof. Dr. Claus Braxmaier) and is group leader for 'System Enabling Technologies'. He is working on several projects in the field of optical metrology, especially optical frequency references and high-sensitivity laser interferometry, with focus on technology development for space operation of the optical systems.

## Dr Nandini Bhattacharya

Optics Research Group, Faculty of Applied Sciences, Delft University of Technology, Netherlands

Nandini Bhattacharya did her PhD (1998) from the T.I.F.R. in Mumbai, India. Her thesis work was based on the spectroscopy and lifetime studies of highly charged ions. She did her post-doctoral work in the Van der Waals-Zeeman Instituut in the Universiteit van Amsterdam. In this period she worked on experiments which use light to manipulate atoms and on classical wave implementations of elements in quantum information theory. After starting in the Delft university of Technology from 2002 she worked on several projects on distance metrology and frequency comb based spectroscopy. Besides this her interests are in mid infrared spectroscopy and study of fluid dynamics using speckle metrology.

## Dr Steven van den Berg

Dutch Metrology Institute (VSL), Delft, Netherlands

Steven van den Berg is principal scientist at VSL, the national metrology institute of the Netherlands. He obtained his master's degree in physics in 1997 at Leiden university with a thesis on second-harmonic generation in a monolayer of liquid crystals. This was followed by a PhD project in physics at Leiden University on the ultrafast dynamics of lasers based on light-emitting polymers. After his graduation in 2002, he joined the dimensional metrology (Length) group at VSL as a scientist. He was responsible for the development and maintainence of optical frequency standards. His initial research work focussed on the development of a home-built optical frequency comb at VSL, which serves as a tool for calibration of optical frequency standards against the VSL atomic clock. The work on the frequency comb was extended within a national research programme with research on absolute distance measurements and fundamental constants which took place in close collaboration with the Technical University of Delft and the Vrije Universiteit Amsterdam, with three PhD students involved. Steven van den Berg supervised the experimental work on distance measurement at VSL and was a member of several PhD committees. He was also involved in several collaborative European projects (iMera+ and EMRP) on absolute distance measurements with frequency combs. Next to his scientific work, Steven was the team leader of the Length group for several years. In 2012 he expanded his scope to the field of photometry and radiometry, with a research focus on calibrations for Space and Earth Observation applications.

## Professor Seung-Woo Kim

Department of Mechanical Engineering, Korea Advanced Institute of Science and Technology (KAIST), South Korea

Seung-Woo Kim is a professor at Korea Advanced Institute of Science and Technology (KAIST). His research interests include precision engineering, optical metrology, and ultrafast optics. During the last three decades he has published ~160 technical papers in peer-reviewed journals, ~ 240 presentations in conferences, and ~50 patents. He has been working as principal investigator for many national and industrial research projects and is currently involved in an important national creative research initiative project for the development of next generation precision engineering key technologies using femtosecond pulse lasers. He was president of the Korea Society of Precision Engineering (KSPE) and is currently a member of OSA (Optical Society of America), SPIE (International Society of Optical Engineering), CIRP (International Academy for Production Engineering), and euspen (European Society Precision Engineering).

## Dr Yoon-Soo Jang

Department of Mechanical Engineering, Korea Advanced Institute of Science and Technology (KAIST), South Korea

Yoon-Soo Jang is a postdoctoral researcher in the Department of Mechanical Engineering at Korea Advanced Institute of Science and Technology (KAIST), Daejeon, Republic of Korea. His current research interests include precision LIDAR using mode-locked lasers.

## Professor Armin Reichhold

Department of Physics, The University of Oxford, Oxford, United Kingdom

Prof. Dr. rer. nat. Armin Reichold is associate Professor of Physics in the sub-department of particle physics at the University of Oxford where he has been working as an academic since 1998. Born in 1966 in Dortmund, Germany he obtained his Diploma in physics from the University of Dortmund in 1992. Following a two-year EEC research fellowship at the University of Oxford he obtained his doctorate in physics July 1996. His thesis focussed on silicon tracking detector development for the ATLAS experiment at the LHC as well as a search for super-symmetric gluinos and was recognised with the universities dissertation prize. During his doctorate he made first contact with precision optical measurement techniques such as electronic speckle pattern interferometry (ESPI) and frequency scanning interferometry (FSI). The strong application focus and problem driven approach of particle physics instrumentation research and its duality with the fundamental nature of the physical phenomena which these instruments seek to observe guide and motivate his research until today. After a two year post doctoral period at NIKHEF in Amsterdam, continuing his work on ATLAS, he became a senior research officer at the University of Oxford and started his first academic teaching role at St John's College Oxford. Three and a half years and several high energy physics projects later he became a University lecturer in Physics and tutorial fellow of Balliol College firmly embedding him into the dual roles of academic researcher and university teacher. Over the years his titles changed via Reader to Associate Professor and research continued further into non-contact optical measurement techniques such as white light interferometric profiling, infra-red lock-in thermography, straightness monitoring and a new generation of FSI techniques. The latter two techniques were developed as part of a robotic survey system for the ILC, a future electron position collider. In 2007 this project led him and his research group to DESY in Hamburg to prototype this survey robot for a full year. The 2007/8 banking crisis abruptly ended the UK's involvement in the ILC and Dr Reichold transformed the metrology side of his research into a collaboration with NPL and Etalon AG to develop FSI into a commercial instrument. In 2008, his particle physics interests found an entirely new home in the SNO+ collaboration and their search for neutrino-less double beta decay using the upgraded SNO detector situated 2 km underground at SNOLAB near Sudbury in northern Ontario. Together with the Oxford SNO+ group he designed, built and operates an optical calibration instrument to measure the scattering properties of many kilo-tons of water and scintillator which form the detection medium of the experiment. In 2015, he spent three months as a guest researcher at PTB in Braunschweig in the groups of Florian Pollinger and Uwe Sterr who introduced him to many of the more fundamental aspects of precision metrology which continue to fascinate him to this day.

## Dr Birk Andreas

Physikalisch-Technische Bundesanstalt (PTB), Braunschweig, Germany

Birk Andreas was born in 1975 in Bramsche (Osnabrück), Germany. He studied physics at the University of Osnabrück and graduated in 2001. Then, he moved to the University of Bonn where he received his Doctor of Science in 2005 for his thesis: 'Modifikation des Brechnungsindexes von Dielektrika mit Hilfe ionisierender Strahlung' (modification of the refractive index of dielectrics by means of ionizing radiation). Afterwards, he accepted a position with the optics department of the Physikalisch-Technische Bundesanstalt (PTB), the national metrology institute of Germany, where he resides ever since. He is working in the field of x-ray interferometry, where he is mainly interested in traceable length measurements, diffraction corrections for optical interferometers, optical simulations, crystallography and since recently dynamical diffraction theory.

## Dr Christoph Weichert

Physikalisch-Technische Bundesanstalt (PTB), Braunschweig, Germany

Christoph Weichert was born in 1984 in Halle (Saale), Germany. He studied physics at the Friedrich Schiller University Jena and received his diploma in 2007. Afterwards, he started working at the PTB and received his PhD in mechanical engineering for his thesis: 'Implementation of straightness measurements at the Nanometer Comparator of the Physikalisch-Technische Bundesanstalt' from the technical university of Braunschweig in 2016. Currently, Christoph Weichert continues his work at the department 'Dimensional Nanometrology' at the PTB dealing with the development of displacement interferometer systems, their integration into measuring machines as well as the development of new measurement methods and evaluation algorithms.

# Editor biography

## Dr René Schödel

René Schödel is head of the 'Interferometry on Material Measures' department of the Physikalisch-Technische Bundesanstalt (PTB), the German national metrology institute, where he has been working since 2000. Born in 1966 in Ludwigsfelde, Germany, he obtained both his diploma in physics 1993 and his doctorate (Dr. rer. nat.) in 1999 from the Humboldt-University of Berlin where he was employed as a scientist in the field of nonlinear laser spectroscopy on constituents of the photosynthetic apparatus of higher plants. During these years he recognised that fundamental improvements require careful analysis and characterization of the measurement process and as a consequence published several successful papers in this field. Joining the PTB and working in the field of optical interferometry for the realization of the SI unit for the length was a great chance to continue persistent engagements in improving device accuracy. His first project involved the redesign and refinement of PTB's precision interferometer for highly accurate measurements of the coefficient of thermal expansion (CTE) based on absolute lengths of material measures in vacuum. In 2004 he successfully demonstrated the achieved improvements within an international comparison. Since then many ultra-precise CTE measurements on new materials have been demanded, mainly from the semiconductor industry. 2005 he and his working group started building a new Ultra-Precision Interferometer (UPI). Demands from the European Space Agency for ultra-precise CTE measurements lead to the extension of the UPI for empowering cryogenic temperatures. This project was successfully finished in 2011 by enabling the first absolute length measurements from room temperature down to 7 K. These activities and the necessary research have been documented in many publications. In addition to his current role as department head, he remains active as a scientist supporting the activities of the working groups of the department 'Interferometry on Material Measures'. Further, he is engaged in the international conference series MacroScale: Recent Developments in Traceable Dimensional Measurements, which he co-organises every three years, and as well as other committee work.

**IOP** Publishing

Modern Interferometry for Length Metrology
Exploring limits and novel techniques
**René Schödel (Editor)**

# Chapter 1

# Practical realisation of the length by interferometry—general principles and limitations

**René Schödel and Florian Pollinger**

## 1.1 A short history of the metre and the present definition

The search for a universal measure of length goes back many centuries. In the Middle Ages, many different measures of length had been established; some of them are still commonly used today. This diversity was an obstacle to trade which was becoming increasingly supraregional. One of the first ideas for a universal measure of length was the seconds pendulum which was suggested by Jean-Félix Picard in 1668. This suggestion was based on the physical connection between the length of a pendulum and its oscillation period. However, it turned out that due to regional differences in the gravitational field of the Earth, the oscillation period of such a pendulum exhibited considerable differences [1]. This definition of length was thus not universal enough. After the French revolution, the concept of a metric system of units started establishing itself, among these the metre as the unit for the length. It was decided that the new measure should be equal to one ten-millionth of the distance from the North Pole to the Equator (the quadrant of the Earth's circumference), measured along the meridian passing through Paris [2]. The task of surveying the meridian arc fell to Pierre Méchain and Jean-Baptiste Delambre, and took more than six years (1792–98) [3]. In 1799, several platinum bars were made, the bar whose length was found to be closest to the meridional definition of the metre was selected and placed in the National Archives on 22 June 1799. This standard metre bar became known as the mètre des Archives[1].

[1] Soon it became apparent that Méchain and Delambre's result was slightly too short for the meridional definition of the metre [2]. The modern value for the 'World Geodetic System 1984' reference spheroid is 1.000 196 57 m (see NIMA Technical Report 8350.2 'Department of Defense, World Geodetic System 1984, Its Definition and Relationships with Local Geodetic Systems').

doi:10.1088/2053-2563/aadddcch1

The metre was officially established as an international measurement unit by the Metre Convention of 1875. In 1889, the International Bureau of Weights and Measures (BIPM) introduced the International Prototype of the Metre for the unit 'metre'; it was a rod made of platinum/iridium alloy with a cross-shaped section, and the length of the metre was defined as the distance between two centre lines, each belonging to a group of lines, at a temperature of 0 °C, the melting point of ice. In 1927 the 7th Conférence Générale des Poids et Mesures (CGPM) defined that this platinum–iridium bar is under atmospheric pressure and supported by two rollers [4].

The first suggestion that a light wave could serve as means for the realisation of the length was probably made by the French natural philosopher Babinet as early as 1827 [5] (see [6]). In 1893, Albert A Michelson declared in a review article that 'light waves [were] now the most convenient and universally employed means we [possessed] for making accurate measurements' [7]. It was only in 1960 that the definition of the metre stopped relying on a material measure which needed to be maintained (or its copies). At the 11th CGPM, the metre was defined as the length equal to 1 650 763.73 wavelengths in vacuum of the light radiated at the transition between the levels $5d_5$ and $2p_{10}$ of $^{86}$Kr. This definition was decisively rooted in a krypton-86 spectral lamp with a wavelength of approximately 606 nm which was developed at the Physikalisch-Technische Bundesanstalt (PTB), the German national metrology institute, by Johann Georg Ernst Engelhard.

In the definition from 1983 'The metre is the length of the path travelled by light in vacuum during a time interval of 1/299 792 458 of a second' light is understood as a plane electromagnetic wave or a superposition of plane waves and the length is considered along the propagating direction of the light wave. Under vacuum conditions there is a strict relationship between the frequency $\nu$ and the wavelength $\lambda_0$ of monochromatic light:

$$c_0 = \lambda_0 \cdot \nu = 299\ 792\ 458\ \mathrm{ms}^{-1}, \tag{1.1}$$

in which $c_0$ represents the speed of light. It was set to a fixed value by the metre definition in 1983. Besides perceptible light intensity and colour—the latter is an indication of the wavelength—a single monochromatic light wave does not reveal its parameters to any kind of detector. The realisation of a length based on the 1983 definition requires superposition of at least two light waves.

While the definition for the metre refers to vacuum conditions, in most cases the realisation of the length is performed under atmospheric conditions. Then, the exact quantification of the influence of the air on the speed of light is of major importance. Under atmospheric conditions, the air refractive index $n$ downscales the speed of light ($c = c_0/n$) as well as the wavelength. The relative effect of $n$ is of the order of $3 \times 10^{-4}$ corresponding to 0.3 mm per metre of measured length under standard air conditions. Further, in the case of light pulses travelling along a pathway of a length that should be realised, it is highly important to consider the group refractive index of air $n_g$ instead of the (phase-) refractive index $n$. For example, for green light ($\lambda \approx 520$ nm) $n_g - n$ is $\approx 10^{-5}$, i.e. considering $n_g$ instead of $n$ causes a considerable

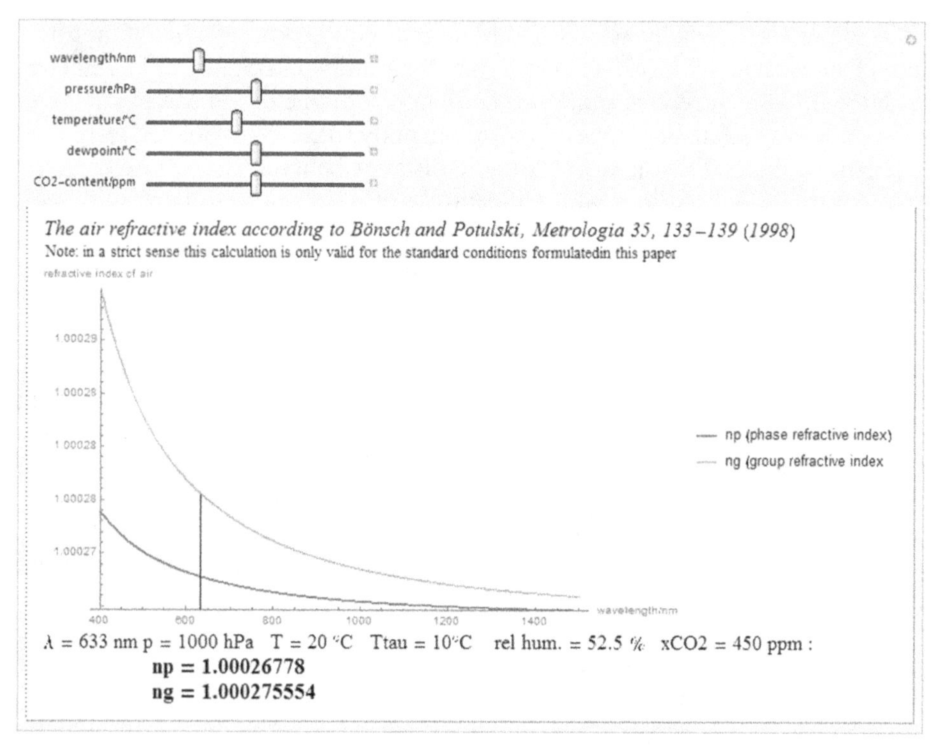


**Figure 1.1.** Interactive diagram for the calculation of the air refractive index (np: phase refractive index, ng: group refractive index) as a function of the light wavelength and the environmental parameters pressure, temperature, humidity and $CO_2$-content according to [8].

additional effect of 10 μm per each metre of measured length. It is noted that the magnitude of this difference is comparable to the variation of the phase refractive indices within the entire range of visible light is: $n(380\ \text{nm}) - n(780\ \text{nm}) \approx 9 \times 10^{-6}$ (see figure 1.1). The relevance and the compensation of the air refractive index are discussed in more depth in chapter 5 'Interferometry in air with refractive index compensation'.

Realisation of the path length travelled by light requires a measurement method. The following subsections describe the two basic ideas of such measurements that are related to either direct or indirect measurements of the travelling time of the light.

The SI definition for the metre presumes a measuring method allowing the length to be realised along a spatial dimension or along material measures for the length. The accuracy of the realisation of lengths by means of suitable length measurement methods is in perpetual development, triggered by new requirements placed on their accuracy.

## 1.2 Realisation of the length by direct measurement of the light travelling time (time-of-flight measurement)

As a single plane wave is not locatable, a direct measurement of the travelling time of light requires modulation. Any kind of modulation applied to monochromatic light, including application of simple shutters, generates a superposition of light waves forming a wave packet, e.g. light pulses. The path length of propagating light pulses can be determined as it is indicated in figure 1.2. A light pulse is split in two parts so that two pulses are generated, one of which travelling a short reference pathway, the other the measurement pathway. The reflectors in both pathways are arranged such that the light is retro-reflected. After the second passage of the beam splitter the light pulse originating from the reference pathway first hits a light detector which sets a first trigger at a defined threshold defining a reference point of the time. A second trigger is generated by the delayed light pulse originating from the measurement pathway.

Measurement of the time delay $\Delta t$ between both detector signals allows determination of the length difference $\Delta z$ between measurement and reference pathways which represents a length $l$:

$$l = \Delta z = \frac{1}{2} c_g \cdot \Delta t, \tag{1.2}$$

in which $c_g$ is the group velocity of the wave packet. While under vacuum conditions $c_g$ is identical to $c_0$, under the influence of the atmosphere $c_g$ is obtained from $c_g = c_0 / n_g$, in which $n_g$ is the above mentioned wavelength dependent group refractive index of air: $n_g = n(\lambda) - \lambda \cdot dn/d\lambda$.

A typical application of this measurement technique is the determination of long distances. A prominent example is the measurement of the distance from the Earth to the Moon. Here the length of the reference pathway can be neglected and the major part of the measurement pathway is in space (vacuum), i.e. a relative error of less than $10^{-8}$ is caused by usage of $c_0$ as the speed of light in equation (1.2) for this example.

The above principle for the direct measurement of light travelling time is mainly used for long distance measurements on the Earth under the influence of the atmosphere in which the air refractive index, its homogeneity and invariance are limiting the attainable measurement uncertainty (besides the accuracy of the

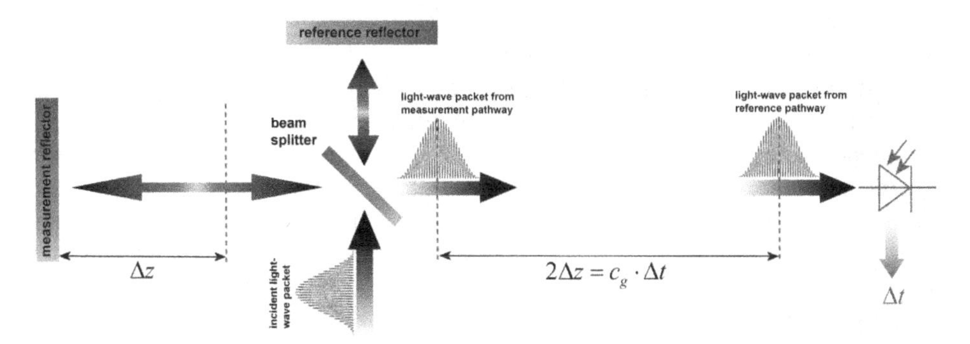


**Figure 1.2.** Primary realisation of the length by direct measurement of time delay between light pulses travelling pathways of different lengths before reaching a detector. Light pulses are indicated by fast oscillations of a temporal light intensity (black curves). The green curves inside indicate the light intensity averaged over several oscillation periods that is sensitised by the detector.

electronics and clocks used). The relationship between time delay and length differences can also be easily demonstrated with standard equipment (oscilloscope) for short distances. However, since the time delay is as small as approximately 3 ns per metre, the attainable accuracy of length measurements for short distances is limited by the electronics. For example, measurement of 1 m on a level of only 1 mm requires an accuracy of 3 ps for the measurement of the travelling time.

More indirect approaches to the time-of-flight measurement can be used to achieve higher resolutions. In most commercial electronic distance meters, for example, amplitude modulation is used. The intensity of a CW diode laser is modulated in the radio frequency regime. By comparison of the phase shift of the intensity of the beam before and after traversing the distance to be measured, the time-of-flight can be derived from the modulation frequency. Amplitude modulation of the order of a few megahertz can be obtained by direct modulation of the diode driving current. Higher modulation frequencies of the order of several Gigahertz can be obtained by external electro-optic [9] or electro-absorption modulators [10]. Multiple highly stable modulation frequencies from several hundred megahertz up to several tens of gigahertz can be generated in parallel by intermode beats of femtosecond lasers [11]. Using gigahertz modulation and rapid phase meters, micrometre resolution can be achieved.

Finally, more sophisticated detection schemes can increase the temporal resolution even further. Using nonlinear optics, for example, the temporal overlap between two femtosecond pulses can be determined in balanced cross-correlation to fractions of the pulse width [12]. This signal can be used to optimise the temporal overlap between two beams of a reference and a measurement pulse by tuning the pulse repetition rate. The pulse repetition rate can be measured to $10^{-12}$ relative uncertainty. For very short distances over a 3 μm, accuracy of 20 nm has been demonstrated for this type of time-of-flight measurement [13].

## 1.3 The basic concept of length measurement by interferometry

For the realisation of lengths below a few metres, but also for the most accurate realisation of the length in general, interferometric techniques are preferable. In the following, the basic idea for the realisation of the length by interferometry and its relationship with the SI definition for the metre is explained.

Optical interferometry is a measurement method basing on the superposition (interference) of light waves. Assuming light as an electromagnetic wave, the electric field of a single light wave propagating along the $z$-direction is given by $E(z, t) = A\ \cos[\varphi] = A\ \cos[\omega t - kz + \delta]$ in which $A$ is the wave amplitude, $\varphi$ the phase, $t$ the time, $\omega$ the angular frequency, $k$ the wave number and $\delta$ the initial phase. The relationship between the parameters $\omega$ and $k$ with wavelength $\lambda$ and frequency $\nu$ is given by $k = 2\pi/\lambda$ and $\omega = 2\pi\nu$. Wavefronts travel the distance of a single wavelength during one oscillation period $T$ ($T = 1/\nu$). Consequently, the speed of a monochromatic light wave, $c$, the so-called phase velocity, is given by $c = \lambda \cdot \nu$. The frequency of visible light waves lies in the range from 300 THz to approximately 600 THz. The respective oscillation period is therefore extremely small and could not be acquired directly with any detector. Thus, the only measurable parameter of a single light wave is its mean intensity which is essentially proportional to the square of the wave amplitude[2]:

$$I = \langle E^2 \rangle_t = \lim_{t\to\infty} \frac{\int_0^t (E(t, z))^2 \, dt}{t} = \frac{A^2}{2}. \tag{1.3}$$

Therefore, a single light wave does not disclose its wave properties, apart from the 'colour' property. Only when several waves overlap, access to parameters such as the wavelength is obtained, as shown in the following.

Assuming the simple case of two identical light waves spreading along the same axis (in the $z$-direction), then the resulting interference of the waves is given by:

$$\underbrace{E_1 = A\cos[\omega t - kz + \delta_1]\ , \quad E_2 = A\cos[\omega t - kz + \delta_2]}_{\Downarrow} \\ E_1 + E_2 = 2A\cos[\omega t - kz + (\delta_1 + \delta_2)/2] \cdot \cos[(\delta_1 - \delta_2)/2]. \tag{1.4}$$

As equation (1.4) reveals, the overlain wave is not a 'standing wave', as is often misleadingly stated, it spreads in the same direction as the single waves, with an amplitude which is affected by the shift of the two waves in relation to each other, $\delta_1 - \delta_2$.

The more general case of different amplitudes for the interfering light waves is shown in the interactive figure 1.3.

Here the intensity of the interference at any point is given by:

$$\underbrace{E_1 = A_1\cos[\omega t - kz + \delta_1]\ , \quad E_2 = A_2\cos[\omega t - kz + \delta_2]}_{\Downarrow} \\ I = \langle (E_1 + E_2)^2 \rangle_t = \frac{A_1{}^2}{2} + \frac{A_2{}^2}{2} + A_1 A_2 \cos[\delta_1 - \delta_2] \\ = I_1 + I_2 + 2\sqrt{I_1 I_2}\cos[\delta_1 - \delta_2] \\ = I_0(1 + \gamma\cos[\delta_1 - \delta_2]), \tag{1.5}$$

in which $I_0 = I_1 + I_2$ denotes the maximum intensity, and $\gamma$ the interference contrast that is given by $\gamma = 2\sqrt{I_1\ I_2}/(I_1 + I_2) = (I_{\max} - I_{\min})/(I_{\max} + I_{\min})$.

For the practical realisation of the length, interfering waves are generated by means of optical interferometers, the simplest arrangement of which is shown in figure 1.4, left.

In analogy to the situation in figure 1.2, the length of the reference pathway is assumed to be unchanged, while the length of the measurement pathway is variable. In such arrangement, the initial phase is same ($\delta_1 = \delta_2$), but phase differences are

[2] In a strict sense the intensity of an electromagnetic wave, i.e. its power density, is defined as the temporal average value of the Poynting Vectors $\vec{S} = \vec{E} \times \vec{H}$. The density of the electric field, $\vec{E}$, is proportional to the density of the magnetic field $\vec{H}$. For simplicity, all constants of proportionality are set to unity here.

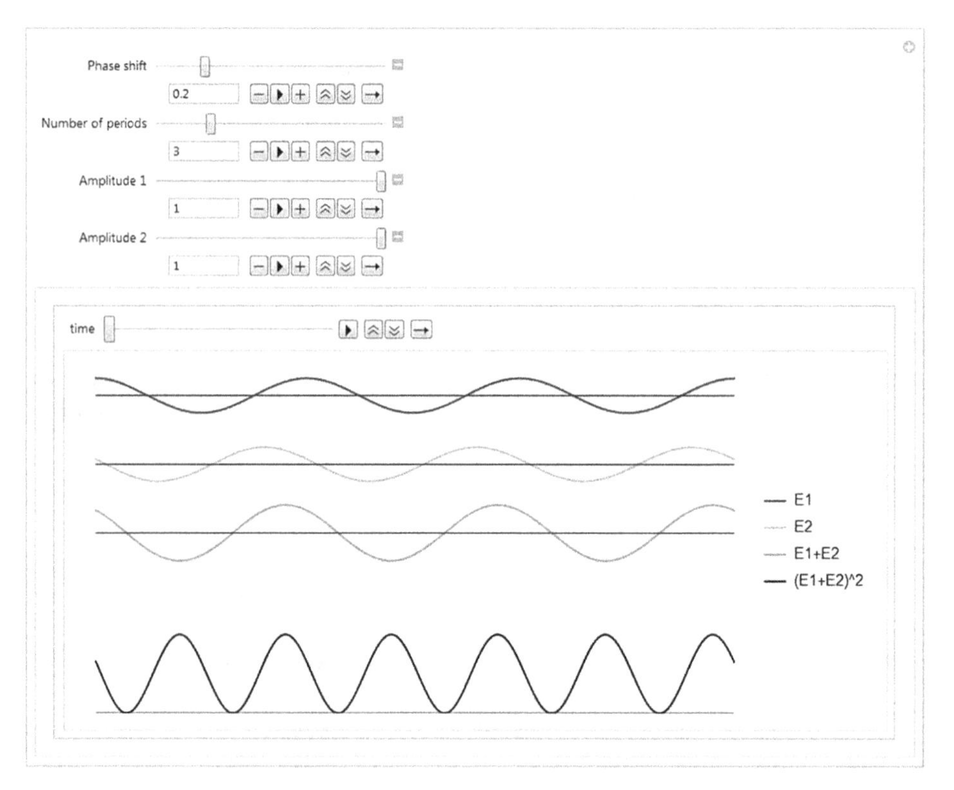


**Figure 1.3.** Interactive diagram for illustrating the interfering waves when setting values for phase shift $\delta_1 - \delta_2$ between the waves and individual amplitudes. Movement of the waves can be initiated by pressing 'play' in the 'time'-field.

generated by length differences $2\Delta z = 2(z_1 - z_2)$ between measurement path and reference path:

$$\underbrace{E_1 = A_1 \cos\left[\omega t - k(z_0 + 2z_1) + \delta\right], \quad E_2 = A_2 \cos\left[\omega t - k(z_0 + 2z_2) + \delta\right]}$$
$$\Downarrow$$
$$I = I_1 + I_2 + 2\sqrt{I_1 I_2}\cos\left[2k(z_1 - z_2)\right] = I_0\left(1 + \gamma \cos\left[2\pi \frac{\Delta z}{\lambda/2}\right]\right), \tag{1.6}$$

where $z_0$ indicates common pathways of the two light waves. Figure 1.4, right, shows the resulting Intensities according to equation (1.6) for different values of the interference contrast. It becomes evident that even for $I_1/I_2 = 0.001$ a remarkable modulation of the interference signal is obtained, i.e. interference signals are well detectable.

According to the relationship given in equation (1.6), a length along a measurement pathway can be determined from a phase difference $\Delta\varphi$. In the simplest case, this is made by counting the number of intensity periods (interference order, $\Delta\varphi/2\pi$ in integer) while continuously shifting the measurement mirror. Modern AD

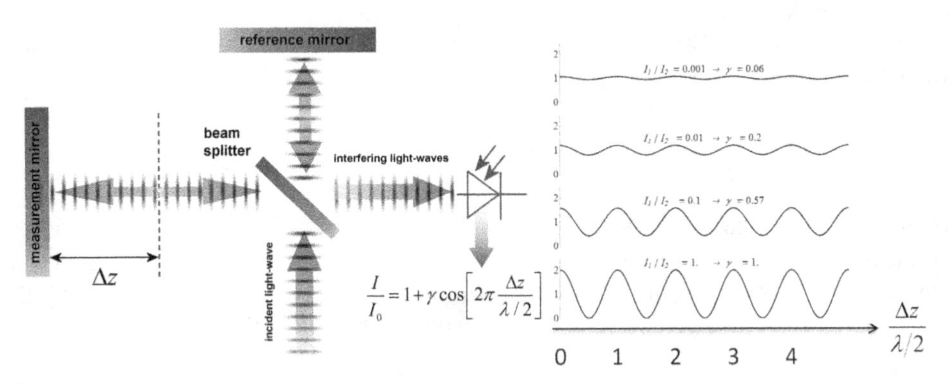


**Figure 1.4.** Realisation of a length distance $\Delta z$ by interferometry, i.e. indirect measurement of the time delay between monochromatic light waves travelling pathways of different lengths before reaching a detector. The blue dashed line indicates the position of equal path length of the measuring and reference pathways.

conversion enables phase resolution to $2\pi/100$ or even better. The amount of the shift along the length axis is therefore given as a multiple of the half wavelength:

$$l = \Delta z = \frac{\lambda}{2}\frac{\Delta\varphi}{2\pi} = \frac{1}{2}\underbrace{c}_{\text{speed of light (phase velocity)}} \cdot \overbrace{\Delta t}^{\text{delay time between wave fronts originating from measurement beam with respect to reference beam}} \quad , \tag{1.7}$$

which can be written as the product of the speed of light and the time delay. Accordingly, the travelling time of the light waves amounts to

$$\Delta t = \frac{1}{\underbrace{\nu}_{\text{frequency of the light}}} \frac{\overbrace{\Delta\varphi}^{\text{measured phase change by means of interferometry}}}{2\pi} \quad . \tag{1.8}$$

Equation (1.7) demonstrates the relationship between a length and the travelling time, as formulated in the SI definition of the metre. Equation (1.8) indicates what is required and what is the measurand in the realisation of the length:

(1) the frequency $\nu$ of the light (required input)
(2) the phase difference between the two interfering waves resulting from the observation of the intensity of interference using an interferometer (measurand).

Length measuring interferometers are mainly used to realise distances in air. Like the direct measurement of time delays (see section 1.2), the presence of air reduces the speed of the light and the wavelength due to the air refractive index $n$ ($c = c_0/n$, $\lambda = \lambda_0/n$). A reduction of the speed of light is compensated for by an increased phase difference to be measured (see equation (1.7)), so that a unique length is derived from measurement of $\Delta\varphi$, irrespective of whether such measurement is performed in vacuum or in air:

$$l = \Delta z = \frac{\lambda_0}{2n}\frac{\Delta\varphi}{2\pi}, \tag{1.9}$$

where $\lambda_0$ is the size of the so-called 'vacuum wavelength' ($\lambda_0 = c_0/\nu$). The actual amount of the air refractive index is dependent on the thermodynamic air parameters, its composition and the frequency of the light. For simplicity, the terminus '$\Delta\varphi$' in equation (1.9) is mostly replaced by '$\varphi$' in the following.

## 1.4 Optical frequency standards

As explained above, knowledge of the light frequency $\nu$ is an essential requirement since it provides the scaling factor between a measured phase difference and the length that is realised by interferometry (see equation (1.8)). Often, the value of the so-called 'vacuum wavelength', which describes the distance between the wavefronts in vacuum under idealised conditions ($\lambda_0 = c_0/\nu$), is stated instead of the frequency.

When two light waves of different frequency are overlapped, then the resulting interference of the waves is given by[3]:

$$\underbrace{E_1 = A\cos\left[2\pi\nu_1 t - k_1 z\right], \quad E_2 = A\cos\left[2\pi\nu_2 t - k_2 z\right]}$$

$$\Downarrow$$

$$\begin{aligned} E_1 + E_2 &= 2A\cos\left[\pi(\nu_1+\nu_2)\cdot\left(t-\frac{z}{c}\right)\right]\cdot\cos\left[\pi(\nu_1-\nu_2)\cdot\left(t-\frac{z}{c}\right)\right] \\ (E_1+E_2)^2 &= 4A^2\left(\cos\left[\pi(\nu_1+\nu_2)\cdot\left(t-\frac{z}{c}\right)\right]\right)^2\cdot\left(1+\cos\left[2\pi(\nu_1-\nu_2)\cdot\left(t-\frac{z}{c}\right)\right]\right) \\ \langle(E_1+E_2)^2\rangle_{t\gg 1/\nu_{1,2}} &= 2A^2\cdot\left(1+\cos\left[2\pi(\nu_1-\nu_2)\cdot\left(t-\frac{z}{c}\right)\right]\right), \end{aligned} \tag{1.10}$$

in which $\langle(E_1+E_2)^2\rangle_{t\gg 1/\nu_{1,2}}$ denotes a temporal average over the fast term, comprising $\nu_1+\nu_2$. Consequently, at a constant point of $z$ ($z$ may be set to zero in equation (1.10)) the interference intensity oscillates with the frequency difference $\nu_1-\nu_2$, the so-called 'beat frequency'. The beat frequency can be sufficiently small to be detectable with a fast light detector. In this way, an unknown frequency can be compared with a standard frequency.

Figure 1.5 illustrates the result of the interference of two waves—whose frequencies differ from each other—in a fixed place, as a function of time according to equation (1.10). The intensity is shown in the lower part of the figure.

The green curve in figure 1.5 shows the mean intensity which oscillates with the beat frequency $\nu_{\text{Beat}} = |\nu_1-\nu_2|$. It is noted that in figure 1.5, unrealistic assumptions were made about the single waves ($\lambda_1 = 630$ nm, $\lambda_2 = 670$ nm, $\nu_{1/2} = c/\lambda_{1/2}$), to

[3] $\cos x + \cos y = 2\cos\frac{x+y}{2}\cos\frac{x-y}{2}$, $(\cos x)^2 = \frac{1}{2}(1+\cos 2x)$.

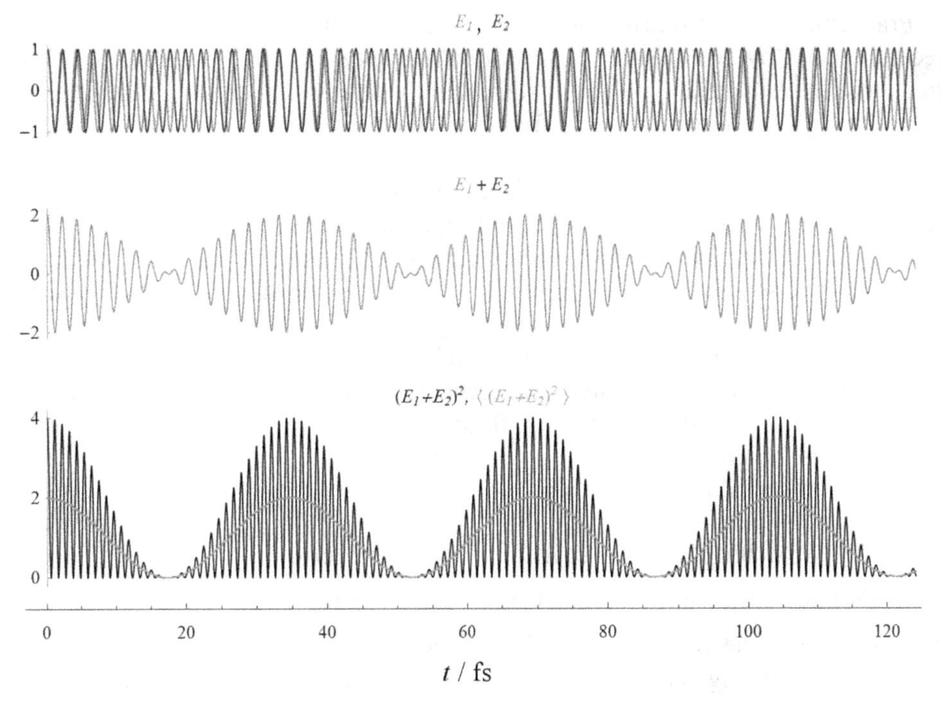


**Figure 1.5.** Overlapping of two light waves whose frequencies differ from each other, considered at a given point in space. To illustrate both the fundamental mode (high-frequency fraction) and the beating waves (low-frequency fraction) of a light wave, the following parameters were used: $\lambda_1 = 630$ nm, $\lambda_2 = 670$ nm, $\nu_{1/2} = c/\lambda_{1/2}$.

clearly demonstrate this relationship. In real frequency comparisons, the frequencies lie much closer to each other, so that beat frequencies in the Hz, kHz or MHz regime are observed. The frequency of a light source to be calibrated ($\nu_{\mathrm{EUT}}$) is therefore related to the frequency of a standard ($\nu_{\mathrm{Standard}}$): $\nu_{\mathrm{EUT}} = \nu_{\mathrm{Standard}} \pm \nu_{\mathrm{Beat}}$. The ambiguity ($\pm$) comes from the fact that the beat frequency only indicates the amount of the difference $\nu_{\mathrm{Beat}} = |\nu_{\mathrm{Standard}} - \nu_{\mathrm{EUT}}|$. To achieve unambiguity, it is therefore necessary to have more than one standard frequency light source available. Frequencies of suitable light sources can be obtained by comparison with the primary Cs atomic clocks at 9.2 GHz. Respective frequency chains used to be highly complex [14]. Today, this is done by means of frequency comb generators. By control of the carrier envelope offset frequency $f_{\mathrm{CEO}}$ and the repetition rate frequency $f_{\mathrm{REP}}$, both typically in the radio frequency regime, optical frequencies $\nu$

$$\nu = f_{\mathrm{CEO}} + n f_{\mathrm{REP}} \tag{1.11}$$

can be generated with $n$ representing a (large) integer number [15]. In this approach so-called 'optical frequency standards' can be realised with relative uncertainties of $10^{-15}$ and better [16]. With the above method, a secondary frequency standard (such as a laser for use in interferometric length measurements) can be calibrated by using a suitable mode of the synchronised frequency comb.

Practicable light sources for use in length measurements by interferometry are those recommended by the *Comité International des Poids et Mesures* (CIPM) whose frequency is established by laser stabilisation on molecular hyperfine structure transitions [17]. Hereby, relative frequency uncertainties of typically $<10^{-11}$ are achieved [18].

Today, optical frequency standards are mainly further developed for the future realisation of the time unit by so-called 'optical clocks' that promise to be more precise than the current caesium atomic clocks. At present, it is already possible to attain relative frequency uncertainties of $<10^{-16}$ [19], depending on the averaging period considered. It is noted that, contrary to the situation before the 1980s, the uncertainty of the length measurement is no longer determined by the uncertainty of the frequency of the light source. The benefit of even more stable laser sources is hence negligible for the practical realisation of the metre. On the other hand, improvement of practical light sources with respect to frequency range, intensity stability and availability are highly desired and need technological improvements.

## 1.5 Types of length measuring interferometers

### 1.5.1 Distance-scanning by fringe counting interferometers

The basic principle of distance-scanning interferometry is the observation of the periodically changing detector signal of the interference intensity during a continuous change of the distance of a measuring mirror with respect to a reference mirror, considered stable (see figure 1.4). Each period corresponds to a single interference order, i.e. to a change in distance by half the wavelength of the light used (equation (1.9)).

In such interferometers, the mirrors are positioned on carriages. To cope with angular changes during the movement, retroreflectors are typically used instead of flat mirrors which makes the observed interference signal insensitive to small tilts. Over the distance to be measured a huge number of interference orders is counted, e.g. approx. 3 million oscillations per metre when using a typical red laser (wavelength 633 nm, $\Delta\varphi/2\pi = 1\text{ m}/(633/2\text{ nm})$). In principle, with the laser light sources available today (coherence length in the kilometre range), it is theoretically possible to measure very large distances. However, air turbulences can make counting of interference orders impossible at larger distances, especially in an uncontrolled environment. A further limit is the mechanical stability of the interferometer.

Since the interference is decisively determined by the cosine function (see equation (1.5)) the periodical signal measured by a detector cannot, however, indicate the direction in which the measuring mirror is moved. In extreme cases, moving the mirror forwards and backwards can even simulate uniform motion along an axis. For this reason, most counting interferometers are equipped with additional techniques, described in the following, allowing a unique relationship to the position of the measuring mirror.

### 1.5.2 Counting interferometers using the 'quadrature procedure'

In the so-called 'quadrature procedure' (see figure 1.5), a laser beam whose polarisation axis is tilted by 45° hits a polarising beam splitter which splits the incident light into the two polarisation directions that are vertical to each other, i.e. into the polarised component which is parallel to the incident plane (p-beam) which entirely passes through the beam splitter and into the vertically polarised component (s-beam) which is fully reflected. These two beams are reflected by retroreflectors and then led together again at the polarising beam splitter. At the exit of the interferometer, two beams that are polarised vertically to each other are then available[4]. Polarizers are located behind another beam splitter, in front of a detector; their axis is tilted by 45° so that they project both polarisation components (s and p) onto the polarisation axis of the corresponding polarizer. Only in this way is it possible that the beams that are first polarised vertically to each other lead to an observable interference signal. The beam which is reflected at the beam splitter traverses a $\lambda/4$—delay plate before the polarizer and therefore generates a signal at the upper detector which is shifted by $\pi/2$ compared to the signal of the other detector. The quadrature detection thus provides two interference signals that are shifted by $\pi/2$ to each other and are called sine and cosine signals, respectively. Unavoidable imperfections, e.g. unequal detector gains and offsets can be corrected in the post processing by the Heydemann correction [20]. Then, the signals $\tilde{I}_{\sin}$ and $\tilde{I}_{\cos}$ can be represented as shown in figure 1.6 on the right, e.g. by means of an oscilloscope in the $xy$ mode. A shift of the measuring mirror then leads to a circular or elliptic figure being displayed. Depending on the direction of displacement, the corresponding vector rotates in one or the other direction. Thereby, a full revolution corresponds to one interference order ($\Delta\varphi = 2\pi$) and thus to a shifting of the measuring mirror by $\Delta z = \lambda/2$, which can be clearly detected with this procedure.

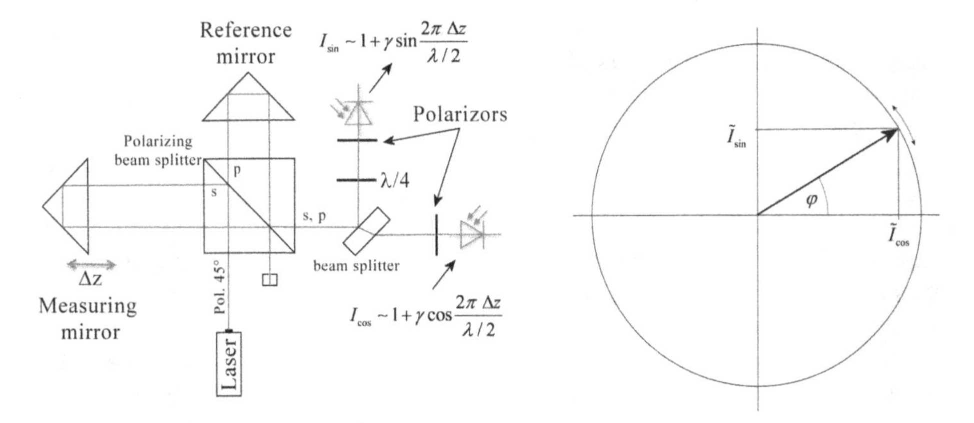


**Figure 1.6.** Extension of an interferometer by polarisation components to generate two interference signals whose phases are shifted by $\pi/2$ (left). On the right, the vector yielded from the signals $\tilde{I}_{\sin}$ and $\tilde{I}_{\cos}$ is plotted.

[4] Superposition of two light waves that are orthogonally polarised to each other, does not lead to an observable interference.

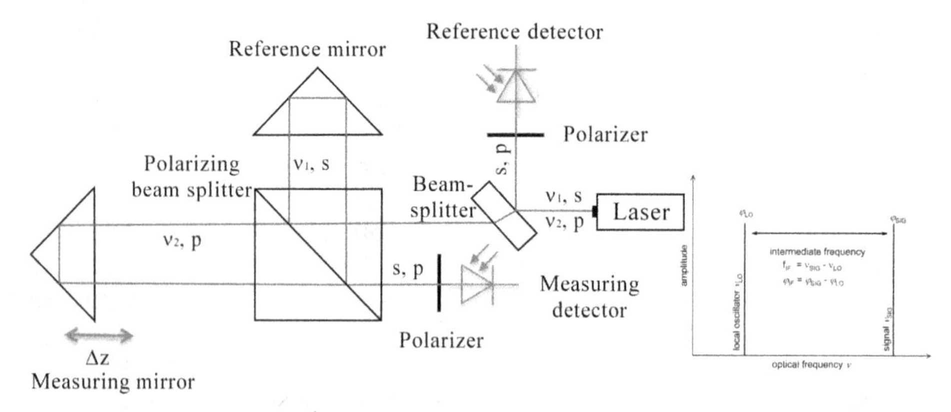


**Figure 1.7.** Scheme of a simple heterodyne interferometer. Two orthogonally polarised laser beams s and p of different laser frequencies ($\nu_1$ and $\nu_2$) are used. The p-beam acts as the signal beam (SIG), the s-beam as the local oscillator (LO). Signals are detected at the intermediate frequency $f_{IF}$, the phase of the $\varphi_{IF}$ corresponds to the difference of the phases of the two optical beams.

### 1.5.3 Heterodyne interferometers

Laser sources can be designed to emit two or more coherent light waves that are polarised vertically to each other and have different frequencies (i.e. two different laser modes). For HeNe lasers, this is often achieved by external magnetic fields inducing the Zeeman effect [21], leading to frequency differences of typically 1 to 3 MHz. Larger frequency differences can be generated by suitable laser resonator designs [22]. Furthermore, it is possible to shift the frequency of a light wave after generation by a defined value, e.g. by means of acousto-optic modulators (AOMs). This can be used to demodulate the interference phase signals at an intermediate frequency in a heterodyne detection scheme. Thus, the typically large low-frequency noise can be avoided. Furthermore, faster moving targets can also be tracked. Finally, multiple interferometers of different wavelengths can be detected in parallel in this detection scheme, using different intermediate frequencies as information carriers [23].

Figure 1.7 shows the typical design of a heterodyne interferometer[5]. The light source used here is a laser with two different frequencies. The light waves whose electric field is polarised vertically to the image plane is labelled ‘s’, the wave which is polarised parallel to the plane is designated as ‘p’. First, both light waves are split by means of a beam splitter. The two reflected waves then hit a polarizer whose axis is tilted by 45° regarding the two polarisation directions.

At the detector located behind these, this leads to a beat interference (see above section 1.4) of the now equally polarised waves of different frequencies ($\nu_1$ und $\nu_2$), the so-called ‘reference signal’. The two light waves transmitted at the beam splitter later hit a polarising beam splitter where the s-polarised wave is reflected and the p-polarised wave is transmitted. After being reflected by the retroreflectors, the two

[5] In more classical interferometers operating with a single frequency for demodulation and measurement beam, the detection scheme is performed at DC. Therefore, these are called homodyne interferometers.

waves are led together again in the polarising beam splitter and pass a polarizer whose axis is, again, tilted by 45° with respect to the two polarisation directions, so that a second beat interference, the so-called 'measuring signal', occurs at the detector located behind it. The p-beam acts as signal beam, while the s-beam plays the role of a classical oscillator.

The phase of the reference signal $\Phi_{ref}$ is equivalent to the phase difference of the individual phases belonging to the different frequencies, $\varphi_{1/2}^{ref} = k_{1/2}\ z^{ref-detector} - \omega_{1/2}\ t + \delta_{1/2}$ , where $z^{ref-detector}$ describes the joint light path of the individual waves up to the reference detector. In contrast, the phase of the measuring signal, $\Phi_{meas}$, is generated by the overlapping of single waves which have covered different distances. If $\Delta z$ is defined as the length difference between the measured length and the distance from the reference mirror $z^{ref-mirror}$, then the measuring signal is obtained from the difference between $\varphi_2^{meas} = k_2\ (z^{ref-mirror} + \Delta z) - \omega_2 t + \delta_2$ and $\varphi_1^{meas} = k_1\ z^{ref-mirror} - \omega_1 t + \delta_1$ . If $z^{ref-detector}$ and $z^{ref-mirror}$ are considered as constant and equal, then the phases of both detectors and their difference can be expressed as follows:

$$
\begin{aligned}
\Phi_{ref} &= (k_2 - k_1)\, z^{ref} - (\omega_2 - \omega_1)t + \delta_2 - \delta_1 \\
\Phi_{meas} &= (k_2 - k_1)\, z^{ref} + k_2 \Delta z - (\omega_2 - \omega_1)t + \delta_2 - \delta_1 \\
\Phi_{meas} - \Phi_{ref} &= k_2 \Delta z.
\end{aligned}
\tag{1.12}
$$

This means that the shift $\Delta z$ of the measuring mirror can be determined from the comparison of the phase position between the beat signals at the reference detector and at the measuring detector. Both signals can be separated from the DC component by capacitive coupling, which makes the procedure insensitive to variations of the light intensity, including influences due to the ambient light. Interpolating an interference signal—as is usually done in homodyne interferometers—becomes superfluous, and reversing the direction is no longer an issue since the phase position of the beat signals always increases and thus the phase difference between the measuring signal and the reference signal can be determined unambiguously. Heterodyne interferometers have been demonstrated to achieve uncertainties down to a level of 10 pm [24].

### 1.5.4 Multi-wavelength interferometry

To construct the full phase change $\Delta\varphi$ for longer lengths, the integer interference order needs to be determined. Classically, this is achieved by moving the measurement path over the length measured while monitoring the phase change continuously (see section 1.5.1). This movement over a distance makes counting interferometry more and more impractical the larger the length to be measured gets. For measurements of larger parts in precision engineering, for example, such a measurement strategy is extremely time-consuming and inflexible. Moreover, beam breaks in more crowded environments require a complete repetition of the counting procedure. The so-called absolute interferometry realises a non-incremental interferometric measurement. There are several approaches for this purpose, all requiring broader spectral sources for the interferometry than the monochromatic discussed so far. For this introduction, the basic principles of synthetic wavelength interferometry, frequency-sweeping interferometry and the interferometric correlation of femtosecond pulses are briefly discussed.

A first approach to absolute interferometry is based on an increase of information. Due to the periodicity of the interference function (see equation (1.6)) a measured length is only non-ambiguous within half of the optical wavelength $\lambda$. When two or more lasers of different frequencies are available for the same length measurement by interferometry, for a given distance $l$ the respective phases $\varphi_1$, $\varphi_2$, ..., $\varphi_k$ can be measured separately. In the simplest case of using two wavelengths $\lambda_1$ and $\lambda_2$, the phase difference $\varphi_2 - \varphi_1$ can be written as

$$\varphi_2 - \varphi_1 = 2\pi\frac{l}{\lambda_2/2} - 2\pi\frac{l}{\lambda_1/2} = 4\pi l\left(\frac{1}{\lambda_2} - \frac{1}{\lambda_1}\right) = 2\pi\frac{l}{\Lambda_{\mathrm{synth}}/2} \equiv \varphi_{\mathrm{synth}}, \tag{1.13}$$

i.e. the measured phase difference $\varphi_2 - \varphi_1$, called the synthetic phase $\varphi_{\mathrm{synth}}$, is periodic with half of the so-called synthetic wavelength:

$$\Lambda_{\mathrm{synth}} = \frac{\lambda_1\lambda_2}{\lambda_1 - \lambda_2}. \tag{1.14}$$

The synthetic phase can be directly measured, for example by a superheterodyne detection scheme at the difference frequency of the two optical wavelengths [25], or constructed in the post processing of measured optical phases. The available additional phase information increases the range of non-ambiguity of the interferometric measurement to half the synthetic wavelength $\Lambda_{\mathrm{synth}}$. Unfortunately, all uncertainties scale in this approach by the factor $\Lambda_{\mathrm{synth}}/\lambda$. This can be circumvented by using multiple wavelengths of different spacing. Then, a series of decreasing synthetic wavelengths can be constructed [26]. The most correct interference measurement of the shortest synthetic wavelength can then be unwrapped by the result derived from the longer synthetic wavelengths. Nevertheless, the range of unambiguity is for this approach still limited to the longest constructible synthetic wavelength.

This limitation is overcome by frequency-sweeping interferometry (FSI). It can be intuitively derived from classical interferometry. Instead of varying the length $l$ to count the phase change $\Delta\varphi$, it is (formally and practically) also possible to vary the laser frequency $\nu$ by $\Delta\nu$ for a fixed distance. The interferometric phase changes by

$$\Delta\varphi(t) = \frac{4\pi l}{c}[n(\nu(t))\nu(t) - n(\nu(t_0))\nu(t_0)] \tag{1.15}$$

when varying the frequency $\nu(t_0)$ to $\nu(t)$ in the time interval $t - t_0$ ($n$ representing the phase refractive index). Typically, sets of data points of $(\nu(t), \Delta\varphi(t))$ are taken during a frequency sweep. The dependence of the phase change $\Delta\varphi(t)$ on the frequency is linear assuming mechanical stability and neglecting refractive index fluctuations during a sweep. Therefore, the absolute distance $l$ can be derived from the gradient of the linear regression by

$$l = \frac{c}{4\pi n_g(\nu(t))} \frac{d\Delta\varphi(t)}{d\nu} \tag{1.16}$$

with the group refractive index

$$n_g(\nu(t)) = n + \nu(t)\frac{dn}{d\nu}. \tag{1.17}$$

In practice, the uncertainty of the absolute distance $l$ is often dominated by the uncertainty of the frequency change $\nu(t) - \nu(t_0)$. Fabry–Perot interferometers [27], highly stable auxiliary interferometers [28], or spectroscopic transitions [29, 30] have been investigated for this purpose. The method and its application is discussed in more detail in chapter 8 of this book.

Finally, the well-defined, sharp temporal and broad spectral properties of femtosecond lasers can also be used for the direct measurement of the absolute distance. In a straightforward approach, one monitors the interferometric cross-correlation signal of two pulse trains in a Michelson-type interferometric arrangement. Only if two pulses temporarily overlap, a cross-correlation signal can be detected. Thus, this can be used to determine lengths as multiples of the pulse separation length $l_{pp}$, resembling a time-of-flight measurement [31, 32].

To achieve higher spatial resolution, one approach is the spectral resolution of the interference pattern. This is referred to as dispersive or spectral interferometry in literature. Here the unwrapped phase $\varphi(\nu)$ can be derived by Fourier transformation from the normalised spectrum. The gradient $d\varphi/d\nu$ is then directly connected with the absolute length by [33]

$$l = \frac{c}{4\pi n_g} \frac{d\varphi}{d\nu}. \tag{1.18}$$

Comparison with equation (1.16) makes clear that dispersive interferometry is formally equivalent to a single-shot, instantaneous FSI measurement.

The ultimate limit of this approach is the case when single femtosecond comb modes can be spectrally resolved and their interference phase directly measured. From the perspective of synthetic wavelength interferometry, this case can also be interpreted as equivalent to thousands of single wavelength interferometers analysed in parallel for the derivation of the absolute length [34]. From this perspective, femtosecond lasers can be considered as ultimate multi-wavelength sources.

Femtosecond laser-based interference technologies are discussed in more depth in chapters 6 and 7.

### 1.5.5 Multiple-beam interferometry

In mulitple-beam interferometry, the light is reflected several times and eventually overlaps. A well-known example of this is the coloured stripes generated by white light falling onto an oily patch on water. This phenomenon is used in its most simple form in surface testing. For this purpose, a nearly perfectly smooth glass plate is laid onto the surface of the test object. Here too, coloured patterns become visible where the surface exhibits a tilt; the number and the direction of these patterns depend on the tilt of the surfaces to each other. The straightness of these patterns is a measure of the flatness of the specimen.

A 'Fizeau interferometer' is a multiple-beam interferometer for which the distance between the two plane surfaces on which the reflection takes place is large compared to the wavelength of the light used. Figure 1.8 shows a Fizeau interferometer, in which the interference of the light is observed in reflection. The light passes through an optical plate with a semi-transparent surface. Part of the light reflected by the mirror is, in turn, reflected by the semi-transparent surface and goes back to the mirror. The higher the reflectivity $R$ of the semi-transparent surface, the more often this sequence repeats itself. Contrary to the two-beam interference, the intensity of the multiple-beam interference observed at the output of the interferometer is not cosine-shaped, but complies with the Airy-formula. This characteristic of the interference is plotted in the inset in figure 1.8 for different reflectivity.

With increasing reflectivity, the structure becomes increasingly sharp. As in the case of the two-beam interferometer, the periodicity of this structure is given by half the wavelength of the light used.

A Fabry–Pérot interferometer is a Fizeau interferometer that is operated with two semi-transparent mirrors while observing the transmitted light; the reflectivity of semi-transparent mirrors is, however, usually higher. A parallel glass plate with reflective coatings is the simplest example of such an interferometer. The resonator thus obtained can be used as a material measure for the optical path length (the product of the refractive index of the glass plate by its thickness) and is also known

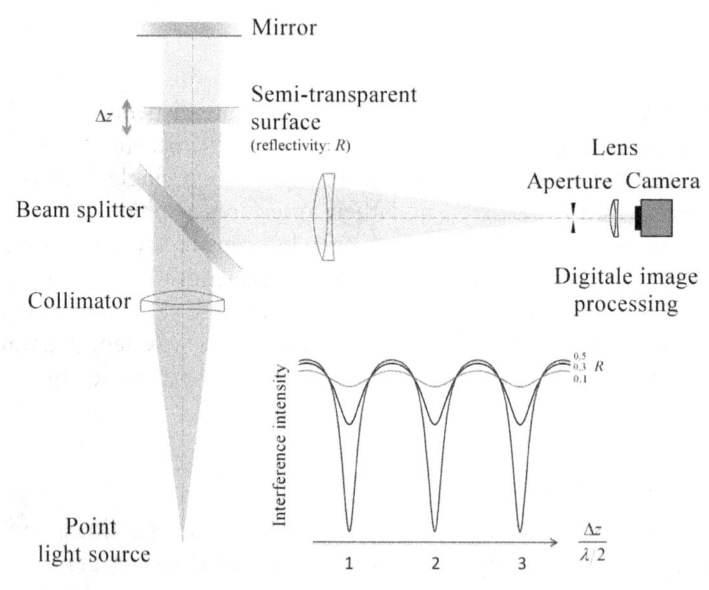


**Figure 1.8.** Fizeau interferometer operated in reflection mode. The characteristics of the interference intensity as a function of $\Delta z$ is dependent on the reflectance of the beam splitter (at any pixel of the camera: dark fringes appear which become tighter with increasing values of $R$).

as a 'Fabry–Pérot etalon'. Since Fabry–Pérot interferometers filter a narrow-band spectrum out of a broad-band radiation, they are often used as optical filters. The so-called 'finesse' serves to characterise the resonator. It is defined as the ratio of the so-called free spectral range $\Delta\lambda$ to the full width at half maximum $\delta\lambda$ of an individual maximum of the interference intensity: $F = \Delta\lambda/\delta\lambda = \pi\sqrt{R}/(1 - R)$. The higher the finesse (i.e. at high reflectivity $R$), the narrower the band of the filtered light appears. Extremely stable resonators of high finesse can be used for short term frequency stabilisation of laser sources. Resonators made of monocrystalline silicon, e.g. enable a frequency stabilisation of commercial laser systems at the inconceivably low-frequency of 0.04 Hz, i.e. better than $10^{-16}$ [35].

### 1.5.6 Large field imaging interferometers

Large field interferometers are equipped with optical components allowing a large diameter, of the order of 50 mm or larger, for the light beam. Besides collimating lenses, exhibiting focal length of typically several hundred millimetres, comparatively large beam splitter/mirrors are necessary. Such interferometers are often combined with an imaging optics allowing objects (e.g. the front face of a gauge block) to be observed by eye or by a camera sensor. These devices are therefore called imaging large field interferometers. They may also be called 'interference macroscopes' to indicate that it is quite the opposite of 'interference microscopes' which are much more common. Imaging interferometers may work as two-beam interferometers (see chapter 2) or Fizeau interferometers (see chapter 3).

The very popular application of large field interferometry in optical shop testing [36], as flatness measurement (e.g. see [37]) or shape measurement (e.g. see [38]) is not addressed in this book.

## 1.6 General requirements and limitations in length measurements by interferometry

In the realisation of the length by interferometry, it is important to consider the physical relationship between the primary result of the measurements, and the geometrical length to be realised. The following important influences lead to limitations in the real world.

1. The frequency of a light source used for the realisation of a length by interferometry must be traceable to the frequency of standard clocks or included in the BIPM-list 'recommended values of standard frequencies'.
2. The direction of wave propagation must coincide with the direction of the length to be realised. This requirement can be satisfied to a certain degree by the appropriate design of the optics (retroreflectors along the measurement pathway) or dedicated adjustment methods (autocollimation adjustment).
3. When the length is realised in air, the presence of air considerably reduces the wavelength of the light used ($\lambda$) compared to the wavelength in vacuum ($\lambda_0$), which is significant for the interferometrically measured length. The refractive index of air $n$ depends both on the wavelength (frequency) itself and on the ambient conditions (the air pressure $p$, the air temperature $t$, the

air humidity $f$, and the $CO_2$ concentration $x$). Since $n$ is close to 1, the relative influence of the refractive index of air on the length is scaled by the so-called 'refractivity', $n - 1$ (approximation: $1/n|_{n\approx1} \simeq 1 - (n - 1)$). Determining the refractive index of air accurately is an important limitation to the precision of a length measurement. When the refractive index of air is determined based on an appropriate empirical equation, input parameters must be measured traceable to the respective unit (mainly pressure, temperature, humidity of the air). Furthermore, distribution of these parameters along the pathway of length must be considered properly. When a vacuum cell is used as refractometer, the effect of the cell windows must be verified (see chapter 2).

4. The finite size of a real 'point light source', positioned in the focal point of a collimating lens, leads to a length proportional aperture correction that must be applied, which is explained in chapter 2.
5. When an extended light beam, covering a certain area within which interferometry is used to realise the length of material measures (see chapter 2) by measuring differences in the phase topography:
   (a) The lateral position of the length measurement must have a clear assignment to the geometry. This can be achieved by using imaging interferometry combined with structure detection algorithms within the resulting topographies for phase and contrast.
   (b) The resulting lengths must be insensitive to the orientation of the phase topography itself. This can be achieved by considering appropriately arranged regions of interest (e.g. for gauge blocks: one region at the front and two regions arranged symmetrically around it).
   (c) The phase change on reflection at the surfaces is 180° for perfect (zero roughness), non-absorbing (zero extinction of the material) surfaces. In length measurements of material measures as gauge blocks these effects must be taken into consideration by appropriate corrections.
6. The shape of the wavefront of real light is not perfectly flat; any deformed wavefront is subject to evolution during propagation along a distance. To keep this effect as small as possible almost ideally flat optical components are necessary. The remaining effect due to wavefront distortion must be treated as a source of measurement uncertainty.
7. Light separation based on polarisation is imperfect in practice. Crosstalk can substantially reduce the achievable measurement uncertainty for example in heterodyne interferometry. The polarisation properties of optical elements are also influenced by the measurement conditions.
8. Unwanted reflections leading to parasitic interferences must be considered as error sources.
9. For incremental and absolute measurements, the stability of the dead path must be ensured. Fluctuations in the air refractive index or mechanical

instability, for example due to thermal expansion, must be considered by design or by proper compensation.

10. In the case of AC detection schemes, the detector can influence the phase measurement. Amplitude to phase-coupling or small beam wandering in the case of local inhomogeneity can increase the uncertainty substantially and must be carefully avoided.
11. Impurity of the light: the light source used in interferometry may contain fractions of light for which the frequency differs from the intended light frequency. Laser light sources are mostly used in length measurements by interferometry. A key component of a laser is an optical resonator. Although in a laser a certain resonator mode is made predominant, the laser light generally contains minor resonator modes, which are called parasitic modes here. When entering an interferometer, the presence of parasitic modes, will affect the length measurement. This error is dependent on the length. It can be shown for a single parasitic mode, that the maximum error is approximately given by $|\delta l|_{\max} = \frac{1}{2\pi} I_{\text{par}}/I_{\text{main}} \cdot \lambda_{\text{main}}/2$ , in which $I_{\text{par}}/I_{\text{main}}$ is the ratio of light intensities of the parasitic and the main mode and $\lambda_{\text{main}}$ the wavelength of the main mode.

This list is not complete. Many of these influences are addressed in the individual chapters of this book. Each of these contributions to the overall measurement uncertainty can only be reduced to a certain level. Hence, it is important to note that the attainable overall measurement uncertainty in the realisation of the length is typically much larger than resolution and repeatability may promise.

## References


[1] Poynting J H and Thompson J J 1907 *A Textbook of Physics: Properties of Matter* 4th edn (London: Charles Griffin) p 20

[2] Larousse P (ed) 1874 *'Métrique', Grand dictionnaire universel du XIXe siècle*vol 11 (Paris: Pierre Larousse) pp 163–4

[3] Alder K 2002 *The Measure of All Things: The Seven-Year Odyssey and Hidden Error that Transformed the World* (New York: The Free Press)

[4] Comptes Rendus de la 7e CGPM 1927, 1928, p 49 https://www.bipm.org/en/CGPM/db/7/1/

[5] Babinet M 1829 Sur les couleurs des réseaux *Ann. Chim. Phys.* **40** 166–77

[6] Meggers W F 1948 A light wave of artificial mercury as the ultimate standard of length *J. Opt. Soc. Am.* **38** 7–14

[7] Michelson A A 1893 Light-waves and their application to metrology *Nature* **49** 56–60

[8] Bönsch G and Potulski E 1998 Measurement of the refractive index of air and comparison with modified Edlen's formulae *Metrologia* **35** 133–9

[9] Fujima I, Iwasaki S and Seta K 1998 High-resolution distance meter using optical intensity modulation at 28 GHz *Meas. Sci. Technol.* **9** 1049–52

[10] Guillory J, Šmid R, Garcia-Marquez J, Truong D, Alexandre C and Wallerand J-P 2016 High resolution kilometric range optical telemetry in air by radio frequency phase measurement *Rev. Sci. Instrum.* **87** 075105

[11] Minoshima K and Matsumoto H 2000 High-accuracy measurement of 240-m distance in an optical tunnel by use of a compact femtosecond laser *Appl. Opt.* **39** 5512–7

[12] Kim J, Chen J, Zhang Z, Wong F N C, Kärtner F X, Loehl F and Schlarb H 2007 Long-term femtosecond timing link stabilization using a single-crystal balanced cross correlator *Opt. Lett.* **32** 1044–6

[13] Lee J, Kim Y-J, Lee K, Lee S and Kim S-W 2010 Time-of-flight measurement with femtosecond light pulses *Nat. Photon.* **4** 716–20

[14] Schnatz H, Lipphardt B, Helmcke J, Riehle F and Zinner G 1996 First phase-coherent frequency measurement of visible radiation *Phys. Rev. Lett.* **76** 18–21

[15] Udem T, Holzwarth R and Hänsch T W 2002 Optical frequency metrology *Nature* **416** 233–7

[16] Huntemann N, Okhapkin M, Lipphardt B, Weyers S, Tamm C and Peik E 2012 High-accuracy optical clock based on the octupole transition in $^{171}Yb^{+}$ *Phys. Rev. Lett.* **108** 090801

[17] Recommended values of standard frequencies http://www.bipm.org/en/publications/mises-en-pratique/standard-frequencies.html [Zugriff am 16.04.2018].

[18] Cordiale P, Galzerano G and Schnatz H 2000 International comparison of two iodine-stabilized frequency-doubled Nd:YAG lasers at 532 nm *Metrologia* **37** 177–82

[19] Chou C W, Hume D B, Rosenband T and Wineland D J 2010 Optical clocks and relativity *Science* **329** 1630–3

[20] Heydemann P L M 1981 Determination and correction of quadrature fringe measurement errors in interferometers *Appl. Opt.* **20** 3382–84

[21] Bobroff N 1993 Recent advances in displacement measuring interferometry *Meas. Sci. Technol.* **4** 907–26

[22] Kim M S and Kim S W 2002 Two-longitudinal-mode He-Ne laser for heterodyne interferometers to measure displacement *Appl. Opt.* **41** 5938–42

[23] Yang R, Pollinger F, Meiners-Hagen K, Tan J and Bosse H 2014 Heterodyne multi-wavelength absolute interferometry based on a cavity-enhanced electro-optic frequency comb pair *Opt. Lett.* **39** 5834–7

[24] Weichert C, Köchert P, Köning R, Flügge J, Andreas B, Kuetgens U and Yacoot A 2012 A heterodyne interferometer with periodic nonlinearities smaller than ±10 pm *Meas. Sci. Technol.* **23** 094005

[25] Dändliker R, Thalmann R and Prongué D 1988 Two-wavelength laser interferometry using superheterodyne detection *Opt. Lett.* **13** 339–41

[26] Dändliker R, Salvadé Y and Zimmermann E 1998 Distance measurement by multiple-wavelength interferometry *J. Opt.* **29** 105–14

[27] Barwood G P, Gill P and Rowley W R C 1998 High-accuracy length metrology using multiple-stage swept-frequency interferometry with laser diodes *Meas. Sci. Technol.* **9** 1036–41

[28] Bechstein K-H and Fuchs W 1998 Absolute interferometric distance measurements applying a variable synthetic wavelength *J. Opt.* **29** 179–82

[29] Dale J, Hughes B, Lancaster A J, Lewis A J, Reichold A J H and Warden M S 2014 Multi-channel absolute distance measurement system with sub ppm-accuracy and 20 m range using frequency scanning interferometry and gas absorption cells *Opt. Express* **22** 24869–93

[30] Prellinger G, Meiners-Hagen K and Pollinger F 2015 Spectroscopically *in situ* traceable heterodyne frequency-scanning interferometry for distances up to 50 m *Meas. Sci. Technol.* **26** 084003

[31] Ye J 2004 Absolute measurement of a long, arbitrary distance to less than anoptical fringe *Opt. Lett.* **29** 1153–5

[32] Cui M, Zeitouny M G, Bhattacharya N, van den Berg S A, Urbach H P and Braat J J M 2009 High-accuracy long-distance measurements in air with a frequency comb laser *Opt. Lett.* **34** 1982–4

[33] Joo K-N and Kim S-W 2006 Absolute distance measurement by dispersive interferometry using a femtosecond pulse laser *Opt. Express* **14** 5954–60

[34] van den Berg S A, Persijn S T, Kok G J P, Zeitouny M G and Bhattacharya N 2012 Many-wavelength interferometry with thousands of lasers for absolute distance measurement *Phys. Rev. Lett.* **108** 183901

[35] Kessler T, Hagemann C, Grebing C, Legero T, Sterr U, Riehle F, Martin M J, Chen L and Ye J 2012 A sub-40-mHz linewidth laser based on a silicon single-crystal optical cavity *Nat. Photon.* **6** 687–92

[36] Malacara D (ed) 2007 *Optical Shop Testing* 3rd edn (New York: Wiley)

[37] Dew G D 1966 The measurement of optical flatness *J. Sci. Instrum.* **43** 409–15

[38] Garbusi E, Pruss C and Osten W 2008 Interferometer for precise and flexible asphere testing *Opt. Lett.* **33** 2973–75

**IOP** Publishing

Modern Interferometry for Length Metrology
Exploring limits and novel techniques
**René Schödel (Editor)**

# Chapter 2

# Large field imaging interferometry for the measurement of the length of bar shaped material measures

**René Schödel**

## 2.1 Introduction

In contrast to the widespread commercially available distance measuring interferometers, introduced in chapter 1 in this book, large field interferometers make use of beam diameters typically bigger than 40 mm. A common type of large field interferometer is the Fizeau interferometer shown in figure 1.8 of chapter 1. Another type of large field interferometer is a Michelson interferometer operated with a collimated beam of light, originally suggested by Twymen and Green [1]. Both types of interferometers are ideal for measurements of surface shape and surface texture. With very little effort these interferometers can be set up in an optical workshop for routine testing of optical components [2, 3]. Besides that, such interferometers are also suitable for the measurement of the length of bar shaped material measures, which is explained in this chapter.

Interchangeable manufacturing was already state of the art at the beginning of the 20th century. Gauge blocks with parallel end faces (representing a prismatic body) made of hardened steel were first produced by the Swedish engineer Johannson around 1904. Equipped with faces nearly as flat and accurate as optical glass, gauge blocks could satisfy the increasing need for accuracy in tool-rooms and inspection departments at that time. Today, gauge blocks are indispensable material measures for length and appear in different grades which define their geometrical and surface finish properties [4]. Among these, grade K represents the highest grade and refers to gauge blocks which are suitable to be primarily calibrated by interferometry. Two or more gauge blocks can be wrung together when their faces are brought into close contact, forming a single gauge for the overall length. With a set of steel gauge blocks consisting of 103 pieces, more than 20 000 measurments between 1 mm and 201 mm can be realised with a gradation of 0.005 mm by combining them.

doi:10.1088/2053-2563/aadddcch2

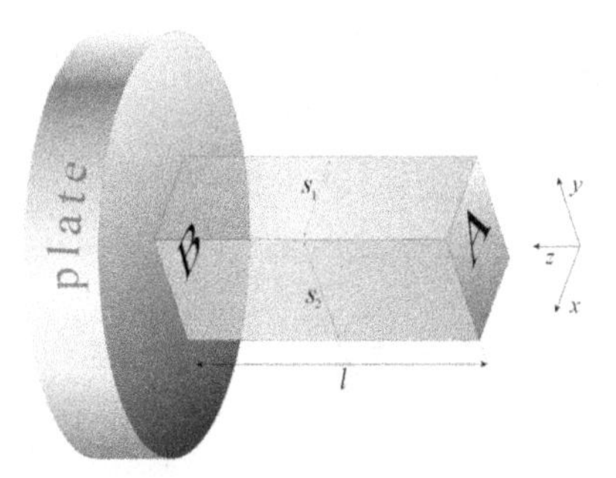


**Figure 2.1.** Prismatic body with end faces A and B that are parallel to each other, wrung to an end plate. 

Figure 2.1 shows the classic situation where one face of a prismatic body is wrung to a flat platen. If made properly, the body and the platen hereby adhere strongly together due to their close contact. Such assembly represents a step height which can be measured interferometrically as described below.

Types of refined interferometers for measuring the length gauge blocks have been suggested and developed in the first half of the 20th century at PTR by Kösters [5, 6] and at NPL [7] and further developed in the second half of the century (e.g. [8–10]).

Although the basic physical principle of modern large field interferometers is unchanged, today's interferometers benefit from the revolutionary technical developments of the 20th century as highly monochromatic stabilised lasers, high dynamic range cameras and computer controlled equipment. This also includes modern light sources that can be stabilized to standard clocks by using optical frequency comb techniques (e.g. see [11]).

This chapter describes, besides principles, the state of the art of such interferometer systems and general limitations.

The large field of such interferometers allows an assembly as shown in figure 2.1 covered by the large beam diameter within an interferometer as shown in figure 2.2. A point light source, set in the focus of a collimating lens, generates a large bundle of parallel beams which covers the prismatic body together with its end platen. The light, after having been reflected and having returned to the plane of the beam splitter, follows a 'quasi-joint' path to a camera which records the interferogram. In this way, the interference intensities for each partial beam of the bundle are resolved laterally[1]:

$$I = I_0 \left\{ 1 - \gamma \cos\left[ \overbrace{\frac{2\pi}{\lambda/2}(Z_1(x, y) - Z_2(x, y))}^{\varphi(x,y)} \right] \right\} \tag{2.1}$$

[1] Since the reference beam is reflected once more by the optically denser medium (the outer surface of the beam splitter) than the measuring beam, the sign of the interference term is reversed compared to the basic 2-beam interference equation (equation (1.5) of chapter 1).

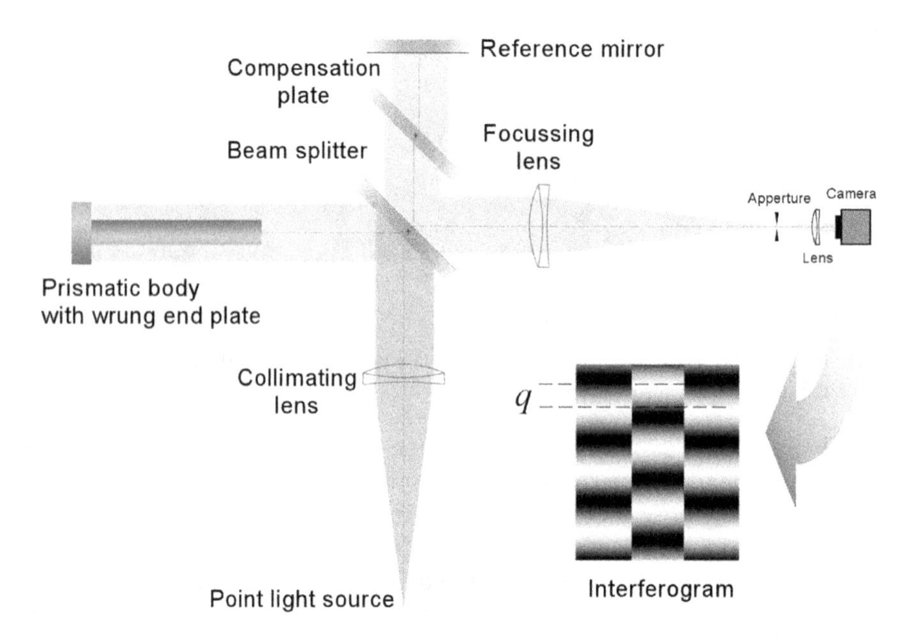


**Figure 2.2.** Interferometer for the realisation of the length of prismatic bodies. The interferogram, shown as the inset, symbolises the situation when the reference mirror is slightly tilted. In such a case the fractional order of interference can be directly read from the fringe mismatch.

where $z_{1/2}(x, y)$ represents the distribution of the geometric paths (measuring: 1, reference: 2) vertically to the optical axis. If the reflecting surfaces were perfectly flat, the path difference $z_1(x, y) - z_2(x, y)$ would describe a plane whose tilt depends on the orientation of the surfaces to each other. The cosine function is then responsible for the typical fringe pattern with maxima for path differences of $m\ \lambda/2$ and minima for $(m + \frac{1}{2})\lambda/2$, where $m$ is an integer. The prismatic body in the measuring arm of the interferometer represented in figure 2.2 causes the appearance of two mismatched fringe systems. The value of this mismatch corresponds to the fractional order of interference $q$ and represents the direct result of the interferometric measurement by large field imaging interferometry. Together with the integer order of interference $i$, which must be determined separately (shown below) the length of a prismatic body can then be obtained according to

$$l = \frac{\lambda_0}{2n}\frac{\varphi}{2\pi} = \frac{\lambda}{2} \cdot (i + q). \tag{2.2}$$

For the setup shown in figure 2.2, at the output of the interferometer, diffraction of light at any aperture will become visible within the cross section of the collimated beam. The effect of diffraction caused by the edges of the prismatic body itself can be managed by using an appropriate lens in front of the camera. Such a lens, together with the focussing lens at the interferometer's output, represents the imaging system of the interferometer which is dimensioned in such a way that the front face of the body is imaged 'sharply'. The sharpness of the interferograms is however affected by the finite size of the aperture positioned in the focal plane of the focussing lens. This aperture serves to suppress disturbing secondary reflections. The smaller the aperture, the less light deflected at the edges can reach the camera and contribute to 'sharpness'. This generally known fundamental principle of optical imaging is illustrated in figure 2.3. For the sake of convenience, the optical system is replaced by a single lens (with an effective focal length $f$). It appears that diffraction orders outside the angular sections, represented in green, are blocked at the aperture.

The effective angular section of the diffracted light finally also influences the measured topography of the interference phase. The larger the aperture, the less the phase topography is distorted around the edges. The smaller the aperture the more diffraction is present near the edges of the body which manifest themselves both as blur and, finally, as phase error.

## 2.2 Topography-based measurement of the interference phase and extraction of the fractional order of interference

At the time when suitable large field interferometers were developed, visual observation of interference fringes, as indicated in figure 2.2, was the only way to go for the extraction of the length of prismatic material measures, e.g. gauge blocks. This situation changed dramatically when CCD cameras became available. Moreover, the application of digital processing techniques [12, 13] to phase shifting interferometry has greatly improved measurement efficiency and accuracy. The fundamental concept of phase shifting interferometry is that one can determine the phase modulo $2\pi$ over a beam diameter by acquiring multiple interferograms, each phase shifted by a certain amount. Substantial effort was made in effective and error-compensating phase extraction algorithms [14–21]. Among these the five-frame algorithm proposed by Tang [21] is even able to obtain an accurate phase at every pixel point over a wide range of equally spaced phase shifts $\alpha$. The theoretical relation

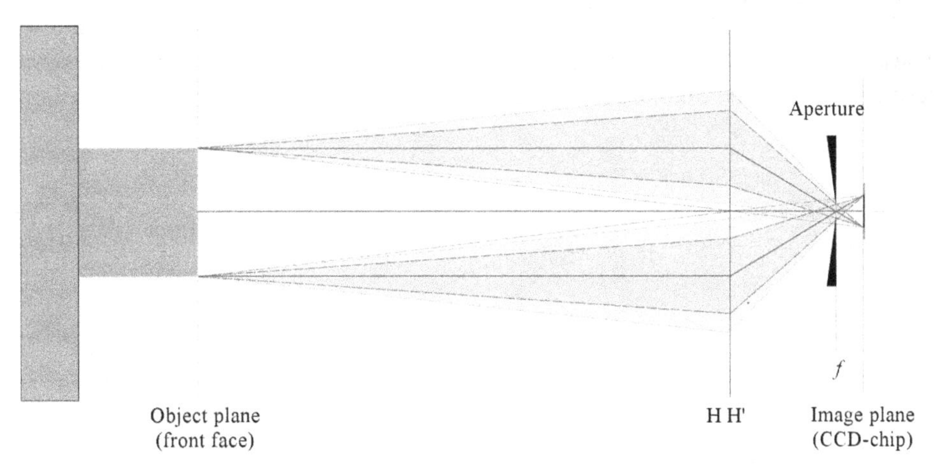


**Figure 2.3.** Schematic diagram of light beams deflected at the edges of a gauge block shaped body attached to an end plate. Only the beams within the green marked region will pass the apperture and reach the image plane of the camera. 

$I_k = I_0\,\{1 - \gamma\cos[\varphi + (k-3)\,\alpha]\}$ results—according to the Tang algorithm—to the following relation for the interference phase:

$$\tan\varphi = \frac{\sqrt{[(I_2 - I_3) + (I_1 - I_4)][3(I_2 - I_3) - (I_1 - I_4)]}}{I_2 + I_3 - I_1 - I_4}. \tag{2.3}$$

This way the interference phase can be calculated from a set of interference intensities as illustrated in figure 2.4.

In large field interferometers, phase-stepping interferometry is realised by a subsequent stepwise equidistant shift of the length of the interferometer's reference arm. Any algorithm, e.g. the Tang algorithm via equation (2.3), can be applied at each existing pixel position within the interferogram's recordimg. An example for a phase topography resulting from Tang's five interferograms is shown in figure 2.5.

From such a phase topography $\varphi(x, y)$ the fractional order of interference, $q$, can be determined much more accurately than from the above mentioned fringe mismatch method, indicated in the inset of figure 2.2.

Before regions of interest can be defined, it is first necessary to identify the body's lateral position within the phase topography. The centre coordinate of the body's front face can be determined highly accurately from the coordinates of the body's edges and pixel-wise observation of the transition *interference → no interference*. Defined areas of *no interference* around the edges can e.g. be made by mounting small brackets, as shown in figure 2.6, left, or by a compact piece that tightly matches the edges of a body (figure 2.6, right).

At pixels related to these 'mask pieces' intensity variation due to interference is absent. Assuming, an expected interference contrast, $\gamma = (I_{\max} - I_{\min})/(I_{\max} + I_{\min})$ as a threshold for a 'mask criterion', a 'mask array' can be obtained. Figure 2.7, left, shows the mask array for the prismatic body shown in figure 2.6, right. Figure 2.7, right, shows a map of pixel coordinates for transitions *interference → no interference* within the mask array for the cylindrical cross section. The coordinates left/right and top/bottom are each averaged resulting in the lines which cross near the centre. These pixel coordinates are again averaged, separately for the two pixel directions, resulting in a single highly accurate sub-pixel coordinate for the central position of the front face.

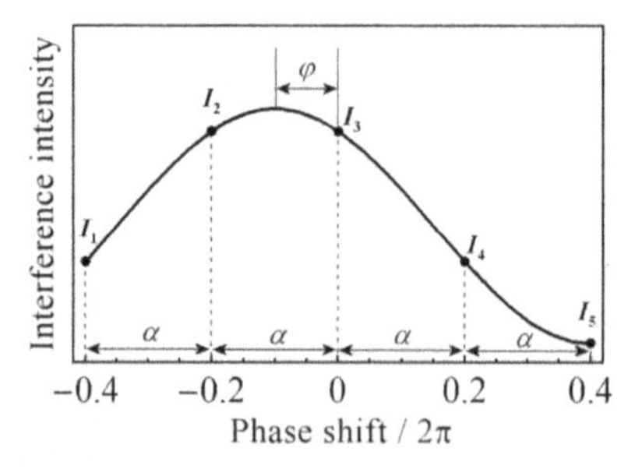


**Figure 2.4.** Illustration for a phase value obtained from a series of intensities using a five-frame phase-stepping algorithm.

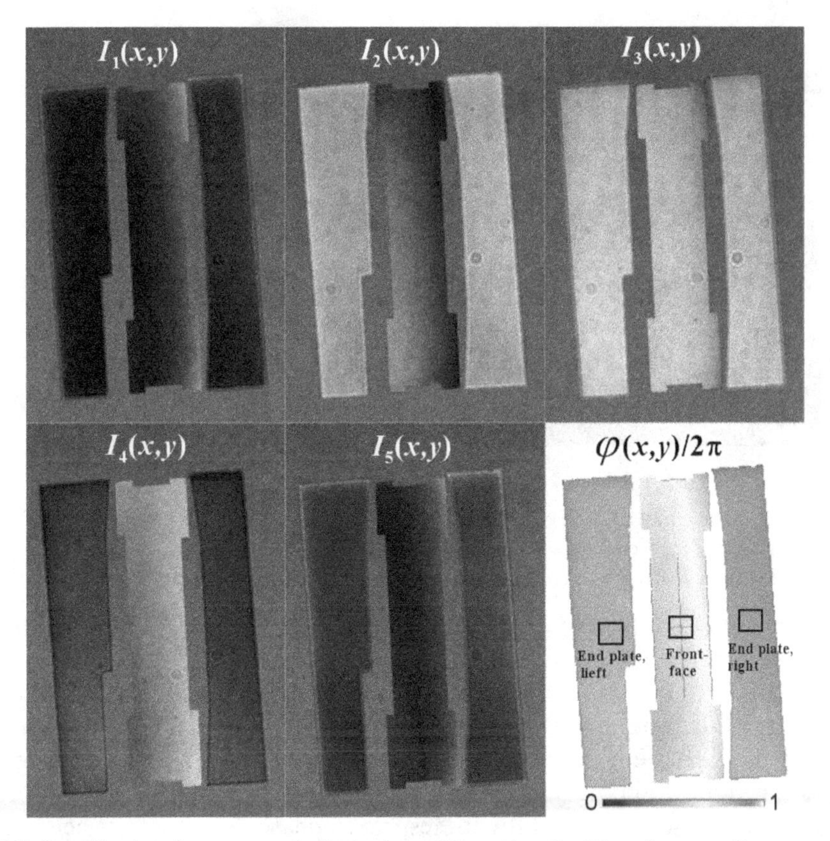


**Figure 2.5.** Set of five interferograms each obtained at a different length of the reference pathway, resulting in a phase topography. The grey pixels around the front face indicate the transitions *interference* → *no interference*, the black squares show ROIs in which unwrapped phase values are averaged.

**Figure 2.6.** Preparation of edges around the front face of prismatic bodies as necessary for highly accurate determination of the centre pixel coordinate of the body's front face. Left: brackets held in contact against the body's side edges by a rubber band. Right: a metal ring matching the edges of a cylindrically body.

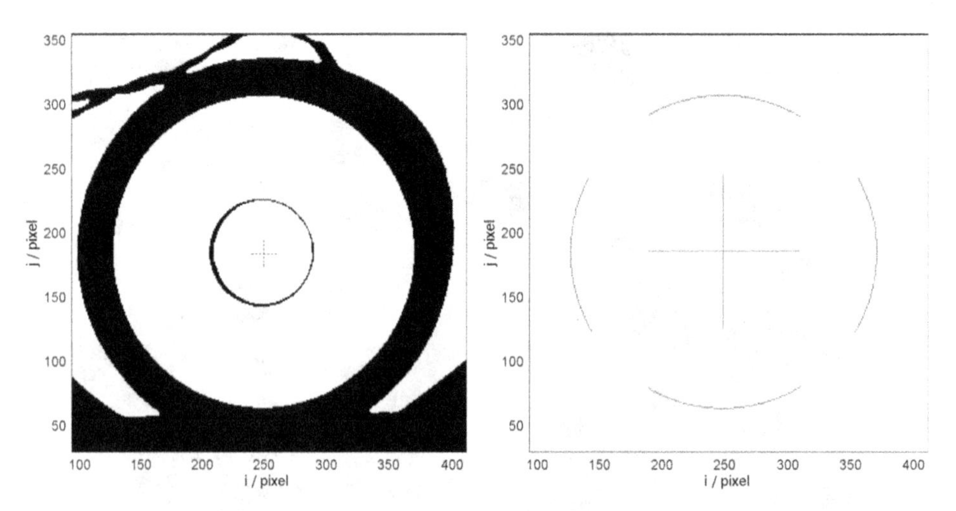


**Figure 2.7.** Left: Mask array for the prismatic body shown in figure 2.6, right, indicating 'no interference' as black pixels. Right: $(i, j)$-coordinates resulting from the mask array, based on the appearance and disappearance of the interference.

Independent of the type of cross section, the central pixel coordinate of the front face is used to define the position for the regions of interest (ROIs) at the front face and the platen, respectively. For a gauge block shaped body, figure 2.5 shows three rectangular ROIs, one at the front face and two symmetrically arranged at the platen. In the next step, $2\pi$ discontinuities existing within the ROIs are removed which is called 'unwrapping'. This procedure can easily be performed by processing the phase values within each ROI with respect to phase differences at neighbouring pixel positions. The unwrapped phase values may, e.g. be obtained from

$$\varphi^{\mathrm{uw}}(x, y) = \varphi(x, y) - \mathrm{Round}\,[\varphi(x, y) - \varphi(x - 1, y)]. \tag{2.4}$$

For a rectangular ROI the averaged unwrapped phase values are then simply obtained from

$$\overline{\varphi} = \frac{1}{(x_{\mathrm{f}} - x_{\mathrm{i}} + 1)\left(y_{\mathrm{f}} - y_{\mathrm{i}} + 1\right)} \sum_{\substack{x=x_{\mathrm{i}}..x_{\mathrm{f}} \\ y=y_{\mathrm{i}}..y_{\mathrm{f}}}} \varphi^{\mathrm{uw}}(x, y), \tag{2.5}$$

in which $(x_{\mathrm{i}}, y_{\mathrm{i}})$ and $(x_{\mathrm{f}}, y_{\mathrm{f}})$ indicate the initial and final coordinates within a ROI. For highest accuracy, the exact sub-pixel coordinate of the front face centre, $(x_{\mathrm{c}}, y_{\mathrm{c}})$, can be incorporated into the calculation by using

$$\overline{\varphi} = \left( \frac{1}{(x_{\mathrm{f}} - x_{\mathrm{i}} + 1)\left(y_{\mathrm{f}} - y_{\mathrm{i}} + 1\right)} \sum_{\substack{x=x_{\mathrm{i}}..x_{\mathrm{f}} \\ y=y_{\mathrm{i}}..y_{\mathrm{f}}}} \varphi^{\mathrm{uw}}(x, y) \right) + \overline{\varphi}^{x} \Delta x_{\mathrm{c}} + \overline{\varphi}^{y} \Delta y_{\mathrm{c}} \tag{2.6}$$

instead of equation (2.5). Here $\overline{\varphi}^{x}$ and $\overline{\varphi}^{y}$ indicate the slopes within a ROI along the respective pixel directions, and the $\Delta x_{\mathrm{c}} = x_{\mathrm{c}} - \mathrm{Round}\,[x_{\mathrm{c}}]$, $\Delta y_{\mathrm{c}} = y_{\mathrm{c}} - \mathrm{Round}\,[y_{\mathrm{c}}]$ take into account the sub-pixel residuals of the centre position of the front face.

For the extraction of the correct fractional order of interference it is further necessary to remove possible $2\pi$ discontinuities between the ROIs located on the platen. In the example shown in figure 2.5, the average phases of ROIs on the platens left-hand side, $\overline{\varphi}_{\text{paten, left}}$, and right-hand side, $\overline{\varphi}_{\text{platen, right}}$ may differ by multiples of $2\pi$, which must be checked and removed. This can be done by 'left-right unwrapping' so that the left value, $\overline{\varphi}_{\text{platen, left}}$, is extrapolated to the right-hand side and compared with the value of $\overline{\varphi}_{\text{platen, right}}$, i.e.:

$$\overline{\varphi}^{\text{uw}}_{\text{platen, right}} = \overline{\varphi}_{\text{platen, right}} - \text{Round}\left(\frac{1}{2\pi}\begin{bmatrix} \overline{\varphi}_{\text{platen, left}} - \overline{\varphi}_{\text{platen, right}} \\ + \overline{\varphi}^{x}_{\text{platen}}(x_{\text{right}} - x_{\text{left}}) \\ + \overline{\varphi}^{y}_{\text{platen}}(y_{\text{right}} - y_{\text{left}}) \end{bmatrix}\right), \tag{2.7}$$

where $x_{\text{right}} - x_{\text{left}}$ and $y_{\text{right}} - y_{\text{left}}$ are the pixel distances between the ROIs on the left-hand and the right-hand sides, $\overline{\varphi}^{x}_{\text{platen}}$ and $\overline{\varphi}^{y}_{\text{platen}}$ are the average phase slopes within the ROIs on the platen, i.e. $\overline{\varphi}^{x,\,y}_{\text{platen}} = \frac{1}{2}\left(\frac{\partial}{\partial x,\, y}\overline{\varphi}_{\text{platen, left}} + \frac{\partial}{\partial x,\, y}\overline{\varphi}_{\text{platen, right}}\right)$. Finally, for the example phase topography shown in figure 2.5, from the averaged phase values, the fractional order of interference is obtained:

$$q = \frac{1}{2\pi}\left[\frac{1}{2}\left(\overline{\varphi}_{\text{platen, left}} + \overline{\varphi}^{\text{uw}}_{\text{platen, right}}\right) - \overline{\varphi}_{\text{front face}}\right]. \tag{2.8}$$

The availability of a phase topography covering the front face of a prismatic body together with the platen can further be exploited to correct potential platen flexing. Figure 2.8 schematically shows a side view for a sample attached to a flexed platen in which mean phase values and detected slopes within the platen's ROIs are indicated.

The slopes $\frac{\partial}{\partial x}\overline{\varphi}$ can be used to calculate extrapolated values for $\overline{\varphi}^{\text{E}}_{\text{platen, left}}$ and $\overline{\varphi}^{\text{E}}_{\text{platen, right}}$, as suggested in [22]:

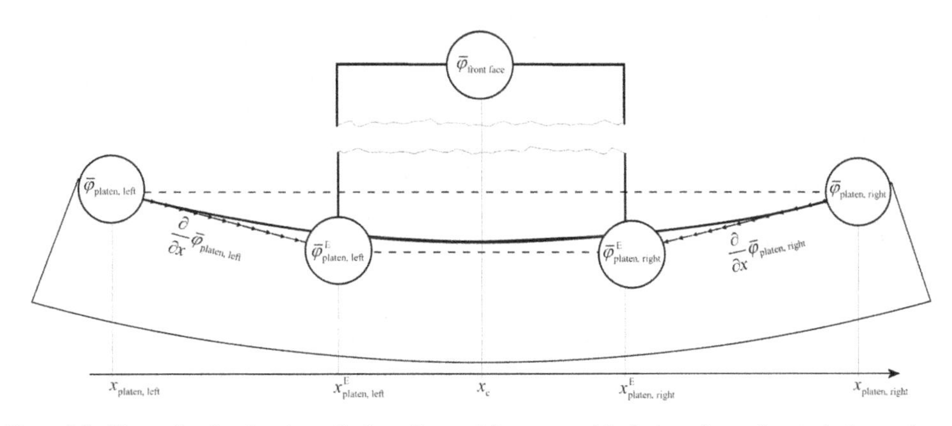


**Figure 2.8.** Illustration for the platen flexing effect and for a gauge block shaped sample attached to a platen.

$$\overline{\varphi}^{E}_{\text{platen, left}} = \overline{\varphi}_{\text{platen, left}} + \frac{\partial}{\partial x}\overline{\varphi}_{\text{platen, left}} \cdot \left(x^{E}_{\text{platen, left}} - x_{\text{platen, left}}\right)$$
$$\overline{\varphi}^{E}_{\text{platen, right}} = \overline{\varphi}_{\text{platen, right}} + \frac{\partial}{\partial x}\overline{\varphi}_{\text{platen, right}} \cdot \left(x^{E}_{\text{platen, right}} - x_{\text{platen, right}}\right). \tag{2.9}$$

Alternatively, a parabolic deformation of the platen can be assumed [23], resulting in the phase correction:

$$\delta\overline{\varphi}_{\text{platen}} = \left(\frac{\partial}{\partial x}\overline{\varphi}_{\text{platen, right}} - \frac{\partial}{\partial x}\overline{\varphi}_{\text{platen, left}}\right)\frac{\left(x_{\text{c}} - x_{\text{platen, left}}\right)}{4}. \tag{2.10}$$

## 2.3 Determination of the integer order of interference

In large field interferometry, the fractional order interference, $q$, is the primary measurement result. Accordingly, due to the relation $l = (i + q)\lambda/2$, for a given value of $q$ there exist many possible lengths, separated by $\lambda/2$. On the other hand, the integer order of interference, $i$, existing between the two faces of a body, cannot be accessed directly. An estimate for the number $i$ can be derived from a pre-value of the length $l_{\text{est}}$, e.g. from a given nominal length:

$$i_{\text{est}} = \text{Round}\left[\frac{l_{\text{est}}}{\frac{1}{2}\lambda}\right]. \tag{2.11}$$

If only a single wavelength is available as a light source, the unambiguous determination of $i$ requires prior knowledge of $l_{\text{est}}$ within $\pm\lambda/4$. This accuracy is far from what can be achieved with conventional methods. Moreover, the length of a body is sensitive to the temperature by the thermal expansion property of the material, e.g. for 1 m steel a temperature difference of 10 mK corresponds to a length difference of approximately 130 nm which is comparatively large as $\lambda/4$. Often the thermal expansion coefficient of the body is not even accurately known, i.e. even small deviations from the target temperature (usually 20 °C), to which $l_{\text{est}}$ refers, lead to changes in the length which cannot be accounted for with sufficient accuracy. Consequently, the risk of 'miscounting' the number $i$ is relatively high when only a single wavelength is available.

If at least two wavelengths are available, each related to respective measurement results for $q$, the corresponding integer orders of interference can be determined by the so-called method of exact fractions which goes back to Benoît [24]. The basic idea of this method consists of the comparison of lengths, obtained at the different wavelengths according to equation (2.2), i.e. $l_k = (i_k + q_k)\lambda_k/2$, under variation of the respective integer order of interference and using the coincidence between these lengths as a criterion (see also [25–30]). Generalised analytical expressions have been developed to determine the minimum achievable unambiguous measurement range, which can be applied to arbitrary measurement wavelengths [31] and optimum wavelength selection strategies were presented [32]. However, in the context of primary realization of the metre, the choice of possible wavelengths is restricted, i.e. so many suitable light sources do not exist (see [33]). Assuming $N$ different wavelengths $\{\lambda_1, .., \lambda_N\}$, the general expectation is that coinciding lengths are obtained for each wavelength. For this purpose, the integer interference orders $\{i_1, .., i_N\}$ are each split into $i_k = i_k^{\mathrm{est}} + \delta_k$, i.e. the lengths

$$l_k = \left(i_k^{\mathrm{est}} + \delta_k + q_k\right)\frac{\lambda_k}{2} \tag{2.12}$$

are considered in which $i_k^{\mathrm{est}}$ represents the estimate integer orders according to equation (2.11) and $\delta_k$ variation integers. For the set of $N$ different wavelengths the mean length, $\bar{l}$, and the average deviation from the mean length, $\Delta$, are then calculated according to:

$$\bar{l} = \frac{1}{N}\sum_{k=1}^{N} l_k, \quad \Delta = \frac{1}{N}\sum_{k=1}^{N} |\bar{l} - l_k|. \tag{2.13}$$

The resulting set $\{\bar{l}, \ \Delta\}$ can be displayed as a scatter-plot which represents a so-called coincidence pattern.

In the case of two wavelengths, the value of $\Delta$ is exactly given by $\frac{1}{2}|l_1 - l_2|$. Assuming that in this case $i_1^{\mathrm{est}}$ and $i_2^{\mathrm{est}}$ correspond to the correct integer orders ($\delta_1 = 0, \delta_2 = 0$), the two lengths, written as $l_1 = (i_1 + q_1)\lambda_1/2$ and $l_2 = (i_2 + q_2)\lambda_2/2$, coincide ($\Delta = 0$) when the fractions $q_1$ and $q_2$ are given 'exactly'. A relation for further variation integers ($\delta_1 \neq 0, \delta_2 \neq 0$) for which coincidence is obtained can be extracted by setting $\Delta = \frac{1}{2}|\tilde{l}_1 - \tilde{l}_2| = \frac{1}{2}|(i_1 + \delta_1 + q_1)\lambda_1/2 - (i_2 + \delta_2 + q_2)\lambda_2/2| = 0$, in which $\tilde{l}_1$ and $\tilde{l}_2$ correspond to 'blunder lengths'. This situation is expressed with equation (2.14):

$$\Delta = 0 = |\lambda_1\,\delta_1 - \lambda_2\,\delta_2| \quad \Rightarrow \quad \Delta = 0 \text{ for } \frac{\delta_2}{\delta_1} = \pm\frac{\lambda_1}{\lambda_2}. \tag{2.14}$$

Therefore, integer interference orders can be wrong when their ratio is equal to the ratio of the wavelengths used in the measurements, i.e. a rational number. In other words, when the first wavelength is given by an integer multiple of the second wavelength, i.e. $\lambda_1 = a\,\lambda_2$ ($a$ is integer), coincidence ($\Delta = 0$) is obtained for each set of $\{\delta_1, \delta_2\}$ for which $\delta_2 = a\,\delta_1$.

Figure 2.9 shows a fictive coincidence pattern for the case of two wavelengths {532.3 nm, 548.6 nm}, ($\lambda_1 \neq a\,\lambda_2$, $a$ is an integer) in which the estimate length is set to 10 mm, the deviation of the length $l$ from the estimate length, $\delta l = l - l_{\mathrm{est}}$, is set to 2.5 μm. For the best illustration of this approach, the fractional interference orders $q_1$ and $q_2$ are fictively set to exact values $q_k = l/\frac{1}{2}\lambda_k - i_k^{\mathrm{est}}$. In figure 2.9, only data points of $\{\bar{l}, \ \Delta\}$ are displayed for which $\Delta$ is smaller than 20 nm.

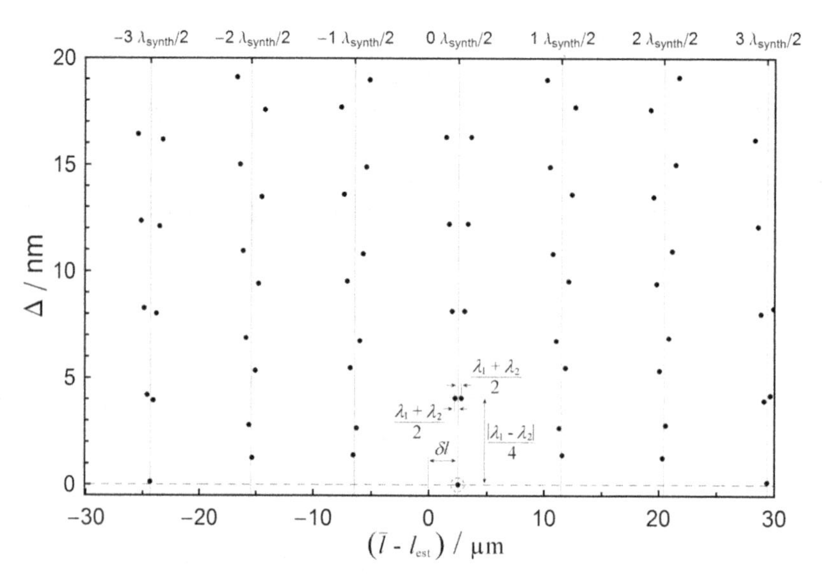


**Figure 2.9.** Fictive coincidence pattern when using two wavelengths. 

With the example coincidence pattern shown in figure 2.9 the following conclusions can be drawn:

(a) At the length deviation $\bar{l} - l_{est} = \delta l$ there exists a minimum of $\Delta$ that is exactly zero (marked by the dotted circle).

(b) Further minima exist which are apart by half of the so-called synthetic wavelength, $\lambda_{synth} = \lambda_1\lambda_2/|\lambda_2 - \lambda_1|$, i.e. at positions $\bar{l} - l_{est} = \delta l + m.\ \lambda_{synth}/2$ in which $m$ is an integer.

(c) The next neighboured minima for which $\Delta$ is zero can be found at specific multiples of $\lambda_{synth}/2$. In this example the multiples are localised at $m = \pm 3$. It is noted that this number is strongly dependent on the amounts of wavelengths $\lambda_1$ and $\lambda_2$ (see equation (2.14)).

(d) The lengths which are closest to the correct minimum $\bar{l} - l_{est} = \delta l$ $(\Delta = 0)$ are separated by a single interference order, i.e. at $\bar{l} - l_{est} = \delta l \pm (\lambda_1 + \lambda_2)/2$. The corresponding values of $\Delta$ amount to $|\lambda_1 - \lambda_2|/4$.

Point (d) reveals a very important requirement for the applicability of the method of exact fractions: the two wavelengths must be clearly separate. Otherwise the value of $\Delta$ for the neighbouring lengths becomes too small. When, e.g. $\lambda_1$ and $\lambda_2$ are just 2 nm apart, for $\{\delta_1 \pm 1, \delta_2 \pm 1\}$ the values of $\Delta$ would result in 0.5 nm which is very close to zero ($\Delta = 0$ is obtained for the correct variation numbers $\{\delta_1, \delta_2\}$). It is important to note that in such a case, despite the small values of $\Delta$, the corresponding lengths are apart by relatively large amounts, namely $\pm(\lambda_1 + \lambda_2)/2$.

Point (d) also illustrates that this variation method requires 'exact fractions'. In the above fictive example this was ensured by setting $i_k^{est} = \text{Round}\,[l/\frac{1}{2}\lambda_k]$ and $q_k = l/\frac{1}{2}\lambda_k - i_k^{est}$. In actual length measurements by interferometry the fractional orders interference $q_k$ are subject to a measurement uncertainty. Only when all $q_k$ can be measured 'exactly', i.e. when $\frac{1}{2}\lambda_k\, u(q_k)$ is considerably smaller than $|\lambda_1 - \lambda_2|/4$, 'blunder lengths' correlated with $\{\delta_1 \pm 1, \delta_2 \pm 1\}$ can be discriminated from the 'true lengths'.

In the case of three wavelengths, the value of $\Delta$ according to equation (2.13) is approximately given by:

$$\Delta = \frac{1}{3}\sum_{k=1}^{3}|\bar{l} - l_k| \approx \frac{1}{3}(|l_1 - l_2| + |l_2 - l_3|) \tag{2.15}$$

The probability distribution of $\Delta$ is a folded normal distribution [34] which is represented by a half-normal distribution in the case of coinciding lengths values (e.g. $l_1 \simeq l_2 \simeq l_3$, $\Delta = \Delta_{\min}$). Therefore, the expected value, $E(\Delta_{\min})$, can be derived from the input uncertainties of the involved fractional interference orders. While when using two wavelengths $E(\Delta_{\min})$ is approximately given by $E(\Delta_{\min}) \simeq \frac{1}{2}\left(\frac{1}{2}\lambda_1 u(q_1) + \frac{1}{2}\lambda_2 u(q_2)\right)$, it is obtained from

$$E(\Delta_{\min}) \simeq \frac{1}{3}\left(\frac{1}{2}\lambda_1 u(q_1) + \frac{1}{2}\lambda_2 u(q_2) + \frac{1}{2}\lambda_3 u(q_3)\right). \tag{2.16}$$

when using three wavelengths. It is noted that $E(\Delta_{\min})$ is somewhat overestimated in equation (2.16) since the actual expected value for the half-normal distribution is smaller by the factor $\sqrt{2/\pi} \approx 0.8$, which is ignored here for simplicity.

In analogy to the above conclusion (d), derived for a two wavelengths coincidence pattern, values of $\Delta$ correlated with integer orders apart by $\{\delta_1 \pm 1, \delta_2 \pm 1, \delta_3 \pm 1\}$ from the correct values result in:

$$\Delta \approx \frac{1}{3}(|\lambda_1 - \lambda_2| + |\lambda_2 - \lambda_3|) \tag{2.17}$$

in the case of using three wavelengths. When these $\Delta$, which are correlated to wrong lengths (blunders), become as small as $E(\Delta_{\min})$ according to in equation (2.16), there does not exist a suitable coincidence criterion, i.e. the method of exact fractions cannot distinguish blunders from the correct length. Therefore, when unambiguous length measurements by interferometry are desired, at least one of the wavelengths must be clearly apart from the others. Ideally, the wavelengths of all light sources are well separated from each other as is e.g. the case in [35].

An additional advantage of the method of exact fractions is that, independent lengths are obtained for each wavelength which should coincide. Therefore, 'coincidence' is also a check for the possibility of the whole length measurement and can help detecting possible errors in the determination of $q_k$ such as an incorrectly set wavelength $\lambda$ of the light source.

It is noted that the above described determination of the integer order of interference can fail in the case of chromatic effects to the measurement of $q_k$, e.g. wavelength dependent phase change on reflection (e.g. see section 2.6) but also optical effects (e.g. from the imaging system as shown in [36]). In the case when the air refractive index is determined interferometrically (e.g. [37]) additional chromatic effects may occur. In all such cases the coincidence criterion can lead to a completely wrong integer order of interference, i.e. a straightforward application of the method of exact fractions can even lead to lengths which are far from the actual length. Therefore, a critical view and understanding of a specific measurement is necessary to avoid blunder lengths. Apart from these restrictions and limitations, the method of exact fractions cannot be replaced.

## 2.4 Measurement of the air refractive index

Measurement of the length by means of interferometry, e.g. for primary calibrations of gauge blocks, is nearly always done in air, since material measures are also used under atmospheric conditions when they serve as secondary standards for the length. The presence of air considerably reduces the wavelength $\lambda$ of the light compared to the wavelength $\lambda_0$ in vacuum:

$$\lambda = \frac{\lambda_0}{n(\nu, p, t, f, x)}. \tag{2.18}$$

The refractive index of air, $n$, depends both on the light frequency $\nu$ and on the ambient conditions (the air pressure $p$, the air temperature $t$, the air humidity $f$, and the $CO_2$ concentration $x$). Among a variety of influences onto a length measurement, the air refractive index is mostly limiting the attainable accuracy. Consequently, the exact quantification of $n$ is of major importance.

Since $n$ is close to 1, the relative influence of the refractive index of air on the length measurement is scaled by the so-called 'refractivity', $n - 1$ (approximation: $1/n|_{n\approx 1} \simeq 1 - (n - 1)$), which is of the order of $3 \times 10^{-4}$ under atmospheric conditions. Thus, dependent on the frequency of the light, the effect of the air is approximately 0.3 mm per metre. Many sources can be found in the literature (e.g. [38–43] and references therein) in which the air refractive index and its dependence on the parameters pressure, temperature, $CO_2$-content, was investigated and formulas have been established. Without these key experiments and the respective considerations accurate realisation of the length by interferometry would not work today.

The Lorentz–Lorenz equation provides the physical relationship between the refractive index of a gas mixture, e.g. air, and the partial densities $\rho_i$ of its constituents:

$$\frac{n^2 - 1}{n^2 + 2} = \sum_i R_i \rho_i, \tag{2.19}$$

where $R_i$ is the specific refraction of a constituent which is related to $R_i = \frac{4}{3}\pi N_A \alpha_i / M_i$, $N_A$ is the Avogadro constant, $M_i$ is the molecular weight and $\alpha_i$ the dynamic polarizability. For atmospheric air, the dispersion curves for $N_2$, $O_2$, and Ar are sufficiently similar and it can therefore be presumed that all the constituents, including $CO_2$, can be combined in a single term on the right-hand side of equation (2.16) (see [41]). Furthermore, at low densities, the functionality of the Lorentz–Lorenz term on the left-hand side of equation (2.19) is approximately given[2] by $\frac{2}{3}(n-1)$. Therefore, the refractivity of dry air is nearly linearly related to the air density $\rho_{\text{air}}$. For an ideal gas the density would be almost proportional to the pressure $p$ divided by the Kelvin temperature $T$. For dry air, at the conditions 20 °C, 100 000 Pa, the relationship between $n-1$, $p$ and $T$ becomes visible when rewriting Bönsch's empirical equation (see [43]) in the form:

$$(n^{Bönsch}-1) \propto \frac{p/\text{Pa}}{273.15+t/°\text{C}} \cdot [1+0.5953\times10^{-8} \times(1-0.01\ 659.\ t/°\text{C})\times p/\text{Pa}], \tag{2.20}$$

in which $t$ is the Celsius temperature.

$n(\nu, p, t, f, x)$ can be determined from the set of highly accurately measured air parameters and applying an empirical formula. It should be noted that the existing formulae are, in a strict sense, only valid for the conditions these are derived from, e.g. nearly standard conditions defined in [43] (air pressure of 100 000 Pa, temperature of 20 °C, $CO_2$-content 0.04%). Moreover, the applicable range of wavelengths is restricted since the original formulae are based on the usage of cadmium lamps, i.e. a restricted range of the visible spectrum.

Therefore, it is advantageous, to measure the actual refractive index of air, which is effective in a length measurement, as described in the following. Figure 2.10 shows an evacuated cell, which is tightly closed at both sides by means of flat cell windows, located inside the measuring arm of an imaging Twyman–Green interferometer. The collimated bundle of light beams traverses the evacuated inner part of the cell, and the windows outside the cell, along the same geometric path. A series of interferograms is recorded at slightly different equidistant tilt angles of the compensation plate resulting in a phase topography by using an appropriate phase-stepping algorithm. Within the phase topography, shown in figure 2.10, circular regions of interest (ROIs) are defined at pixel positions of the vacuum pathway (labelled 'cell') and the air pathway (labelled 'out1' and 'out2').

Within the ROIs, unwrapped phase values are averaged resulting in $\overline{\varphi}_{\text{air}}^{\text{out1}}$, $\overline{\varphi}_{\text{air}}^{\text{out2}}$ and $\overline{\varphi}_{\text{vac}}^{\text{cell}}$ as indicated in figure 2.10, in which the suffix 'vac' indicates the vacuum path and the suffix 'air' indicates an air path. For a specific vacuum wavelength, the fractional interference order is then obtained from

$$q = \frac{1}{2\pi}\left[\frac{1}{2}\left(\overline{\varphi}_{\text{air}}^{\text{out1}} + \overline{\varphi}_{\text{air}}^{\text{out2}}\right) - \overline{\varphi}_{\text{vac}}^{\text{cell}}\right]. \tag{2.21}$$

Like the situation for the fractional interference order of a prismatic body (see equation (2.8) and text above), it is important that the ROIs for the air pathways ('out1' and 'out2'), are arranged symmetrically around the ROI within the vacuum path of the cell. In this way it is achieved that the average $\frac{1}{2}(\overline{\varphi}_{\text{air}}^{\text{out1}} + \overline{\varphi}_{\text{air}}^{\text{out2}})$, which is

[2] For $n = 1.0003$, the approximation is as good as $(n^2-1)/(n^2+2) - \frac{2}{3}(n-1) \approx 10^{-8}$. An even better approximation is given by $(n^2-1)/(n^2+2) \approx \frac{2}{3}(n-1)(1-\frac{1}{6}(n-1))$ for which the respective difference is $\approx 4\times10^{-12}$.

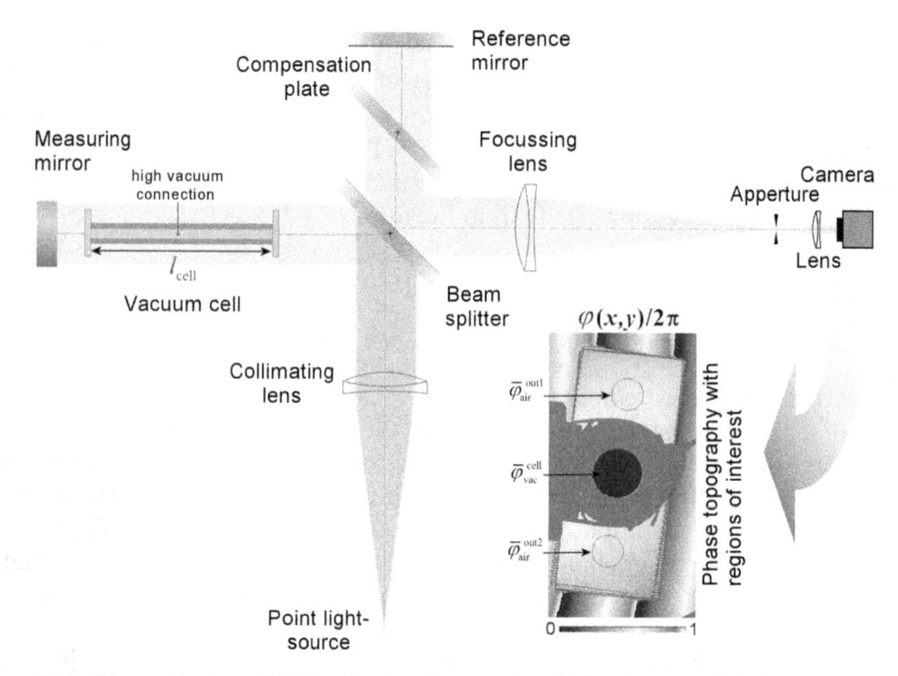


**Figure 2.10.** Scheme of a large field imaging interferometer in which a vacuum cell is installed. From the recorded interferograms a phase topography is obtained by a phase-stepping procedure (figure taken from [44]). © IOP Publishing Ltd. All rights reserved.

related to $\overline{\varphi}_{vac}^{cell}$, is insensitive to tilts in the phase topography caused by the individual adjustment state of the interferometer. For high accuracy, the position of the central ROI within the vacuum pathway should be set near the centre of the vacuum cell which can be identified within the phase topography. The length of the vacuum cell, $l_{cell}$ as indicated in figure 2.10, can be written, in analogy to equation (2.2), in two different ways:

$$l_{cell} = (i_{vac} + q_{vac})\frac{\lambda_0}{2} = (i_{air} + q_{air})\frac{\lambda_0}{2n}, \tag{2.22}$$

From equation (2.22) the air refractivity can be written as:

$$n - 1 = \frac{1}{l_{cell}}\left(\overbrace{i_{air} - i_{vac}}^{i} + \overbrace{q_{air} - q_{vac}}^{q}\right) \cdot \frac{\lambda_0}{2}. \tag{2.23}$$

Thus, measurement of $q$ according to equation (2.21) 'ideally' allows the determination of $n - 1$:

$$(n - 1)^{ideally} = \frac{(i + q)\lambda_0/2}{l_{cell}}. \tag{2.24}$$

In the real world, the cell windows exhibit refractive index inhomogeneities. Further, imperfections of the interferometer optics affect the measurement of the air refractive index. The mentioned effects can be quantified by a measurement of the residual optical path differences assessable when also the pathways 'out1' and 'out2', i.e. when the entire interferometer can be evacuated. From such a measurement, in which the two outer vacuum pathways are compared against the inner vacuum pathway across the cell, an offset in the fractional interference order can be derived:

$$q^{\text{offset}} = \frac{1}{2\pi}\left[\frac{1}{2}\left(\overline{\varphi}_{\text{vac}}^{\text{out1}} + \overline{\varphi}_{\text{vac}}^{\text{out2}}\right) - \overline{\varphi}_{\text{vac}}^{\text{cell}}\right]. \tag{2.25}$$

Since the phase topography includes all imperfections of the optics, the averaged values $\overline{\varphi}_{\text{vac}}^{\text{out1}}$, $\overline{\varphi}_{\text{vac}}^{\text{out2}}$, $\overline{\varphi}_{\text{vac}}^{\text{out2}}$ and therefore $q^{\text{offset}}$ are sensitive to the individual pixel positions within the phase topography. Figure 2.11, left, shows an example phase topography for this offset measurement taken from [44]. At a wavelength of $\lambda$ = 532 nm figure 2.11, right, shows the length equivalent value for the resulting offset, $l^{\text{offset}} = q^{\text{offset}}.\ \lambda_0/2 = -16.73$ nm, and illustrates its sensitivity to variations in the pixel positions of the regions of interest. Accordingly, in this example the variation of $l^{\text{offset}}$ with the pixel positions is well within 0.1 nm when the pixel positions of the ROIs are shifted not more than ± 3 pixel. The complete equation for the air refractivity measurement, including the offset values according to equation (2.25), is:

$$n - 1 = \frac{(i + q - q^{\text{offset}})\lambda_0/2}{l_{\text{cell}}}. \tag{2.26}$$

Additional effects of the cell windows exist which are not just included in $q^{\text{offset}}$. When optical components are subject to pressure variations, anomalous path differences may be induced where the dimensional change in the component is not compensated for by a corresponding change in the refractive index of the material. Such effects, discussed in more detail in e.g. [45], must be taken into consideration for the specific geometry and material of a vacuum cell.

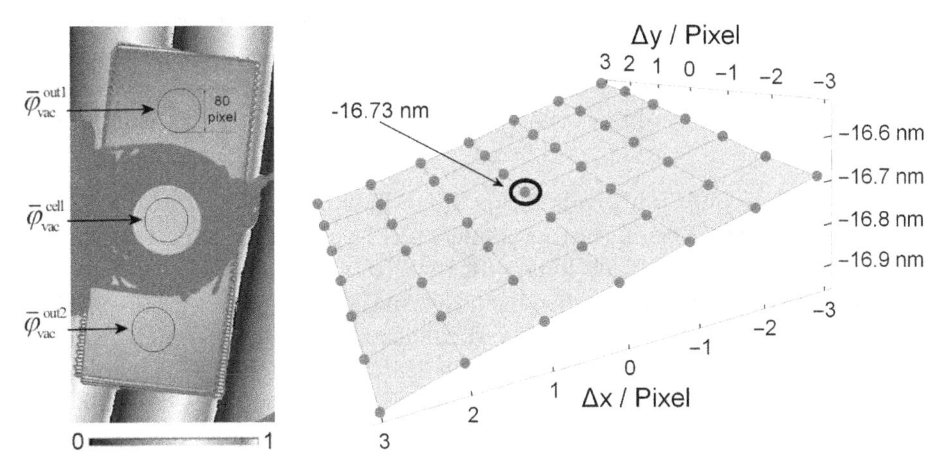


**Figure 2.11.** Left: example phase topography for an offset measurement at $\lambda$ = 532 nm and indication of the circular regions of interest in which the phase values are averaged right: length equivalent value for the resulting offset and its sensitivity to the pixel position of the regions of interest (figure taken from [44]). 

It remains to be clarified in which way the integer order of interference, $i$, in equation (2.26) can be obtained. First, in analogy with a length measurement of a prismatic body, an estimate integer order, $i^{\text{est}}$, can be obtained from an estimate refractive index $n^{\text{est}}$ when setting $q = 0$ in equation (2.26): $i^{\text{est}} = (n^{\text{est}} - 1)\, l_{\text{cell}}/\frac{1}{2}\lambda_0$. Here, $n^{\text{est}}$ is obtained from the air parameters using an empirical formula as mentioned above. When light of multiple sources is available, the air refractivity can thus be written as

$$n_k - 1 = \frac{\left(i_k^{\text{est}} + \delta_k + q_k - q_k^{\text{offset}}\right)\lambda_k/2}{l_{\text{cell}}}, \tag{2.27}$$

in which $\delta_k$ are variation integers and $i_k^{\text{est}} = (n_k^{\text{est}} - 1)\, l_{\text{cell}}\big/\frac{1}{2}\lambda_k$.

Under ideal conditions, $n_k$ determined from equation (2.27) are identical to those obtained from the air parameters, i.e. to $n_k^{\text{est}}$. This can almost be the case when an interferometer is operated at room temperature, a temperature stabilized housing is kept closed for a very long time and the absolute air pressure is measured highly accurately. Then, assuming three available wavelengths as an example, the variation integers, $\delta_k$ in equation (2.27), would result in zero when using the following simple coincidence criterion to be minimum for the correct integer interference order of the refractive index:

$$\Delta^{\text{refrac\_1}} = \tfrac{1}{3}\left(|n_1 - n_1^{\text{est}}| + |n_2 - n_2^{\text{est}}| + |n_3 - n_3^{\text{est}}|\right). \tag{2.28}$$

Using $\Delta^{\text{refrac\_1}}$ as a coincidence criterion for the determination of $n_k$ can however lead to non-physical results for the air refractive index when e.g. temperature gradients exist or the measurement of an air parameter is incorrect. Non-physical refractive indices can be made visible when considering the ratios $n_1/n_2$ and/or $n_2/n_3$ and checking if these ratios are discrepant from the expected ones, i.e. from the dispersion ratios $n_1^{\text{est}}/n_2^{\text{est}}$ and/or $n_2^{\text{est}}/n_3^{\text{est}}$. A criterion, based on a so-called 'overall dispersion value'

$$\text{odv} = \frac{1}{3}\left(\left| n_2 - n_1\frac{n_2^{\text{est}}}{n_1^{\text{est}}}\right| + \left| n_3 - n_2\frac{n_3^{\text{est}}}{n_2^{\text{est}}}\right|\right) \tag{2.29}$$

was suggested in [30]. This value is a quantitative measure for the expected dispersion relations. For illustration, figure 2.12, left, shows an example with possible values $n_1$, $n_2$, $n_3$ (black lines with steps, estimates $n_1^{\text{est}}$, $n_2^{\text{est}}$, $n_2^{\text{est}}$ shown as grey continuous lines) resulting when the traditional coincidence criterion $\Delta^{\text{refrac\_1}}$ via equation (2.28) is used. While the blue vertical line indicates the values for a fictively measured pressure, the red vertical line indicates a situation of a non-physical dispersion relation which becomes apparent by a large 'overall dispersion value' (figure 2.12, right). This demonstrates that non-physical results for the refractive indices can occur when one of the parameters determining the estimates $n_k^{\text{est}}$ is subject to an error. Consequently, even a small pressure error of approximately 30 Pa can falsify the resulting refractive indices, finally leading to a completely wrong result for the length measurement.

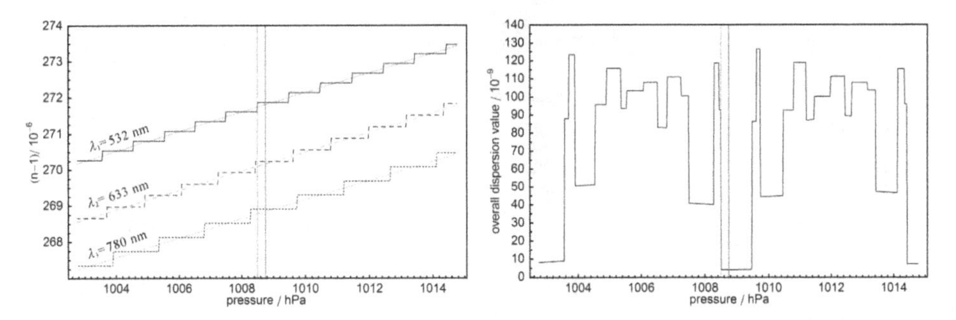


**Figure 2.12.** Example data for the usage of $\Delta^{\text{refrac\_1}}$ as coincidence criterion via equation (2.28). Left: estimates for the air refractive indices for the wavelength 532 nm, 633 nm and 780 nm (grey continuous lines) and resulting values $n_1$, $n_2$, $n_3$ obtained from equation (2.27) (black lines with steps). Right: the respective 'overall dispersion value' according to equation (2.29). The data are shown in a range of air pressures, within which the blue vertical lines indicate fictive actual air pressure and the red vertical lines indicate the closest neighbouring pressure that leads to a non-physical dispersion relation (figure taken from [30]). 

Non-physical results can be avoided by using the 'overall dispersion value' to be minimum as a criterion, while the integers, $\tilde{\delta}_k$, are varied in equation (2.27):

$$\Delta^{\text{refrac\_1}} = \text{odv}. \tag{2.30}$$

Figure 2.13, left, shows the values $n_1$, $n_2$, $n_3$ (black lines with steps, estimates shown as grey continuous lines) resulting when $\Delta^{\text{refrac\_1}}$ is used as coincidence criterion. Again, the blue vertical lines indicate the values obtained for a fictively measured pressure. In contrast to the situation in figure 2.12, the resulting refractive indices are insensitive to large changes of its estimates, e.g. a falsified air pressure does not affect the resulting air refractive indices in a fairly large range (almost ±5 hPa in this example). Accordingly, the 'overall dispersion value', shown in figure 2.13, right, is generally smaller for deviating pressures, indicating the high physical relevance of the results, even when these are based on 'wrong' estimates.

## 2.5 Interferometer adjustment and limitations due to optical components and the light source

### 2.5.1 Alignment basics for large field interferometers

In length measurements by interferometry it must be ensured that the measurement beam travels parallel with the length axes assumed. Compared to distance measuring interferometers, generally equipped with corner cube retroreflectors instead of measurement mirrors, alignment is more complex for large field interferometers.

A first prerequisite for achieving a collimated beam of light, ideally representing even wavefronts, is that the point light source (see e.g. figure 2.10), represented by the fibre end at the entrance of the interferometer, is exactly positioned at the focal point of the collimating lens. This can be ensured when the collimating lens is

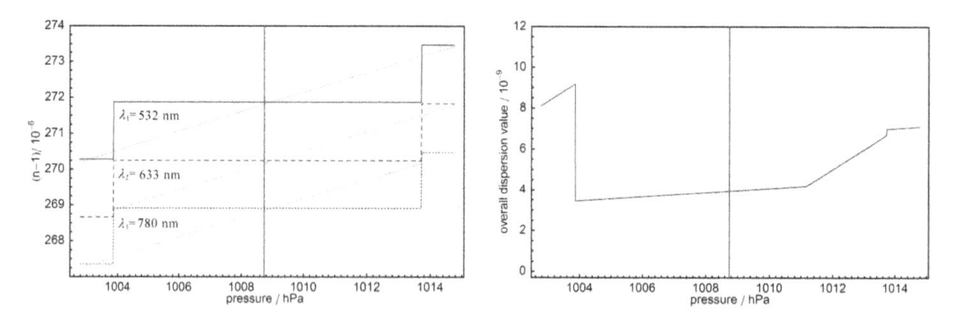


**Figure 2.13.** Example data for the usage of $\Delta^{\text{refrac\_2}}$ as coincidence criterion (equation (2.30)). Left: estimates for the air refractive indices for 532 nm, 633 nm and 780 nm (grey continuous lines) and resulting values $n_1$, $n_2$, $n_3$ obtained from equation (2.27) (black lines with steps). Right: the respective 'overall dispersion value' according to equation (2.29). The data are shown in a range of air pressure, within which the blue vertical lines indicate a fictively correct air pressure (figure taken from [30]). 

mounted highly symmetrically with respect to its optical axes. Then a flat mirror is positioned at the lens mount so that the collimated beam is retro-reflected resulting in a return spot near the fibre end. The position of the fibre end is then adjusted until the return spot is coincident with the source.

The beam splitter of a large field interferometer typically consists of a wedged glass plate with one face coated with a semi-transparent layer (mostly metallic). The faces of this key optical component, particularly the semi-reflective one, must be manufactured as flat as possible. Otherwise, the reflected wavefronts of the measuring and the reference beam will be affected in opposite directions, giving rise to significant error in the length measurement. Figure 2.14 shows the flatness topography of an almost perfect beam splitter plate, 130 mm × 20 mm in size, measured before being installed into the interferometer described in [35].

A wedge angle between the two faces of the beam splitter plate ensures that the beam reflected at the second face travels in a different direction compared to the beam reflected on the semi-reflective face. The respective secondary light spot, which normally would give rise to parasitic interference can therefore be blocked by the aperture in the focal plane of the focusing lens at the output of the interferometer (see figure 2.10). On the other hand, since glass exhibits dispersion, the wedge angle of the beam splitter plate causes a wavelength dependent adjustment state of the interferometer, i.e. the number of fringes visible in the interference pattern can be strongly dependent upon the wavelength of the light source used. This effect can be compensated by inserting a wedged compensation plate into the reference pathway[3],

[3] In the usual case that the wedged compensation plate is installed within the reference pathway of the interferometer, the semi-reflecting surface of the wedged beam splitter must be directed to the reference mirror, as shown in figure 2.10. Otherwise, although the dispersion effect could also be compensated for, but for the opposite orientation of the wedge angles, both plates together would constitute a parallel plate, i.e. the separation of the unwanted secondary reflection from the beam splitter is negated causing parasitic interference.

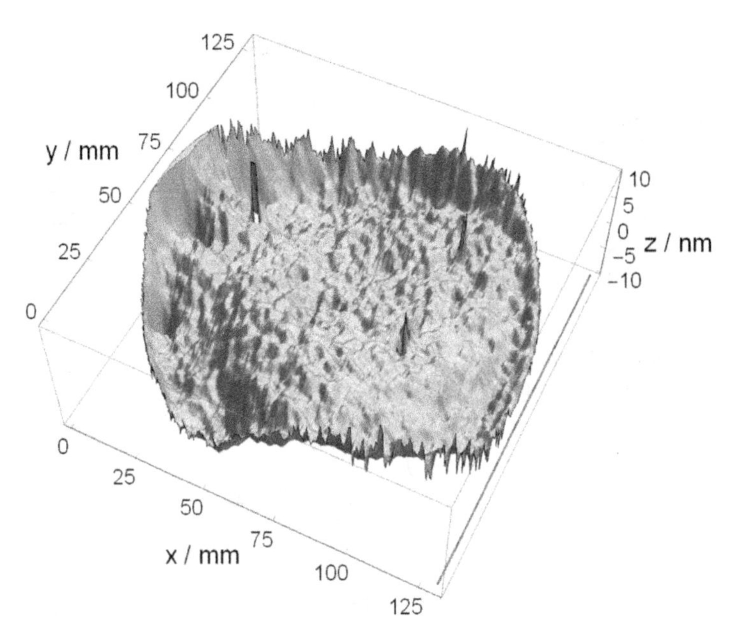


**Figure 2.14.** Flatness topography of the beam splitter plate, measured before being installed into the interferometer described in [35]. 

the wedge angle, its orientation, and the material of which are identical with those of the beam splitter.

### 2.5.2 Autocollimation adjustment of the light beam

A misalignment angle $\alpha$ between the beam direction and the length axes results in a length $\tilde{l}$ which is measured too short by the factor cos $\alpha$:

$$l = \tilde{l} \,.\, \cos\alpha \overset{\alpha << 1}{\cong} \tilde{l} \,.\left(1 - \tfrac{1}{2}\alpha^2\right). \tag{2.31}$$

Accordingly, a lateral misalignment error causes a so-called 'cosine error' $e_\alpha = (l - \tilde{l})/l \cong -\frac{1}{2}\alpha^2$, which therefore must be quantified for an existing adjustment state. In this approach the small angle $\alpha$ can be expressed in terms of the distance of a partial beam from the optical axis in the focal plane of the collimating lens: $\alpha \cong r/f$. Considering a physical light source of certain size, e.g. a multimode fibre end, positioned in the focal plane of the collimating lens, as shown in figure 2.15, a cosine error can be attributed to each partial beam[4].

Considering all the partial beams from the light source, a mean cosine error is caused. For a circular homogeneous light source, e.g. a multimode fibre end, $D$ in diameter, centred at the optical axis, the mean cosine error $e_D$ is derived from:

[4] Since the fibre is in the focal plane, all the beams that pass the collimating lens will have the same directions as the central beam that is shown in figure 2.15.

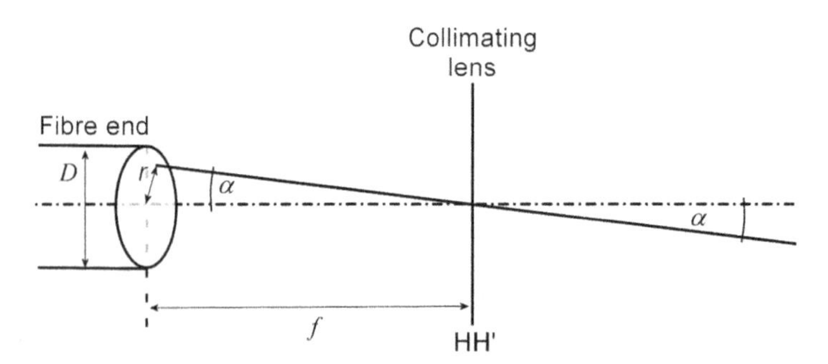


**Figure 2.15.** Partial beam originating from a certain position at a fibre end, acting as light source, located in the focal plane of the collimating lens.

$$e_D = \frac{\int_A -\frac{1}{2}(r(x, y)/f)^2 \, dx \, dy}{\int_A dx \, dy} = \frac{\int_0^{2\pi} \int_0^{D/2} -\frac{1}{2}(r/f)^2 r \, dr \, d\varphi}{\int_0^{2\pi} \int_0^{D/2} r \, dr \, d\varphi} = -\frac{D^2}{16f^2}, \quad (2.32)$$

as was originally shown by Bruce in 1954 [46, 47]. In a length measurement, the size effect of such a light source can therefore be corrected by $\delta l_D = l. \frac{D^2}{16f^2}$. As an example, for $D = 0.2$ mm and $f = 500$ mm, this so-called aperture correction is 10 nm per metre ($10^{-8}$).

For completeness, when the centre of the circular light source is apart from the optical axis by the value $r_0$, the total cosine error amounts to $e_{D,\,\alpha} = -\frac{1}{f^2}(\frac{r_0^2}{2} + \frac{D^2}{16})$.

While the size effect of a circular light source can be corrected as described above, the light source must be brought to the optical axis of the interferometer, ensuring rectangular incidence to the measuring faces. This can be achieved effectively by autocollimation adjustment procedures as suggested in [48, 49]. In such an approach the return spot originating from the reference mirror is brought into coincidence with the position of the light source. During this, the measurement path of the interferometer is closed. Figure 2.16 shows a scheme for the procedure suggested in [49]. The amount of light re-entering the fibre at its output increases, when the separation $\delta$ between the fibre position and its retroreflection reduces. Consequently, the signal of the retroreflection from the interferometer can be optimised. This retroreflection signal is extracted via a beam splitter close to the fibre input, while contributions from internal fibre reflections are measured as an offset when the shutter is closed. A computer controlled translation stage can be used for lateral movement of the fibre to grid points in the $xy$-plane. At each such position of the fibre end, the retroreflection signals are read out (figure 2.17).

For the optimum axial position of the fibre, i.e. at the focal plane of the collimating lens, the distribution of the retroreflection signals has a conical shape, as expected from the overlap of two shifted circles (the fibre and its retroreflection).

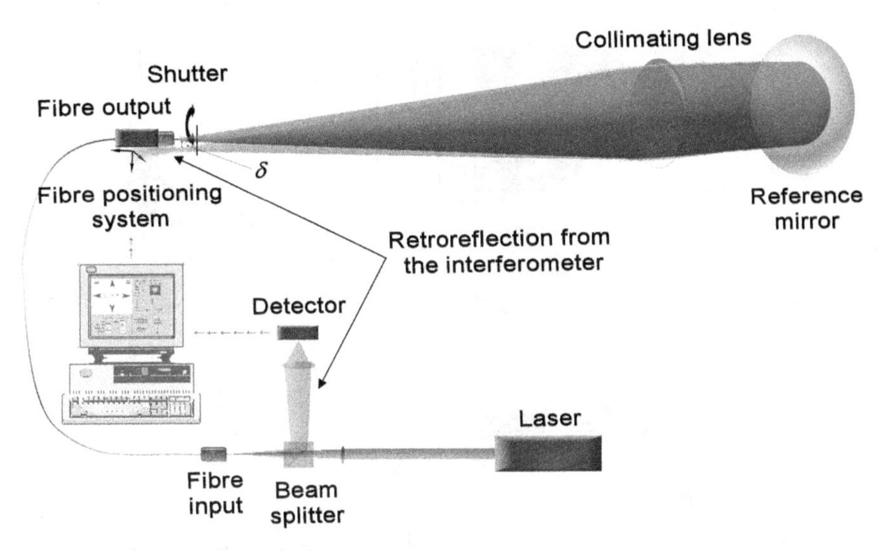


**Figure 2.16.** Scheme for the retroreflection scanning procedure (figure taken from [49]). © 2004 Optical Society of America.

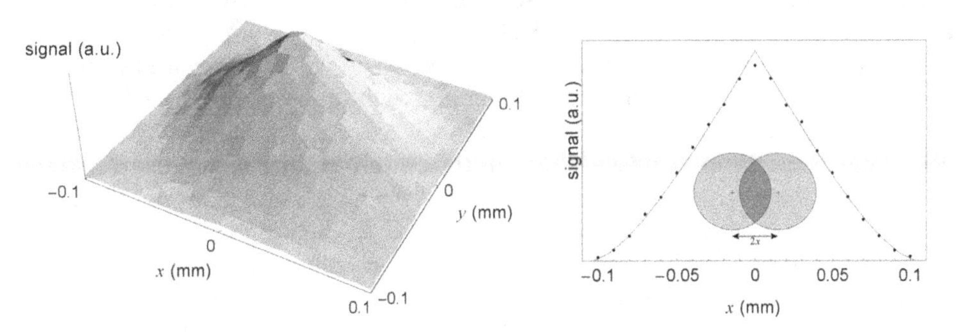


**Figure 2.17.** Left: measured retroreflection signal as a function of the $xy$-fibre position. Right: section in $x$-direction together with a theoretical curve (solid line) obtained from the overlap of two circles as a function of their displacement (figure taken from [49]). © 2004 Optical Society of America.

From the scanned $xy$-distribution of the retroreflection signal the centre coordinate is calculated to which the fibre output is then positioned.

After the autocollimation adjustment, in the next step the measurement path of the interferometer is opened and the interference pattern appearing is observed, while changing the lateral angles within the measurement pathway, e.g. adjustment of the measurement mirror (figure 2.10) or the gauge block shaped body (figure 2.2). Ideally the number of visible fringes can be reduced to zero (the so-called fluffed out mode). In such a case, the two beam directions almost perfectly coincide, i.e. the light reflected at the front face of a prismatic body, e.g. a gauge block with the platen (see figure 2.2) incidents perpendicular to the beam direction, as in the reference pathway adjusted beforehand. Then the length axis coincides with the measurement beam axis.

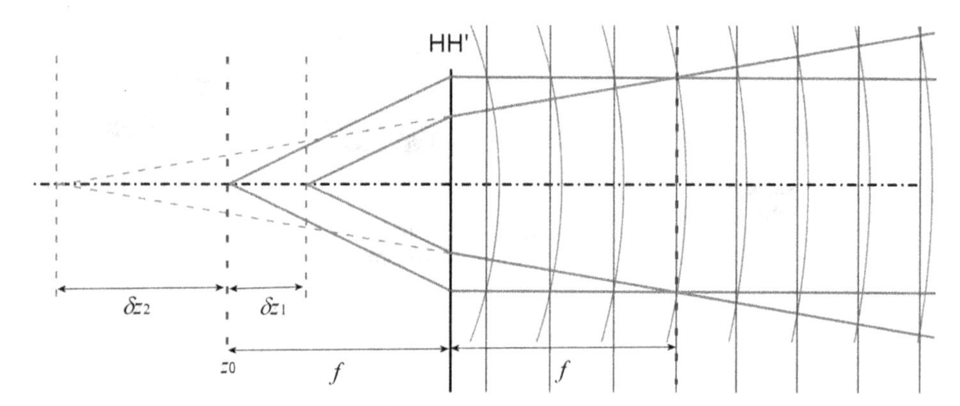


**Figure 2.18.** Progression of partial beams from a fictive point light source, positioned at $z_0$ for the ideal case (green lines) and for the case that the axial position of the light source is shifted $\delta z_1$ from $z_0$ (red lines) which result in curved wavefronts.

In addition to the lateral adjustment of the beam, length measurements by interferometry require axial adjustment of the light source to ensure that all measurement beams across the beam diameter travel parallel with the length axis (except the size effect of the light source which can be corrected as mentioned above). Ideally the axial position of the light source is away from the collimating lens by exactly the focal length. This ideal state is shown with the green lines in figure 2.18. Two partial beams from a fictive point light source, positioned at $z_0$, are shown, illustrating the progression as parallel beams after these have passed the collimating lens, the respective wavefronts (green vertical lines) are shown as straight lines. The red lines represent the situation in which the axial position of the light source is shifted by $\delta z_1$ from $z_0$. After passing the collimating lens these beams are generally not parallel to the optical axis, associated with curved wavefronts. In fact, as is visible in figure 2.18, the red wavefronts are segments of circles with the same origin, namely the axis point which is $\delta z_2$ away from $z_0$[5].

Since the origin of the circular segments is fixed, the corresponding radii of the curved wavefronts are proportional to the propagation length in the misaligned state (red lines in figure 2.18). In the special case when the reference pathway and the measurement pathway of an interferometer is the same, interference would thus happen between wavefronts of the same curvature, i.e. phase differences between such wavefronts will result in the same difference as for flat wavefronts. However, when a length is measured, at least one pathlength of the light in the measuring arm of the interferometer is different from the respective pathlength in the reference arm. Assuming equal pathways for the front face of a gauge block shaped body, the situation for the wave reflected at the end plate is shown in figure 2.19.

The radius of this wave is increased by $2l$. Therefore, at a range $d$ away from the optical axes, the distance between the two wavefronts is increased to $2\tilde{l} = 2l/\cos\beta$. Since the radius $R$ is much larger then $d$, $\beta \cong \sin\beta = d/R$ and

[5] It can be shown easily that $\delta z_2 = f^2/\delta z_1$.

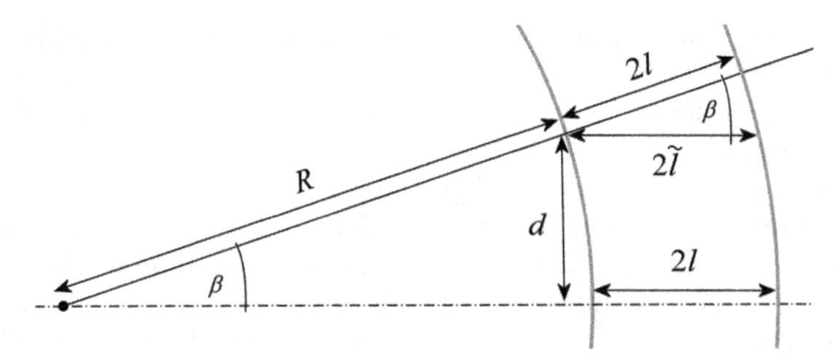


**Figure 2.19.** Wavefronts of two spherical waves with the same source point, one of which travelled the distance $R$, the other $R + 2l$.

$\cos\beta \cong 1 - \frac{1}{2}\beta^2 = 1 - \frac{1}{2}(d/R)^2$ is given in good approximation. Consequently, the relative error in the length measurements can be expressed as:

$$\frac{\delta l}{l} = -\frac{1}{2}\left(\frac{d}{R}\right)^2 \tag{2.33}$$

As an example, at a focal length of 500 mm of the collimating lens, a shift of the light source by $\delta z_1 = 1$ mm away from the focus position (see figure 2.18) leads to a radius of approximately 250 m ($R \approx \delta z_2 = f^2/\delta z_1$) of the wavefronts. Consequently, according to equation (2.33), at a distance $d = 20$ mm away from the optical axis a relative length error of $\approx 3 \times 10^{-9}$ is caused.

It remains to be mentioned that the axial adjustment can be optimized with the same autocollimation alignment procedure as for the above mentioned lateral adjustment. In [49] the relation between a measured length and the adjustment state of an interferometer is shown in more detail.

### 2.5.3 Optimum light sources for large field interferometry

The availability of a monochromatic light source is one of the key points when interferometry is applied for length measurements. Before laser radiation entered laboratories, it was a big challenge to generate monochromatic light from other light sources, e.g. from special spectral lamps. These light sources almost disappeared whilst the list of stabilized laser radiations recommended by the CIPM was growing. One example of a recommended radiation is the light emitted from a frequency-doubled Nd:YAG laser, stabilized with an iodine cell external to the laser, having a cold-finger temperature of −15 °C. Here the absorbing molecule is $^{127}I_2$ and the transition is the $a_{10}$ component, R(56) 32-0 at a frequency of 563 260 223 513 kHz corresponding to a vacuum wavelength of 532.245 036 104 nm with a relative standard uncertainty of $8.9 \times 10^{-12}$.

There is a close link between the spectral linewidth of a light source and temporal coherence. Accordingly, it is not surprising that stabilized laser light sources as mentioned above have superior long coherence lengths. Temporal coherence is an important criterion for the observation of interference in length measurements as the interfering beams travel different path lengths. The size of the light source is also important in length measurements by interferometry. This regards different aspects such as:

(1) The divergence of the beam affects the length measurements as discussed in section 2.5.2.
(2) The emittance of the finite light source is desired to be large. When an aperture or a pinhole is used to reduce the size of the light source, as necessary with spectral lamps, the signal-to-noise ratio of the interference intensity signal is limited by the pinhole size.
(3) The spatial coherence of the light beam. This regards the ability of the beam to interfere with a shifted copy of itself. Spatial coherence is necessary when interference is formed between different parts of beams as in a shearing interferometer. A nearly perfect point source, e.g. monochromatic light coming from a single-mode fibre, will generate maximum spatial coherence. On the other hand, spatial coherence is not necessary when using a non-sheared Twyman–Green arrangement together with rectangular incidence of the beams to the plane reflecting faces. In fact, in the case of the Twyman–Green interferometer, it is even advantageous when spatial coherence is reduced. This way parasitic interferences, generated by any additional reflection within the interferometer's beam path, can be effectively inhibited. A multimode fibre represents such light source. The mode field intensity distribution in such case reveals a characteristic grain as shown in the photograph figure 2.20(a). Figure 2.20(b) shows the intensity distribution observed when the fibre is vibrated at 500 Hz by a speaker's membrane.

In order to illustrate the effect of parasitic interferences, figure 2.21 shows phase topographies resulting from measurements in a Twyman–Green interferometer equipped a 16 bit CCD-camera system applying phase-stepping interferometry.

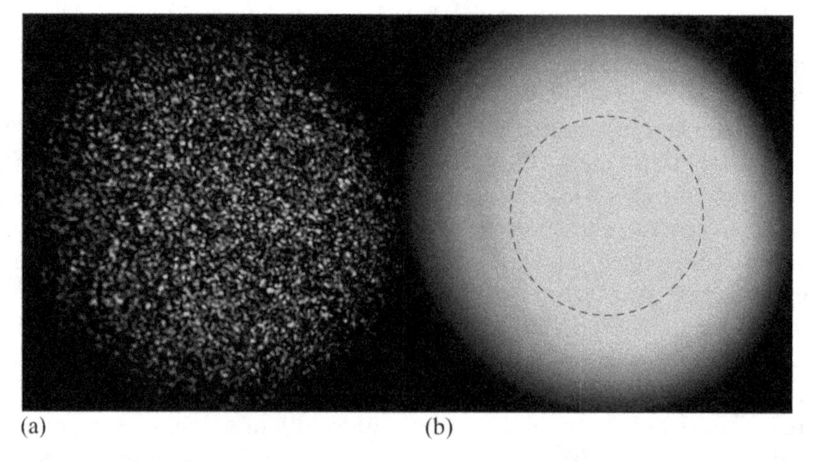


**Figure 2.20.** Photographs of light that passed a multimode fibre before entering the collimating lens of a large field interferometer (exposure time: 50 ms) (a) original grain pattern; (b) mixed grain pattern generated by fibre vibration. The dashed circle indicates the selected beam part entering the collimating lens of a Twyman–Green interferometer.

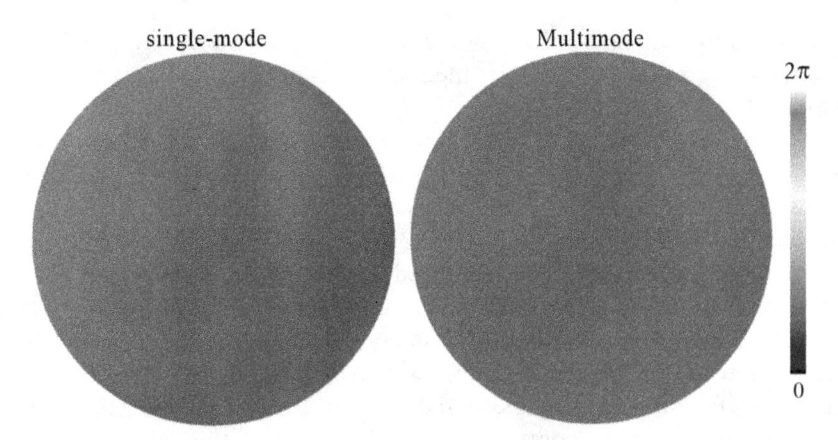


**Figure 2.21.** Phase topographies of a flat mirror measured in a large field interferometer using phase-stepping interferometry with different fibre types, as indicated. In the case of the single-mode fibre, distortions appear with amplitudes is about ± 0.03 interference orders.

The phase topographies each resulting from eight interferograms, recorded at 0.5 s exposure, show the total field of view of approximately 60 mm when an almost ideal optical flat is inserted and aligned so that zero fringes are observed (fluffed out mode).

Figure 2.21, left, shows the interference phase topography when a single-mode fibre is used as the light source. In this a case parasitic interference appears, the amplitude of which is about ± 0.03 interference orders corresponding to about ± 8 nm ($\lambda \cong 532$ nm). Figure 2.21(b) shows the field of view of the same interferometer when the light source is a vibrated multimode fibre. The parasitic interference pattern disappears in the latter case, and the remaining optics error of the interferometer becomes visible. Parasitic interference, affecting the actual phase topography when using single-mode fibres, as shown in this example, are probably caused by reflections at almost flat internal faces of the collimating lens.

### 2.5.4 Optics error correction

With the adjustment procedure described in section 2.5.2, the straightness of the wavefronts of the light can be optimized to a certain degree which is limited by the quality of the optical components of an interferometer. An effective way to check the interferometer's optics is to place an optical flat in the measuring pathway of the interferometer. The phase topography measured with an 'ideal' optical flat represents the sum of optics errors of the interferometer, that is, the imperfectness of the optical components with respect to (1) the flatness of the optical faces and (2) the homogeneity of the optical pathway travelled through the material. This includes the refractive index homogeneity mainly of the beam splitter but also the compensation plate. An example phase topography for the optics error of a Twyman–Green interferometer that is mainly used for primary length calibration of gauge blocks at PTB is shown in figure 2.22. Here a primary calibrated optical flat, 150 mm in diameter, with a maximum flatness deviation of 5 nm was applied. The phase topography shown in figure 2.22 includes the field of view of about 60 mm in diameter ($\lambda \cong 612$ nm). Accordingly, the maximum deviation from a plane is of the order of 20 nm.

The optics error correction to a measured phase topography is then considered as the optics error with opposite sign, related to the same wavelength. For noise reduction, interpolated averaged values, as displayed by the solid lines along the $x$- and $y$-directions in the 2D plots of figure 2.22, are used. The correction for the optics error of the interferometer can be measured at each of the wavelengths used in the measurements. In the interferometer mentioned above, the three wavelengths 612 nm, 633 nm and 780 nm are used alternately. In figure 2.23, the phase difference topographies resulting at the different wavelengths are shown, considering the scaling factors for the wavelengths, as indicated in the figure. Consequently, in this example, the optics error corrections are consistent within about 0.02 radians (corresponding to about 1 nm). Parasitic interference can therefore be excluded in this case.

The above optics error corrections are measured under the condition that both optical pathways within the interferometer are the same. Therefore, wavefront errors of the light waves which might even exist before entering the interferometer, would not become visible. As already mentioned above, length measurements always comprise unbalanced pathways between the measurement pathway and the reference pathway. The effect of wavefront distortion to a measured length, caused by the collimating lens, but also by the optics error, can be investigated by measurement of the optics error's evolution when the optical flat is placed at different axial positions. Figure 2.24 shows the difference of phase topographies measured away from the

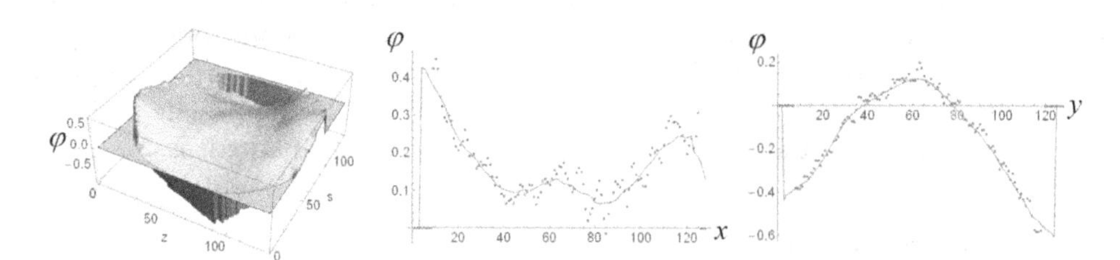


**Figure 2.22.** Phase error due to the interferometer optics measured using an optical flat at a wavelength of 612 nm.

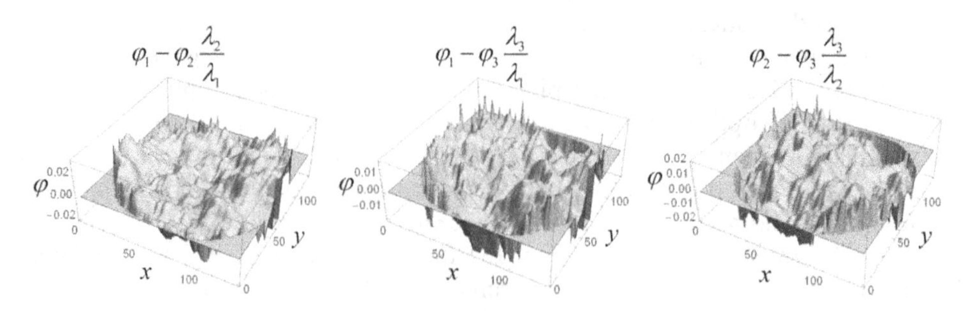


**Figure 2.23.** Consistence check between the optics error corrections obtained at three different wavelengths.

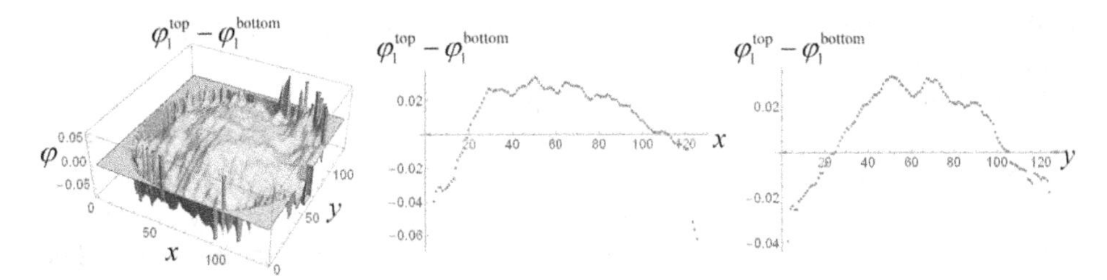


**Figure 2.24.** Difference of phase errors due to the interferometer optics measured at different positions of the optical flat separated by 200 mm (wavelength: 612 nm).

balanced position by −100 mm (labelled 'top') and +100 mm (labelled 'bottom') at 612 nm wavelength. The difference is up to 0.06 radian (about 3 nm) at the borders of the field of view.

Certainly, the specific amount of this uncertainty contribution is dependent upon the geometry of the arrangement, that is, definition of the regions of interest within the phase topography. In the example mentioned above, it was estimated that the optics error correction is related to an uncertainty contribution in the length measurement of 3 nm, independent of the length of gauge blocks to be calibrated.

## 2.6 Effect of surface roughness and phase change on reflection

Even almost 100 years ago surface effects to a length measurement were being discussed [50, 51]. It is generally supposed that when light is reflected in air or vacuum at glass or quartz surfaces a change of phase equal to $\pi$ radians takes place, and that the optical and geometrical surfaces of glass and quartz are coincident. But when light is reflected at a metallic surface the phase change is somewhat fewer $\pi$ radians due to a 'phase loss' $\delta\varphi_p$. This phase loss varies slightly with the wavelength and influences optical measurements, when e.g. a metallic gauge surface is used, which is equivalent to an apparent displacement $\delta l = \frac{1}{2}\lambda \cdot \delta\varphi_p$ of the reflecting surface from the geometrical surface towards the interior of the body. This effect definitely exists when the reflecting medium is a conductor (non-dielectric), i.e. free electrons are present at the surface, and leads to a complex refractive index $\tilde{n}_2$, involving the optical constants $n_2$ and $\kappa_2$ ($\tilde{n}_2 = n_2 + i\kappa_2$) of the material. It can be shown [52] that $\delta\varphi_p$, is given by

$$\tan \delta\varphi_p = \frac{2n_1k_2}{n_1^2 - n_2^2 - k_2^2}, \tag{2.34}$$

where $n_1$ is the refractive index of the surrounding medium and $\kappa_2$ the extinction coefficient of the reflecting material. Measurements of $n_2$ and $\kappa_2$ for a number of steel gauge blocks using ellipsometry indicated that the variation of $\delta\varphi_p$ for a certain steel is small [53]. For a 20% variation in $\kappa_2$, $\delta\varphi_p$ varies by 5°, corresponding to a variation in measured length of 4.4 nm, for $\lambda \cong 633$ nm [54]. Accordingly, it was assumed that this variation in $\kappa_2$ is probably typical for the range of steels used for gauge blocks, length bars and platens with the worst possible difference in material properties causing a length measurement error of ±8.8 nm.

The measured length is further affected by the roughness of the surface at which the light is reflected. This effect cancels out in length measurements in which a platen is attached to the body and both have the same surface roughness. However, this is generally not the case, as e.g., the effect due to surface roughness of steel gauge blocks has been found to vary by several nanometres within a set of gauges. There are several techniques for the measurement of surface roughness. One common technique uses an integrating sphere [8, 52, 55] in which light is reflected and the ratio of the diffuse to directly reflected light is measured. In this approach it is assumed that the roughness of a test surface is proportional to the square-root of this ratio. This technique requires a calibrated surface for reference. The roughness difference between a body and a platen is used as an additional correction, the uncertainty of which is typically 3 nm.

The corrections for the surface roughness and the phase change on reflection are often combined into a single correction termed as a 'phase correction'. The difference in phase change of light on reflection from the gauge and the platen's surface is a significant contributor to measurement uncertainty in primary gauge block calibrations by optical interferometry. Considerable effort is devoted to understanding and evaluating its physical value.

## 2.7 Wringing contact between a body and a platen

Most length measurement techniques involve a platen to be attached to a body (e.g. gauge block). The surfaces of the body and of the platen must be cleaned carefully using a solvent, e.g. acetone. When both surfaces are polished/lapped to a high degree they may represent almost perfect optical flats. Putting together such flat surfaces enables a strong molecular attraction which is also called 'optical contacting.' In cases of increased surface roughness when the wringing of completely dry-cleaned surfaces is difficult, a small quantity of wringing fluid may be used which fills the microscopic gaps. Although in such a case, the wringing force is somewhat smaller compared to the optical contacting, a so-called 'wringing film' contributing to the length measurement can be neglected as long the amount of wringing fluid is small. The gauge block and platen can be separated when the measurement has been completed.

The wringing influences the length measurements by the fact that the surface topography of the body is influenced. The size of this effect is dependent upon any deviation from flatness, mainly of the body's surface, but also, surface flexing caused by thermal expansion, when there exists temperature difference between the two parts while wringing them together. Platen flexing, if not corrected (as described in see section 2.2) is problematic for the application of high-resolution phase-stepping instruments because phase measurement must consider the error contribution from this form variation relative to the ideal plane. The difference of length measurements obtained when the platen is wrung to either one or the other face of the body gives a good indication for these effects. Dependent on the size of the resulting length difference, the wringing influence is considered as an uncertainty contribution, typically ranging from 2 nm to 10 nm for the length measurements.

## 2.8 Double ended interferometry

The primary length calibration of gauge blocks is basically made by single ended large field imaging interferometers in which one of the measuring surfaces is wrung onto a platen, as shown in figure 2.25, left. A first double ended interferometer (DEI) for measurement of gauge blocks was already patented and built by Dowell in 1943 [56]. In this setup two interfering beams traverse the same path in opposite directions. However, Dowell's DEI was made for measuring length differences between two GBs, only. In 1958 Hariharan suggested a DEI in which the beam traverses two triangular pathways [57]. This system is based on the observation of a single interferogram and was designed for absolute length measurements of long GBs. However, the capability of this interferometer was limited due to the existence of multiple interferences.

In 1972 Dorenwendt suggested a double ended design that was based on the extension of a Twyman–Green interferometer [58]. In 1983 Gerasimov proposed a ring optical scheme [59]. Later Khavinson reported a variety of means to control the optical phase differences in such interferometers [60]. In 1993 Lewis reported the setup of a DEI at NPL which was part of 'NPL's Primary Length Bar Interferometer' [61]. Similar designs have been suggested 1998 and 2003 [62, 63]. In 2006 Kuriyama [64] built the first DEI which was equipped with CCD cameras and phase-stepping interferometry. This interferometer was mainly built for quality control of the production process of gauge blocks. In 2012 PTB finished a prototype DEI [65] that was built to study the limitations which were taken into consideration for the final version which is still under development at PTB. The idea of a double ended interferometer seems to have become more and more popular [66] since it

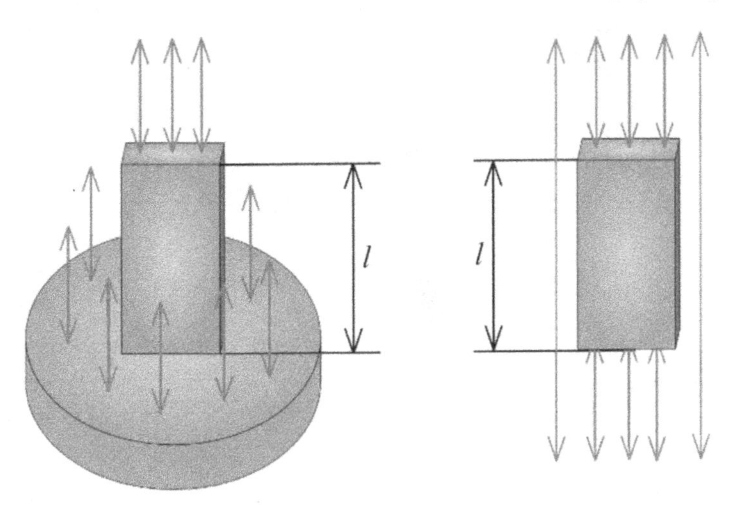


**Figure 2.25.** Left: single ended measurement arrangement for gauge blocks by wringing it to a platen, right: basic idea of a double ended length measurement.

deals with the shortcomings of single ended interferometers in which a single gauge block must be wrung every time it is calibrated which can affect its length in the long-term. Another difficulty is that the wringing can be time consuming and needs experienced and skilled staff, especially for gauge blocks smaller than 5 mm in length.

Figure 2.26 shows a scheme of PTB's final type DEI which is situated in a temperature-controlled vacuum chamber [67]. Here, in contrast to the prototype [65], a shaken multimode fibre is used as the light source to remove parasitic interferences. A large collimated beam of 80 mm in diameter is formed by an achromatic lens with a focal length of 500 mm. This beam is split by a triangular configuration of three wedged beam splitter plates into two separated reference and imaging arms and a common measurement arm, in which a gauge block is placed. At $BS_1$, half of the intensity of the light is reflected to $BS_2$, the other half is transmitted to $BS_3$. The latter two beamsplitters are aligned so that the beams from left and right have opposite directions. For left and right, the inner part of each of the beams reaches the face of the gauge block where it is reflected. The outer parts pass along the gauge block to the other side of the interferometer. The length of the gauge block can be obtained from the differences of the optical path lengths as indicated in figure 2.27.

The difference in the optical path lengths is calculated using the phase differences of the outer part of the beam, which passes along the gauge block, to the inner part of the beam, which has been reflected at the gauge block face. For each arm of the

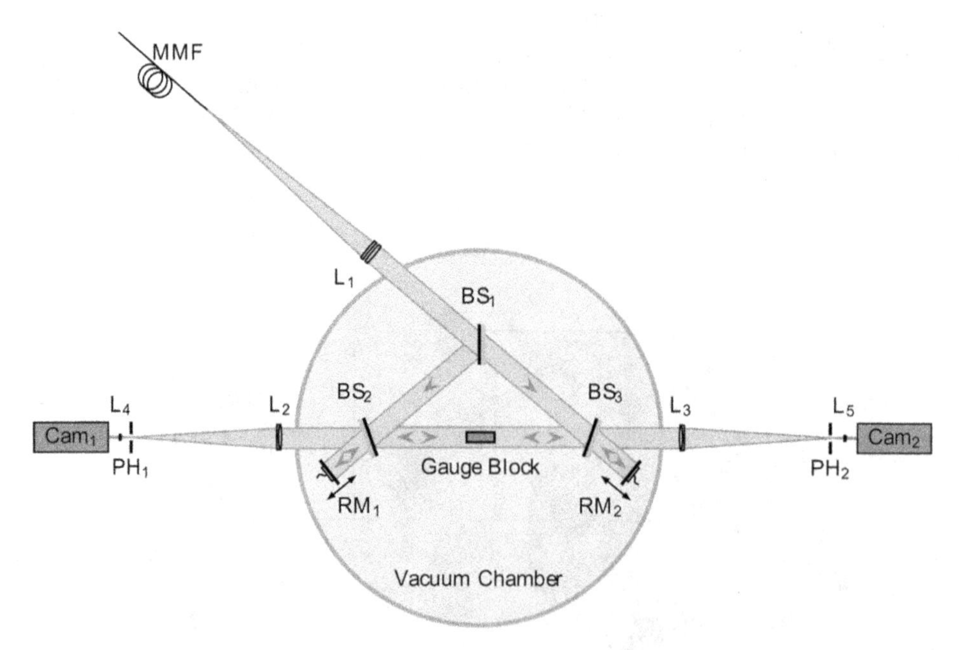


**Figure 2.26.** Setup of PTB's ultimate double ended interferometer with shaken multimode fibre (MMF), wedged beam splitter plates (BS 1,..., BS 3), piezo-actuator controlled reference mirrors ($RM_1$, $RM_2$), apertures ($PH_1$, $PH_2$), achromatic lenses ($L_1$,..., $L_5$) and CCD cameras ($Cam_1$, $Cam_2$).

interferometer, phase topographies are obtained from phase shifted interferograms recorded with the CCDs.

Averaged phase values, as indicated in figure 2.28, are calculated within the respective regions of interest, allowing calculation of fractional orders of interference for the left and the right arm separately. For each wavelength $\lambda_k$, a total fractional order $q_k$ is then obtained from the sum of the individual fractional orders in the left and the right arm:

$$l_{GB} = \frac{\lambda_k}{2}\left\{(i_k + \overbrace{\underbrace{\frac{1}{2\pi}\left(\varphi_{ra} - \varphi_{ra}^{GB}\right)}_{q_k^{ra}} + \underbrace{\frac{1}{2\pi}\left(\varphi_{la} - \varphi_{la}^{GB}\right)}_{q_k^{la}}}^{q_k}\right\} \tag{2.35}$$

The further length evaluation comprises determination of $i_k$, as described in section 2.3.

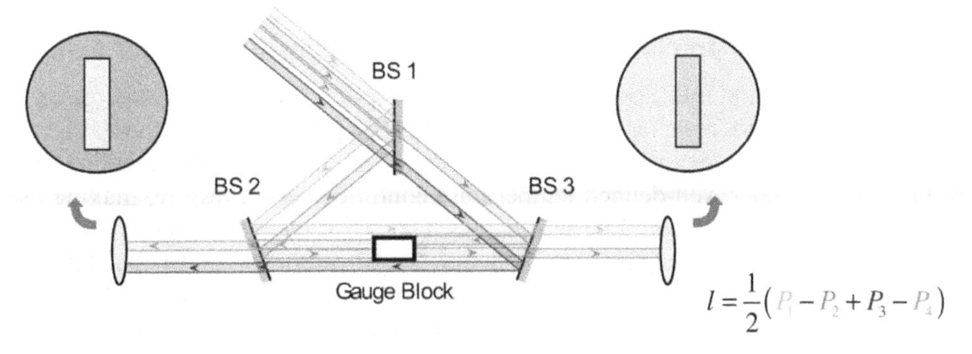


**Figure 2.27.** Scheme for the different pathways from which the length of the gauge block is obtained.

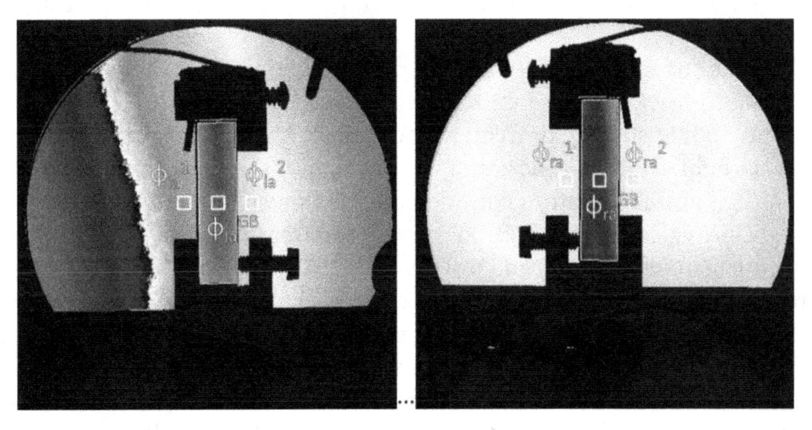

**Figure 2.28.** Phase topographies with regions of interest of left and right interferometer arm measured with PTB's double ended interferometer.

With PTB's DEI [67] the reproducibility of the measured length is well within ± 0.5 nm. Since the influence of wavefront aberrations onto the length measurement is increased compared to single ended interferometers, effective error corrections are necessary. Simulations by a raytracing software deliver insight into potential corrections. A correction for the outer part of the beam can be achieved by a phase measurement of the empty interferometer. However, for wavefront aberrations induced by the beam splitter plates and for long gauge blocks an additional correction is necessary. Such a correction, expected to be rather complex due to varying axial positions of the body's end faces, dependent on the individual length, is to be obtained in future work. Other limitations exist, e.g. only an opaque gauge block can be measured, making a DEI not ready for measurement tasks which are easily made with a single ended interferometer.

While it is not expected that double ended interferometers will soon replace the single ended ones in the field of primary gauge block calibrations, DEIs may show their strength in thermal expansion measurements (absence of temperature dependent platen flexing) and measurement of long-term stability (potential re-wringing to the platens is omitted).

## 2.9 PTB's ultra precision interferometer—design for special tasks

PTB's ultra precision interferometer (UPI) was introduced in 2011 as a successor of the Precision Interferometer described in detail in [35]. In short: the UPI was designed for highly precise absolute length measurements of prismatic bodies, e.g. gauge blocks, under well-defined temperature conditions and pressure, making use of phase-stepping imaging interferometry. The UPI enables several enhanced features, e.g. it is designed for a much better lateral resolution and better temperature stability. In addition to the original concept, the UPI was equipped with an external measurement pathway (EMP) in which a prismatic body can be placed alternatively. The temperature of the EMP can be controlled down to cryogenic temperature by an appropriate cryostat system allowing length measurements for studying thermal expansion of materials from room temperature down to less than 10 K.

A schematic diagram of the UPI in its original concept is shown in figure 2.29. Three stabilized lasers, a Rb-stabilized diode laser @780 nm, a He–Ne laser @633 nm and an iodine-stabilized frequency-doubled Nd:YAG laser @532 nm, are available as light sources. The latter source exhibits a long-term stability better than $10^{-12}$ [68] and is therefore considered as the most stable laser.

The light provided alternatively by the three lasers passes a 200 μm multimode fibre which generates a grainy intensity pattern. Application of a fibre shaker near the fibre end mixes the variety of fibre modes so that excellent homogeneity of the light intensity is achieved while spatial coherence is avoided (see section 2.5.3). The UPI's vacuum chamber, 1.6 m in diameter, represents the housing of the interferometer. Its entire outer surface is covered by water-bearing tubes, arranged in counter flow for optimum temperature homogeneity. With the UPI, a higher lateral resolution was achieved by relatively large wedge angles for the beam splitter and the

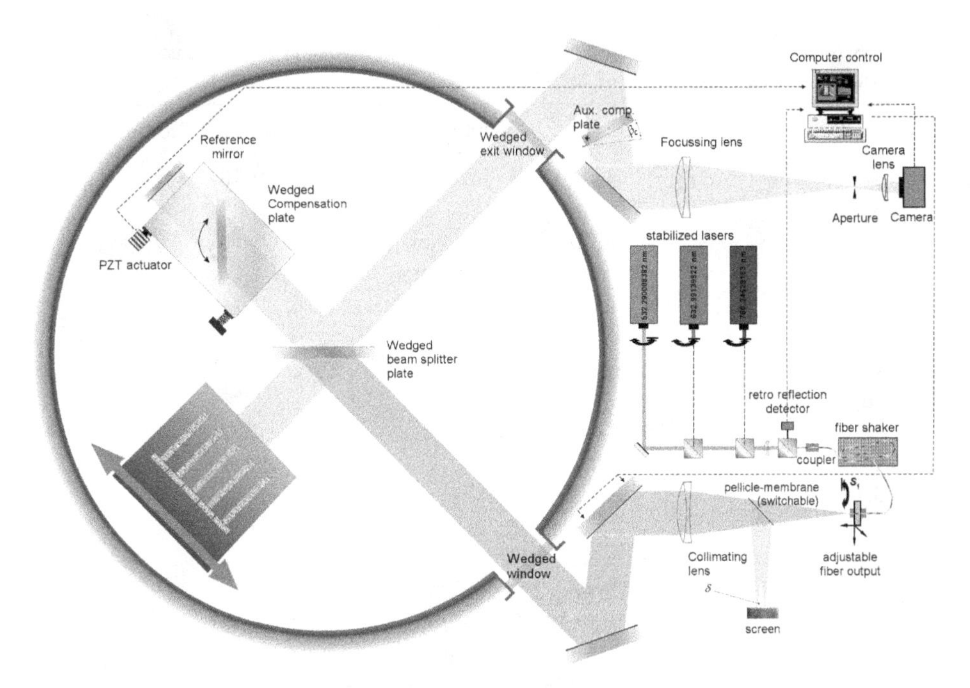


**Figure 2.29.** Configuration scheme of the Ultra Precision Interferometer (figure taken from [35]). © IOP Publishing Ltd. All rights reserved.

compensation plate. The increased image sharpness enables more accurate position detection within a phase topography and less diffraction from a gauge block's edges. Table 2.1 lists the uncertainty contributions of length measurements with the ultra precision interferometer. An additional uncertainty contribution arises from limited parallelism which is individual to a specific gauge block shaped body. Therefore, the total combined uncertainty of the length measurements is evaluated for each sample individually. As an example, a 200 mm long sample with a non-parallelism of 10 μrad (causing 0.03 nm additional uncertainty) can be measured with a total combined standard uncertainty of approximately 0.15 nm under vacuum. It is noted that this value refers to a certain wringing state and to a well-defined position and size of the regions of interest within the phase topographies (see section 2.2). Figure 2.30 shows an example of such an approach in which the method of phase-stepping interferometry (PSI) was validated by comparative measurements of the length of a gauge block shaped sample body made of single crystalline silicon. Here the reference mirror was tilted so that many fringes appeared allowing for phase evaluation by discrete Fourier transformation (DFT) for each of the ten interferograms separately. The baselines (light blue bands) in figure 2.30 are set to the individual mean value of the 'DFT-lengths' (thin blue lines).

The two cases shown represent borderline situations: whereas in figure 2.30(a) the spread between the DFT-lengths is maximum the coincidence with the PSI-length is 3 pm, only. In figure 2.30(b) the situation is reversed. From these and other investigations (e.g. [69]) it was concluded that the contribution of the interference

**Table 2.1.** Specimen independent contributions to the overall uncertainty of length measurements with the ultra precision interferometer (table taken from [35]).

| | Term | Detail | Standard uncertainty | Unit |
|---|---|---|---|---|
| 1 | Laser vacuum wavelengths | He–Ne 633 nm, $J_2$ stab. | $2.5 \times 10^{-11}$ | 1 |
| | | YAG 532 nm, $J_2$ stab. | $3.5 \times 10^{-12}$ [29] | 1 |
| | | Combined uncertainty contribution to the mean length $(l_{532} + l_{633})/2)$ | $1.3 \times 10^{-11}$ | $l$ |
| 2 | Interferometer adjustment | Angle | $2 \times 10^{-06}$ | rad |
| | | Cosine error | $2 \times 10^{-12}$ | $l$ |
| 3 | Interference fraction determination | Phase@633 nm | 0.0003 | $2\pi$ |
| | | Fraction@633 nm | 0.000 37 | 1 |
| | | Fraction@633 nm | 0.11 | nm |
| | | Phase@532 nm | 0.0002 | $2\pi$ |
| | | Fraction@532 nm | 0.000 24 | 1 |
| | | Fraction@532 nm | 0.08 | nm |
| | | Combined uncertainty contribution to the mean length $(l_{532} + l_{633})/2)$ | 0.07 | nm |
| 4 | Platen flexing correction [23] | Uncertainty | 0.05 .. 0.1 | nm |
| 5 | Temperature effect of the interferometer optics | Unspecific influence | 0.05 | nm |
| 6 | Refractive index of residual air | | | |
| | $p = 3$ mPa | Total effect@ $p = 3$ mPa | $9 \times 10^{-12}$ | 1 |
| | (high vacuum) | Uncertainty ($u(p) = 2$ mPa) | $6 \times 10^{-12}$ | 1 |
| | $p = 10$ Pa | Total effect@ $p = 10$ Pa | $3 \times 10^{-7}$ | 1 |
| | | Uncertainty ($u(p) = 0.01$ Pa) | $3 \times 10^{-10}$ | 1 |
| | $p = 50$ Pa | Total effect@ $p = 50$ Pa | $1.2 \times 10^{-6}$ | 1 |
| | | Uncertainty ($u(p) = 0.04$ Pa) | $1.2 \times 10^{-9}$ | 1 |

evaluation to the overall measurement uncertainty is in fact clearly smaller than 0.1 nm (in agreement with the sum of items 1 to 3 of table 2.1).

Figure 2.31 shows measured absolute lengths of a samples made of carbon fibre reinforced silicon carbide material (SiC 100® of Boostec®) as a function of the temperature (data points) together with a polynomial fit of the degree $n = 3$. The drift of the temperature during a single measurement (approximately 3 min) was smaller than 0.5 mK. The sample length at 20 °C (under vacuum conditions) is indicated as 'offset length' in figure 2.31.

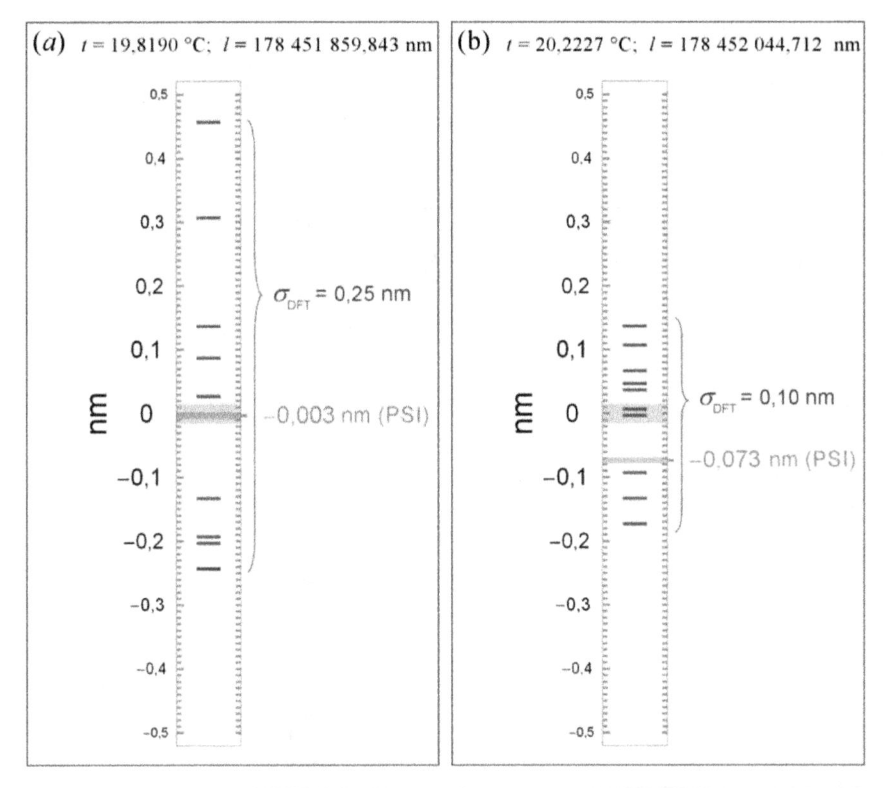


**Figure 2.30.** Two data sets of comparative length measurements at a gauge block of 178 mm in length made of single crystalline silicon. The blue lines indicate length obtained from DFT-based phase evaluation. The red line represents the PSI-based length. The baselines (light blue) indicate the mean of all the DFT-lengths (figure taken from [35]). © IOP Publishing Ltd. All rights reserved.

The top panel of figure 2.31 shows the individual deviation of the data from the fit and the value $\sigma = \sqrt{\sum_{i=0}^{N}(l_i - l^{(n)}(t_i))^2/[N - (n+1)]}$, in which $N$ denotes the number of data points. It is noted that the value of $\sigma = 0.71$ nm is larger than the estimated combined standard uncertainty of the length measurements. This is explained by the uncertainty of the temperature measurement which, in this example, was performed in non-contact mode (giving rise to values ranging from 3.1 mK at 20 °C to 4.1 mK at 50 °C). From the fit polynomial, the temperature dependent coefficient of thermal expansion (CTE) can be obtained (for more details see [35]).

The measurement example shown here aims to indicate achievable measurement uncertainty in absolute length measurements, mainly for vacuum, when these are focused on the influence of temperature (thermal expansion) and time (length relaxation and long-term stability). Effects on the absolute length which can be regarded as constant over the temperature and time are therefore not included in the above measurement uncertainty budget. Primary gauge block calibrations by interferometry more strictly focus on the absolute length in a mechanical sense and are performed under atmospheric pressure. In this case, additional sources of measurement uncertainty arise due to optics error, wringing and phase change on

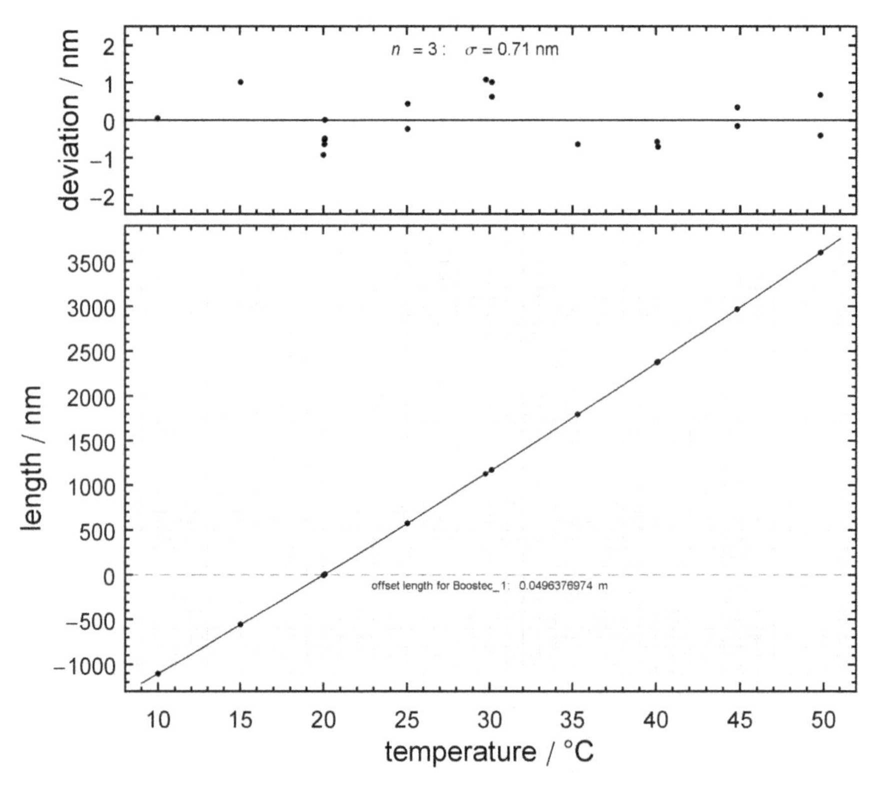


**Figure 2.31.** The length of Boostec_1 as a function of the temperature (data points) and a polynomial fit of the order $n = 3$ (figure taken from [35]). © IOP Publishing Ltd. All rights reserved.

reflection (sections 2.5.4, 2.6. and 2.7) as well as from the air refractive index measurement (section 2.4).

## 2.10 The importance of the temperature measurement

As demonstrate in the above example, the length a material measures is related to the temperature. For this reason, accurate temperature measurement is very important when length measurements are performed. For example, in primary calibrations of gauge blocks by interferometry the quantity to be determined, the 'calibrated length', is the length at a reference temperature $t$, which is mostly 20 °C. Typically, during a length measurement by interferometry the temperature is slightly different from the reference temperature. For this reason, a correction is necessary considering the temperature deviation, resulting in the length at the reference temperature, e.g.:

$$l_{20\ ^\circ\mathrm{C}} = l[1 - \alpha_0(t - 20\ ^\circ\mathrm{C})], \tag{2.36}$$

in which $\alpha_0$ denotes a constant value for the coefficient of thermal expansion. For example, a 1 m gauge block of steel with a typical thermal expansion coefficient of $1.2 \times 10^{-5}\ \mathrm{K}^{-1}$ will be measured about 0.2 µm shorter at $t = 19.985$ °C than at $t =$ 20.000 °C. This is about half of an interference order and illustrates the importance of accurate temperature measurement. Depending on the body's material and the desired uncertainty of the length measurement, it may be necessary to perform the temperature measurements with sub-millikelvin uncertainty, according to the international temperature scale [70].

## 2.11 Primary gauge block calibrations performed today

Gauge blocks are used in the calibration and setting of a huge number of measuring instruments, from micrometers and callipers to the highest accuracy coordinate measuring machines or profilometers. While they may be commonly seen in machine-shop applications, there are grades with optical-quality finish, ranging in nominal length between 0.1 mm and 1 m [4, 71]. The central length is precisely defined as the optical height of the reference point relative to a datum plane (platen), which is parallel to the measuring face of the gauge block. At present gauge blocks are still one of the most accurate transfer standards of the length and they are indispensable within the traceability chain of most dimensional measurements.

Fizeau-type interferometers were at one time popular for gauge block applications, however the potential for unsymmetrical interference makes analysis more complicated than using a two-beam Twyman–Green design where the interference is an exact sinusoid. A Twyman–Green configuration involving a Kösters prism, as shown in figure 2.32, allows a more compact scheme, compared to the rectangular Twyman–Green design, and is therefore especially suitable for long gauge blocks. A refractometer cell placed near the gauge block allows interferometrical measurement of the air refractive index (see section 2.4). A first design for such interferometer was suggested by Kösters around 1938 [6] and further developed by Engelhard [8]. A small series of these instruments was manufactured by Zeiss (Germany) around 1962, and distributed worldwide to several national metrology institutes. PTB's two Kösters interferometers were basically renewed around 2004 [37]. As is visible in figure 2.32, the present setup involves two wedged 8 mm thick plates mounted in one solid frame such that each plate is at a 60° angle to the optical axis. Phase shifting is achieved by slight rotating the mount via leaf springs and piezo-stacks.

Large field interferometers for primary gauge block calibration are typically situated at national metrology institutes where these are often denoted as 'national standards for the length'. More detailed descriptions of upgrades and verifications can be found in the literature (e.g. see [22, 72–74]). Today's measurement capability

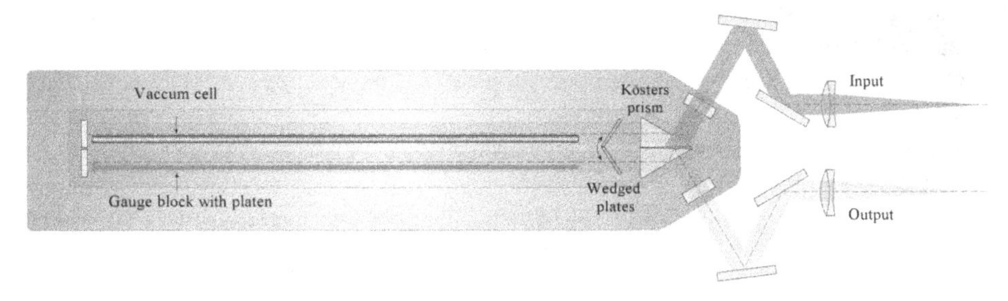


**Figure 2.32.** Diagram of the PTB's Kösters-interferometer.

for primary gauge block length calibration is about $U = 70$ nm expanded uncertainty ($k = 2$ or approximately 95% confidence) for a good quality 1 m gauge block, or $U <$ 20 nm for 25 mm nominal length [75]. International comparisons of gauge block measurements are performed regularly to support measurement capabilities (CMC) claimed individually by the national metrology institutes. Results of such comparisons are listed within the key comparison data base (KCDB) of the BIPM ([76], search item: 'gauge blocks'), each linked to the respective publication.

## References


[1] Twyman F and Green A 1916 *British Patent* 103832

[2] Murty M V R K 1978 Newton, Fizeau, and Haidinger interferometers *Optical Shop Testing* ed D Malacara (New York: Wiley) ch 1, pp 1–45

[3] Malacara D 1978 Twyman-Green interferometer *Optical Shop Testing* ed D Malacara (New York: Wiley) ch 2, pp 47–79

[4] ISO 3650 1998 *International Standard 'Length standards—Gauge Blocks'* (Geneva, Switzerland: International Organization for Standardization)

[5] Kösters W 1926 Ein neuer Interferenzkomparator für unmittelbaren Wellenlängenanschluß *Zeitschrift für Feinmechanik und Präzision* **34** 55–9

[6] Kösters W 1938 Der gegenwärtige Stand der Meter-Definition, des Meteranschlusses und seine internationale Bedeutung für Wissenschaft und Technik *Werkstattstechnik und Werkleiter Bd.* **23** 527–33

[7] Barrell H 1948 Light waves as standards of length *Research* **12** 533–40

[8] Engelhard E 1957 Precise interferometric measurements of gage blocks *Metrology of Gage Blocks Natl. Bur. Stand. Circ.* **581** pp 1–20

[9] Darnedde H 1992 High-precision calibration of long gauge blocks using the vacuum wavelength comparator *Metrologia* **29** 349–59

[10] Lewis A 1994 Measurement of length, surface form and thermal expansion coefficient of length bars up to 1.5 m using multiple wavelength phase-stepping interferometry *Meas. Sci. Technol.* **5** 694–703

[11] Jin J, Kim Y, Kim Y and Kim S 2006 Absolute length calibration of gauge blocks using optical comb of a femtosecond pulse laser *Opt. Express* **14** 5968–74

[12] Bruning J H, Herriott D R, Gallagher J E, Rosenfeld D P, White A D and Brangaccio D J 1974 Digital wave-front measuring interferometer for testing optical surfaces and lenses *Appl. Opt.* **13** 2693–703

[13] Kinnstaetter K, Lohmann A W, Schwider J and Streibel N 1988 Accuracy of phase shifting interferometry *Appl. Opt.* **27** 5082–88

[14] Carre P 1966 Installation et utilisation du comparateur photoelectrique et interferentiel de Bureau International des Poids et Mesures *Metrologia* **2** 13–23

[15] Creath K 1988 Phase-measurement interferometry techniques *Progress in Optics* vol 26 ed E Wolf (New York: Elsevier) pp 349–98

[16] Joenathan C 1994 Phase-measuring interferometry: new methods and error analysis *Appl. Opt.* **33** 4147–55

[17] Hariharan P, Oreb B F and Eiju T 1987 Digital phase-shifting interferometry: a simple error-compensating phase calculation algorithm *Appl. Opt.* **26** 2504–6

[18] Groot P D 1995 Derivation of algorithrns for phase-shifting interferometry using the concept of a data-sampling window *Appl. Opt.* **34** 4723–30

[19] Schwider J, Falkenstörfer O, Schreiber H, Zöller A and Streibl N 1993 New compensation four-phase algorithm for phase-shift interferometry *Opt. Eng.* **32** 1883–85
[20] Zhu Y and Gemma T 2000 Method for designing error-compensating phase calculation alcorythms for phase shifting interferometry *Appl. Opt.* **40** 4540–46
[21] Tang S 1996 Self-calibrating five frame algorithm for phase shifting interferometry *Proc. SPIE* **2860** 91–7
[22] Decker J E, Schödel R and Bönsch G 2004 Considerations for the evaluation of measurement uncertainty in interferometric gauge block calibration applying methods of phase step interferometry *Metrologia* **41** L11–17
[23] Schödel R 2008 Ultra-high accuracy thermal expansion measurements with PTB's precision interferometer *Meas. Sci. Technol.* **19** 084003
[24] Benoît R 1898 Application des phénomènes d'interférence à des déterminations métrologiques *J. Phys. Théor. Appl.* **7** 57–68
[25] Born M and Wolf E 1980 *Principles of Optics* 6th edn (New York: Pergamon) pp 286–306
[26] Tilford C R 1977 Analytical procedure for determining lengths from fractional fringes *Appl. Opt.* **16** 1857–60
[27] Cheng Y and Wyant J C 1985 Multiple-wavelength phase shifting interferometry *Appl. Opt.* **24** 804–7
[28] Creath K 1987 Step-height measurement using two-wavelength phase-shifting interferometry *Appl. Opt.* **26** 2810–16
[29] Decker J E, Siemsen K, Siemsen R, Madej A, Marmet L, Miles J, de Bonth S, Bustraan K, Temple S and Pekelsky J R 2003 Increasing the range of unambiguity in step height measurement using multiple-wavelength interferometry — Application to absolute long gauge block measurement *Appl. Opt.* **42** 5670–78
[30] Schödel R 2015 Utilisation of coincidence criteria in absolute length measurements by optical interferometry in vacuum and in air *Meas. Sci. Technol.* **26** 084007
[31] Falaggis K, Towers D P and Towers C E 2011 Method of excess fractions with application to absolute distance metrology: theoretical analysis *Appl. Opt.* **50** 5484–98
[32] Falaggis K, Towers D P and Towers C E 2012 Method of excess fractions with application to absolute distance metrology: Wavelength selection and the effects of common error sources *Appl. Opt.* **51** 6471–79
[33] BIPM: recommended values of standard frequencies URL: http://bipm.org/en/publications/mises-en-pratique/standard-frequencies.html [accessed 2018/04/01]
[34] Leone F C, Nottingham R B and Nelson L S 1961 The folded normal distribution *Technometrics* **3** 543–50
[35] Schödel R, Walkov A, Zenker M, Bartl G, Meeß R, Hagedorn D, Gaiser D, Thummes G and Heltzel S 2012 A new Ultra Precision Interferometer for absolute length measurements down to cryogenic temperatures *Meas. Sci. Technol.* **23** 094004
[36] Schödel R 2007 Compensation of wavelength dependent image shifts in imaging optical interferometry *Appl. Opt.* **46** 7464–68
[37] Decker J E, Schödel R and Bönsch G 2003 Next generation Kösters Interferometer *Proc. SPIE* **5190** 14–23
[38] Barrell H and Sears J E 1939 The refraction and dispersion of air for the visible spectrum *Philos. Trans. R. Soc. Lond. Ser.* **A238** 1–64
[39] Edlén B 1966 The refractive index of air *Metrologia* **2** 71–80

[40] Owens J C 1967 Optical refractive index of air: Dependence upon pressure, temperature and composition *Appl. Opt.* **6** 51–9
[41] Birch K P and Downs M J 1994 Correction to the updated Edlén equation for the refractive index of air *Metrologia* **31** 315–6
[42] Ciddor P E 1996 Refractive index of air: new equations for the visible and near infrared *Appl. Opt.* **35** 1566–73
[43] Bönsch G and Potulski E 1998 Measurement of the refractive index of air and comparison with modified Edlen's formulae *Metrologia* **35** 133–9
[44] Schödel R, Walkov A, Voigt M and Bartl G 2018 Measurement of the refractive index of air in a low-pressure regime and the applicability of traditional empirical formulae *Meas. Sci. Technol.* **29** 064002
[45] Birch K P, Downs M J and Ferriss D H 1988 Optical path length changes induced in cell windows and solid etalons by evacuation *J. Phys. E: Sci. Instrum.* **21** 690–92
[46] Bruce C F 1955 The effects of collimation and oblique incidence in length interferometers I *Aust. J. Phys.* **8** 224–40
[47] Bruce C F 1957 Obliquity effects in interferometry *Opt. Acta* **4** 127–35
[48] Lewis A and Pugh D J 1992 Interferometer light source and alignment aid using single-mode optical fibres *Meas. Sci. Technol.* **3** 929–30
[49] Schödel R and Bönsch G 2004 Highest accuracy interferometer alignment by retroreflection scanning *Appl. Opt.* **43** 5738–43
[50] Rolt F H and Barrell H 1929 The difference between the mechanical and optical length of a steel end-gauge *Proc. R. Soc. Lond. Ser.* A **122** 122–33
[51] Bennet J 1964 Precise method for measuring the absolute phase change on reflection *J. Opt. Soc. Am.* **54** 612–22
[52] Leach R K 1993 Investigation into the measurement of the wringing effect and the phase shift at reflection applied to the accurate measurement of end standards *MSc feasibility study* Brunel University
[53] Leach R K 1998 Measurement of a correction for the phase change on reflection due to surface roughness *Proc. SPIE* **3477** 138–51
[54] Lewis A J 1993 Absolute length measurement using multiple-wavelength phase-stepping interferometry, Thesis Unpublished for the degree of Doctor of Philosophy of the University of London and for the Diploma of Membership of Imperial College Nov 93
[55] Bönsch G 1998 Interferometric calibration of an integrating sphere for determination of the roughness correction of gauge blocks *Proc. SPIE* **3477** 152–60
[56] Dowell J H 1943 *British Patent* No. 555672
[57] Hariharan P and Sen D 1959 New gauge interferometer *J. Opt. Soc. Am.* **49** 232–34
[58] Dorenwendt K 1973 Interferentielle Messung von nicht angeschobenen Endmaßen *PTB Annual Report 1972* p 121
[59] Gerasimov N P 1983 Investigations in the field of length and angle measurements *Proceedings of the D. I. Mendeleyev Institute for Metrology (Leningrad)* pp 14–8
[60] Khavinson V M 1999 Ring interferometer for two-sided measurement of the absolute lengths of end standards *Appl. Opt.* **38** 126–35
[61] Lewis A J 1993 Absolute length measurement using multiple wavelength phase stepping interferometry *PhD Thesis* London University
[62] Ishii Y and Seino S 1998 New method for interferometric measurement of gauge blocks without wringing onto a platen *Metrologia* **35** 67–73

[63] Lu S H, Chiueh C I and Lee C 2003 Measuring the thickness of opaque plane-parallel parts using an external cavity diode laser and a double-ended interferometer *Opt. Commun.* **226** 7–13
[64] Kuriyama Y, Yokoyama Y, Ishii Y and Ishikawa J 2006 Development of a new interferometric measurement system for determining the main characteristics of gauge blocks *Ann. CIRP* 55/1/(2006)
[65] Abdelaty A, Walkov A, Franke P and Schödel R 2012 Challenges on double ended gauge block interferometry unveiled by the study of a prototype at PTB *Metrologia* **49** 307–14
[66] Kruger O, Hungwe F, Farid N and Schreve K 2014 The design of a double ended interferometer (DEI) *Int. J. Metrol. Qual. Eng.* **5** 408
[67] Rau K, Mai T and Schödel R 2014 Absolute length measurement of prismatic bodies with PTB's new double-ended interferometer under the influence of wavefront aberrations *Website paper of 'MacroScale 2014'*
[68] Cordiale P, Galzerano G and Schnatz H 2000 International comparison of two iodine-stabilized frequency-doubled Nd:YAG lasers at 532 nm *Metrologia* **37** 177–82
[69] Schödel R 2009 Limiting aspects in length measurements by interferometry *Fringe 2009: the 6th Int. Workshop on Advanced Optical Metrology* pp 256–62
[70] Preston-Thomas H 1990 The international temperature scale of 1990 (ITS-90) *Metrologia* **77** 3–10
[71] Doiron T and Beers J S 1995 *The Gage Block Handbook, NIST Monograph 180* (Gaithersburg: US Department of Commerce, National Institute of Standards and Technology (NIST)) http://patapsco.nist.gov/mel/div821/
[72] Bönsch G 2001 Automatic gauge block measurement by phase stepping interferometry with three laser wavelength *Proc. SPIE* **4401** 1–10
[73] Lewis A J, Hughes B and Aldred P J E 2010 Long-term study of gauge block interferometer performance and gauge block stability *Metrologia* **47** 473–86
[74] Byman and Lassila A 2015 MIKES' primary phase stepping gauge block interferometer *Meas. Sci. Technol.* **26** 084009
[75] Decker J E and Pekelsky J R 1997 Uncertainty evaluation for the measurement of gauge blocks by optical interferometry *Metrologia* **34** 479–93
[76] The BIPM key comparison database URL: https://kcdb.bipm.org/ [accessed 2018/06/05]

# Chapter 3

# Fizeau interferometry for the sub-nm accurate realisation of sphere radii

**Arnold Nicolaus and Guido Bartl**

## 3.1 A brief history

A special type of interferometry was established and spawned by an exceedingly challenging and demanding task in the field of the precise determination of a fundamental constant: interferometry on spheres. Initially the goal was to determine the Avogadro constant $N_A$—the number of atoms or molecules in one mole of a substance—with an unprecedentedly small measurement uncertainty. However, the success of the experiments that were further developed in this context have contributed to the fact that the Planck constant $h$, ($h$ and $N_A$ can be directly converted into each other without significant loss of accuracy) has become the basic constant for redefining the 'kilogram', the SI unit for mass [1]. The path pursued for the determination of $N_A$ (or $h$) requires the volume of a test object—a sphere in this case—to be measured. Silicon spheres of 1 kg in weight were firstly manufactured with high optical quality at the Commonwealth Scientific and Industrial Research Organisation (CSIRO) in Australia [2] and after that they were produced at the Physikalisch-Technische Bundesanstalt (PTB) in Germany [3]. Typical properties of the spheres are a diameter of 93.7 mm, sphericity deviations of just a few tens of nanometres and a surface roughness below 0.1 nm. Interferometry for the determination of a volume is by no means new, as the most precise determination of the density of a body is using its definition $\rho = m/V$. Regularly shaped bodies with simple geometric forms were selected over 100 years ago, when the first measurements were carried out on bronze cylinders [4] and glass cubes [5]. With higher demands on volume determination, the simple determination of the linear dimensions of such bodies, for example the edge lengths of a cube, was less important than the requirement for greater form stability. Corner fractures occurred even when the cubes were properly handled. The former National Bureau of Standards (NBS), today the National Institute of Standards and Technology (NIST), USA, therefore

doi:10.1088/2053-2563/aadddcch3

decided to use steel spheres as density reference bodies in the 1960s [6], with the advantage that they were already commercially available at an acceptable quality. But the stability of these steel spheres was found to be insufficient, as they corroded and were often mechanically damaged so that spheres of Zerodur and silicon were selected instead. To determine the volume of highly accurate spheres at the level of 0.1 ppm, an absolute determination of the diameter with an uncertainty of about 1 nm is required. Saunders [7] was the first to construct an interferometer for measuring steel spheres at the former NBS. Following that, special interferometers were developed at the former IMGC [8], now the Istituto Nazionale di Ricerca Metrologica (INRiM), at the National Metrology Institute of Japan (NMIJ) [9], at CSIRO [10] and at PTB [11, 12].

## 3.2 The measurement principle

In contrast to typical optical measurands, in which spherical caps are also called 'spheres', here we only talk about whole spheres, for which the determination of the volume is most important. Hence, typical access to the volume is given by the diameter so that additional properties such as the curvature or distribution of radii are the result of advanced measurement and evaluation techniques (cf section 3.4.2). To determine the diameter of a sphere by means of an interferometer, a difference measurement is applied, which is traced back to Saunders [7]. The sphere to be measured is positioned between two reference faces, and the distances between the two reference faces as well as the distances between the sphere and the reference faces are measured. In the case of the interferometers of NBS, IMGC, NMIJ and CSIRO, plane reference faces and wavefronts are used (figure 3.1), forming a set of Newton's rings between the sphere and the corresponding flat reference face, the centre of which has to be determined. In contrast to that, a spherical geometry is used at PTB. The former only permit a pointwise determination of single diameters, whereas the latter allows diameter determinations across the entire area within the field of view of the lens. By appropriately repositioning the sphere, complete sphere topographies can thus be obtained. Figure 3.2 shows the principle of PTB's Fizeau interferometer. In this interferometer, the sphere is positioned between two lenses

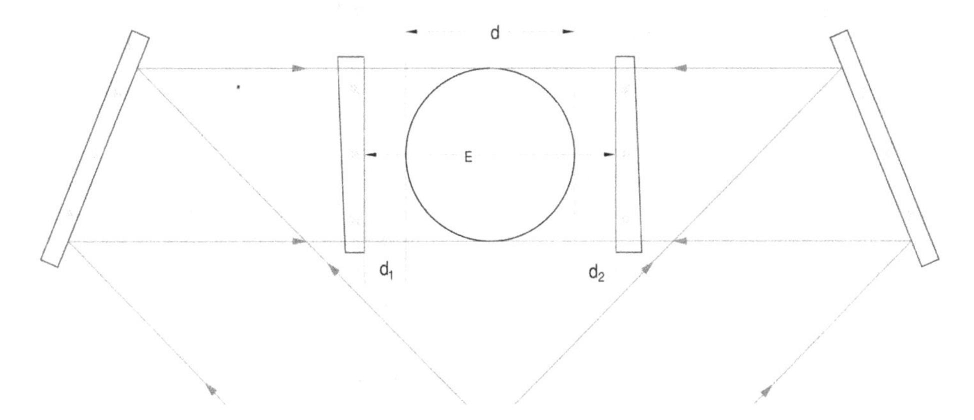


**Figure 3.1.** Principle of determining one diameter by Saunders' idea.

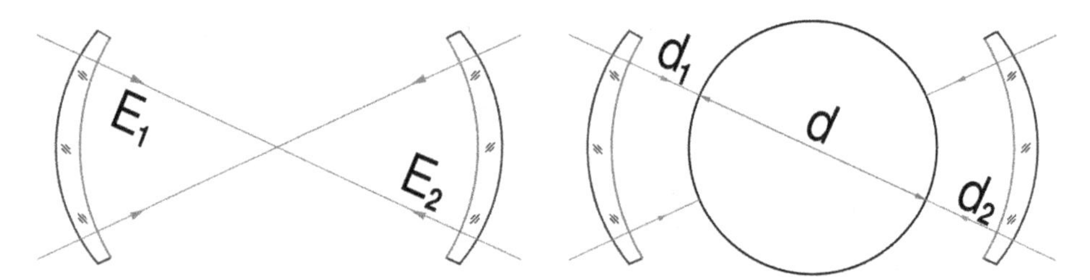


**Figure 3.2.** Principle of the diameter measurements in the spherical interferometer.

which form a spherical etalon with their integrated spherical reference faces. If the centre of the sphere coincides with the curve centres of the reference faces, the—equally spherical—wavefronts that are formed by the lenses reach the surface of the sphere at all points in the perpendicular direction. The diameter of the sphere $d$ can then be determined by measuring, on the one hand, the distance between the lens reference surfaces $E$ in the empty etalon and, on the other hand, the distances $d_1$ and $d_2$ between the sphere and the reference faces. In this way, the following equation is obtained:

$$d = E - d_1 - d_2. \tag{3.1}$$

In practice, however, the test objects are not mathematically exact spheres with one single diameter. Depending on their coordinates $(\vartheta, \varphi)$, different diameters $d(\vartheta, \varphi)$ are obtained which represent a specific topography for each sphere. In principle, the inner diameter of the empty etalon $E$ is expected to be one quantity but, due to possible optical imperfections, it is measured from both sides, yielding $E_1$ and $E_2$, which are averaged to $(E_1 + E_2)/2$. Thus, the final determination formula of the spherical interferometer of PTB for any diameter in the field of view is

$$d(\vartheta, \varphi) = (E_1(\vartheta, \varphi) + E_2(\vartheta, \varphi))/2 - d_1(\vartheta, \varphi) - d_2(\vartheta, \varphi). \tag{3.2}$$

## 3.3 Optical interference

In contrast to two-beam interferences, the interfering waves in the Fizeau interferometer are not so evident. In the case of two-beam interferences, the beam splitter (cube or plate) divides a beam into two differently running paths. In the case here (figure 3.3), the beam splitter is only to separate the observation from the illuminating direction. The beam splitting for the interference occurs at the second face of the semi-transparent mirror $S_1$. The first part is reflected back to the input direction at this surface. The second part is transmitted, impinges on the first face of the second mirror $S_2$ and is reflected. When the light reaches the face of $S_1$ again, it is both transmitted and reflected. The reflected part runs into the gap where the process is repeated several times depending on the reflectances of the surfaces $S_1$ and $S_2$. All transmitted waves of these runs superimpose with the first, directly reflected wave, so that the interference is a finite superposition of waves and can be described by an Airy function. As the Airy function needs an additional parameter to describe the slope of the curve, the evaluation requires a slightly different algorithm [13] than that for the harmonic curves in a two-beam interferometer. Typically phase-stepping algorithms were used for the evaluation of the optical phase. This requires the distance of the optical interference to be changed, often by steps of a quarter of an interference order. For the distance $d$ in figure 3.3, this can be carried out: e.g. by mechanically changing the distance $d$; in air by changing the refractive index $n$ of the matter in the gap; or in vacuum by changing the wavelength of the laser used.

## 3.4 Experimental implementation at PTB

At PTB, two interferometer setups for the measurement of spheres have been developed [14, 15]. Both are based on the principle of spherical reference faces as described in figure 3.2, mainly differing in size, aperture ratio and the quality of the optics. The entire setup (figure 3.4) is defined by the centre of the sphere to be measured. The sphere is placed on a three-point support that is tightly fixed to the monolithic steel frame of the interferometer. On opposite sides of the sphere, two Fizeau lenses with spherical reference faces are collinearly attached to the steel frame so that, in the ideal case, the curve centres of the reference faces coincide with the centre of the sphere. In this way, the lenses form a spherical etalon with the sphere in its centre. The sphere can be lifted from the three-point support with the aid of a displacement apparatus. In this upper position, an aperture in the lift rod clears the

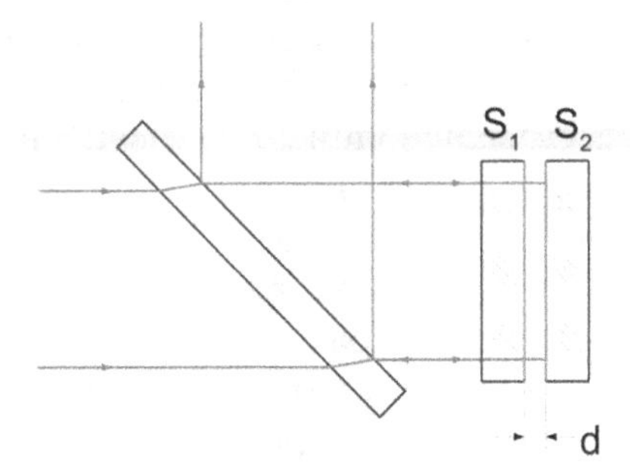


**Figure 3.3.** Setup of a Fizeau interferometer.

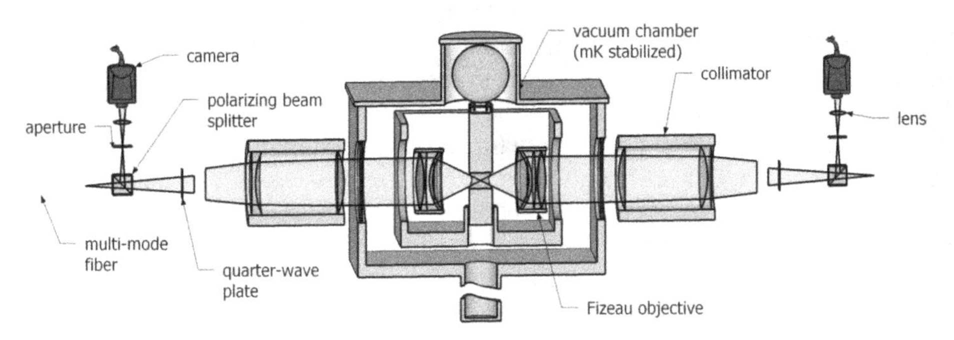


**Figure 3.4.** Sketch of the sphere interferometer setup. The sketch is not to scale, but depicts the symmetrical optical path which spans approximately 3 m. In the situation shown in the figure, the sphere under test is lifted up out of the beam path.

beam path so that the measurements of the empty etalon can be carried out. The sphere can be turned through two axes arranged perpendicularly to each other and is then lowered again. Depending on the aperture-to-diameter ratio of the Fizeau lenses (here 1:1), the sphere can be completely covered with 13 different orientations, thus allowing the sphere to be characterised by about 500 000 to 2 million diameters. The 90° phase steps needed for the Fizeau evaluation algorithm [13] could be obtained by changing the optical path length of the gaps $E$ or $d_1$ and $d_2$. Because of the spherical geometry, a displacement of the optical plates cannot work properly. The measurement evaluations are only correct for the diameters along the moving direction, all other directions will show increasing deviations. For measurements in air, the refractive index of air in the interferometer can be varied by changing the air pressure. As the interferometer is located in a pressure tight housing, the pressure inside can be controlled with a refractometer. This is a modified Jamin interferometer [16], which compares the optical path length in the surrounding air with that inside a reference vacuum chamber. The accuracy reached $5 \times 10^{-9}$ of the refractive index. For the measurement, the pressure is changed in four steps and stabilised by means of a computer-controlled motor-driven bellows. The necessary pressure changes depend on the separation of the reflecting surfaces and amount to about 170 Pa per step for the empty etalon. Due to the adiabatic change, the temperature of the air changes considerably. Therefore, any free volume of the interferometer was filled with a narrow grid of copper fins. Eventually, however, small deformations of the optical system due to the pressure changes began limiting the accuracy. For the highest accuracy the entire interferometer frame—including the three-point support and the lenses—is evacuated until a fine vacuum (0.1 Pa) is reached. Then, the influence of the refractive index on the measurement result can be neglected. The optical parameters of the objectives had to be changed to vacuum conditions so that the position of focus and the wavefront are optimised only for a vacuum. In vacuum, the deviation of the reference faces from a perfect sphere only amounts to a few nanometres. A further correction relates to the fact that the glass fibre ends are not ideal spot light sources, but expand in the perpendicular direction to the beam axis. A beam that does not leave the fibre exactly from the optical axis but a little bit apart, is not continued by the collimator in the exact parallel direction to the optical axis and, consequently, is not focused in the focal point of the lens reference faces. In the case of the sphere interferometer, the impact of beams running to some extent parallel to the optical axis is accounted for by means of an aperture correction [17]. For the absolute determination of the diameter, the temperature needs to be precisely controlled. Given a sphere diameter of approximately 100 mm, a temperature variation of e.g. 4 mK results in a 1 nm change in length. For this reason, the vacuum chamber is stabilised to the reference temperature of 20 °C by means of a thermostat. The interferometer frame is only in contact with the vacuum chamber with its three-point support. Due to the thermal inertia of the interferometer frame, the temperature varies by just a few parts of a millikelvin. To be able to correct the measured lengths to the target value of 20 ° C, the temperature should be determined in the order of 0.1 mK because of the targeted uncertainty. To achieve this, a copper block is used that has thermal contact with the interferometer frame and which serves as a temperature reference. The temperature difference between the sphere, the objectives and other relevant objects to this copper reference block is measured by means of thermocouple pairs [18]. The temperature of the copper block is traced back to the International Temperature Scale ITS-90 by means of a Pt 25 platinum resistor thermometer. The relative temperature measurement of the thermocouples is calibrated with a second, thermally insulated copper block with a second Pt 25 sensor in steps of some 10 mK.

### 3.4.1 Increment control and laser stabilisation

Nowadays, imaging interferometers are evaluated with phase shifting. Due to the spherical geometry, this is realised via a wavelength variation by means of a tuneable diode laser. For this purpose, a three-tiered laser setup is used. The traceability of the sphere diameter to the SI unit of length, the metre, is realised by using an iodine-stabilised He–Ne laser with a wavelength of $\lambda = 633$ nm. A broadly tuneable external cavity diode laser (ECL) [19] that also has a medium wavelength of 633 nm is coupled in two stages via an unmodulated offset laser. To realise this, a photodetector measures the beat signal between the diode laser and the offset laser or between the offset laser and the iodine-stabilised He–Ne laser. A detailed diagram of the laser setup is shown in figure 3.5. The offset laser is tightly coupled to the iodine-stabilised He–Ne laser via a frequency offset lock [20] and provides a stable, unmodulated light source. In the second stage, the frequency of the tuneable diode laser is compared to the offset laser by means of a phase-locked loop (PLL) circuit. The diode laser, together with the PLL circuit, forms a frequency synthesiser—in figure 3.5, the diode laser is to be seen as a voltage-controlled oscillator. A total frequency range between 0.5–15 GHz was realised that can currently be spanned in a

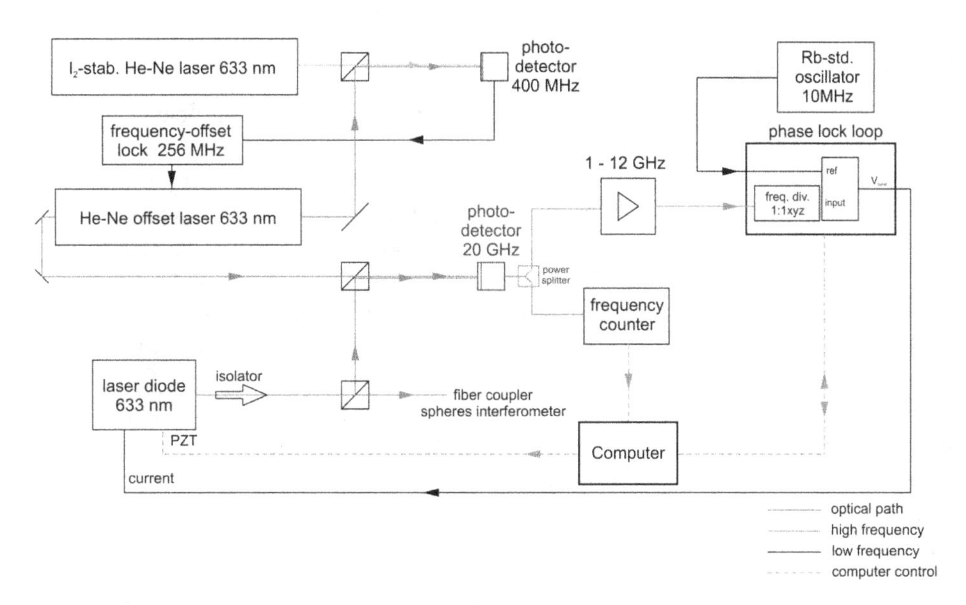


**Figure 3.5.** Diagram of the laser stabilisation system.

grid of 100 kHz. Through the computer-controlled positioning of the diffraction grating, the frequency of the diode laser (Littman–Metcalf arrangement) is brought close to the nominal frequency by means of a piezoelectric actuator (PZT). As soon as the frequency is within the pull-in range of the PLL, the PLL will coordinate the frequency stabilisation. The beat frequency between the diode laser and the offset laser is fractionally divided and compared to a fixed frequency reference oscillator. The phase difference between the reference and the appropriate divided beat frequency is the control signal for the laser diode current modulation. As the mechanical system of the mirror/PZT exhibits a certain inertia, the frequency range would be limited to approximately 1 kHz if the PZT only was used for frequency stabilisation. A faster frequency stabilisation is obtained via the modulation of the laser diode current. At a bandwidth of some 100 kHz, the control via the laser diode current is distinctly faster; on the other hand, the lock-in range with approximately 200 MHz is too small for the total coverage of the operating range. To prevent the frequency from leaving the lock-in range of the PLL, a slow frequency drift is again compensated for by means of the PZT, so that the PLL is always in the centre of the lock-in range. For the interference analysis, the interference must be modified in four steps by one complete interference order in total. A distinction must be made between the two cases of the 'empty etalon' and the 'interferometer with sphere'. The distance between the reference faces in the empty etalon is approximately 150 mm, so that a frequency step width of about 250 MHz results for a step of 1/4 of an interference order. In the second case, the distance between the sphere and the reference faces is approximately 30 mm so that, in this case, frequency steps of 1300 MHz each are necessary. An exact determination of the necessary increments can be made via self-calibration, which is possible by using the phase-shifting procedure itself [21].

### 3.4.2 Reconstruction of radii

According to [22, 23], the volume of a sphere can be determined sufficiently precisely from its mean diameter, if there are only small deviations from the form of an ideal sphere compared to its basic diameter. As the measurements from the spherical interferometer originally aimed at determining the volume of the silicon spheres of the Avogadro project, the datasets of the diameter were analysed correspondingly, which means that the topographic information from both sides of the sphere was mixed by taking the difference of the measured distances. A graphical representation of this kind of result can be realised as a diameter topography which is point symmetric to the centre of the sphere.

When the real form of the sphere, however, is the targeted quantity, which is most likely not point symmetric, the data has to be evaluated in a way that the topographic information of both sides of the sphere is treated separately. The basic idea of the applied stitching approach [24] is to extract the information about the topography of the sphere from the measured data and separate it from systematic effects by concatenating the measured segments of the surface (illustrated in figure 3.6). Therefore, the basic requirement for the reconstruction of the sphere's form is a

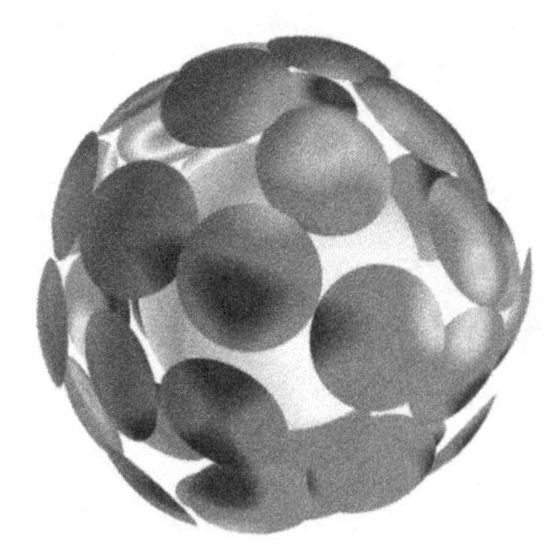

**Figure 3.6.** Illustration of the stitching principle.

set of measured segments of the surface with overlapping areas which cover the entire sphere.

The corresponding implementation is based on a mathematical model which is represented by the following system of linear equations:

$$\left.\begin{aligned} {}^{S1}\mathfrak{m}_{(\vartheta,\,\varphi,\,p,\,s)} &= -\delta\mathcal{R}_{(\vartheta,\,\varphi)}/K_{T_{(s)}} + {}^{S1}\bar{\mathfrak{m}}_{(s)} + {}^{1}\mathcal{C}_{(p,\,s)} + {}^{1}\mathcal{O}_{(p)} \\ {}^{S2}\mathfrak{m}_{(\check{\vartheta},\,\check{\varphi},\,p,\,s)} &= -\delta\mathcal{R}_{(\check{\vartheta},\,\check{\varphi})}/K_{T_{(s)}} + {}^{S2}\bar{\mathfrak{m}}_{(s)} + {}^{2}\mathcal{C}_{(p,\,s)} + {}^{2}\mathcal{O}_{(p)} \\ {}^{E1}\mathfrak{m}_{(p,\,s)} &= {}^{E}\bar{\mathfrak{m}}_{(s)} + {}^{1}\mathcal{C}_{(p,\,s)} + {}^{2}\mathcal{C}_{(p,\,s)} + {}^{1}\mathcal{O}_{(p)} + {}^{2}\mathcal{O}_{(p)} \\ {}^{E2}\mathfrak{m}_{(p,\,s)} &= {}^{E}\bar{\mathfrak{m}}_{(s)} + {}^{1}\mathcal{C}_{(p,\,s)} + {}^{2}\mathcal{C}_{(p,\,s)} + {}^{1}\mathcal{O}_{(p)} + {}^{2}\mathcal{O}_{(p)} \end{aligned}\right|_{(p,\,s)} \tag{3.3}$$

$${}^{\mathrm{bias}}\mathfrak{m}_{(s)} = 2\,\bar{r}/K_{T_{(s)}} - {}^{E}\bar{\mathfrak{m}}_{(s)} + {}^{S1}\bar{\mathfrak{m}}_{(s)} + {}^{S2}\bar{\mathfrak{m}}_{(s)}\big|_{(s)}.$$

In these equations, ${}^{S1}\mathfrak{m}_{(\vartheta,\,\varphi,\,p,\,s)}$ and ${}^{S2}\mathfrak{m}_{(\check{\vartheta},\,\check{\varphi},\,p,\,s)}$ are the measured values (i.e. the phase values converted to length values via the laser wavelength used) from both sides of the sphere associated with the number of the measured segment $s$, the position on the surface of the sphere in spherical coordinates $(\vartheta, \varphi)$ and the pixel number $p$ in the field of view of the camera. $\check{\vartheta} = \pi - \vartheta$ and $\check{\varphi} = (\pi + \varphi) \bmod 2\pi$ specify the opposite side on the sphere's surface. ${}^{E1}\mathfrak{m}_{(p,\,s)}$ and ${}^{E2}\mathfrak{m}_{(p,\,s)}$ are the measured values of the empty etalon taken from both directions and ${}^{\mathrm{bias}}\mathfrak{m}_{(s)}$ are the bias values of the diameter of each segment $s$.

The measured data contain the information about the form deviations of the sphere $\delta\mathcal{R}$, the form deviations from a perfect spherical shape of the reference surfaces ${}^{1}\mathcal{O}$ and ${}^{2}\mathcal{O}$ (Figure 3.7(a)) and systematic corrections ${}^{1}\mathcal{C}$ and ${}^{2}\mathcal{C}$ for the effect of residual misalignments of the objectives (figure 3.7(b)). The terms ${}^{S1}\bar{\mathfrak{m}}_{(s)}$, ${}^{S2}\bar{\mathfrak{m}}_{(s)}$ and ${}^{E}\bar{\mathfrak{m}}_{(s)}$ have to be introduced as the mean values of all the pixels of the respective partial measurement $(s)$. To make the averaging work correctly, also the effect of different temperatures during the sequence of measurements has to be considered due to the thermal expansion of the material of the sphere. This is achieved by introducing the correction factor $K_{T_{(s)}} = 1/(1 + \alpha \cdot \Delta T_{(s)})$ with $\alpha$ being the coefficient of thermal expansion of the respective material and $\Delta T_{(s)}$ the deviation from a

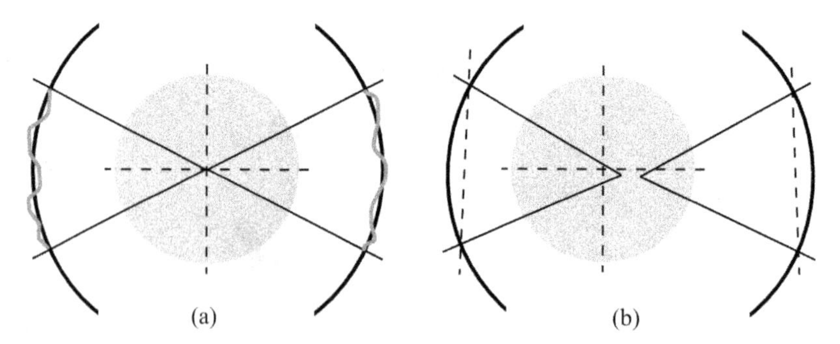

**Figure 3.7.** Effects to be considered in the stitching model: (a) form deviations of the reference faces of the objectives, (b) misalignment of the objectives.

reference temperature during the measurement. The calculation of the mean radius $\bar{r}$ averaged over all contributing measured values is also included.

As regards the parameterisation of the underlying aspects, a convenient parameterisation of the topography of the sphere can be realised by real spherical harmonics which form an orthonormal basis on a sphere:

$$Y_{\ell m}(\vartheta, \varphi) = \begin{cases} \frac{1}{\sqrt{2}}(Y_\ell^m(\vartheta, \varphi) + (-1)^m Y_\ell^{-m}(\vartheta, \varphi)) & : m > 0 \\ Y_\ell^0 & : m = 0 \\ \frac{1}{i\sqrt{2}}(Y_\ell^{-m}(\vartheta, \varphi) - (-1)^m \; Y_\ell^m(\vartheta, \varphi)) & : m < 0 \end{cases} \quad (3.4)$$

where $\ell \in \mathbb{N}$, $m \in \{-\ell, -\ell + 1, \dots, \ell\}$, $(\vartheta, \varphi)$ are spherical coordinates and $Y_\ell^m(\vartheta, \varphi)$ are spherical harmonics as defined in [25]. The form deviations of the sphere related to the mean radius can then be described by

$$\delta\mathcal{R}_{(\vartheta, \varphi)} = \sum_{\ell=2}^{\ell_{\max}} \sum_{m=-\ell}^{\ell} k_{\ell m} \;\; Y_{\ell m}(\vartheta, \varphi). \quad (3.5)$$

The constraint $\ell \geqslant 2$ is motivated by the choice of a coordinate system whose origin coincides with the centre of the sphere which is convenient, as a translational movement of the sphere cannot be separated from a misalignment of the objectives. In practice, the number of coefficients $k_{\ell m}$ is limited by $\ell_{\max}$.

A widely used representation of form deviations in the field of optics is given by Zernike polynomials $Z_n^m(\rho, \varphi)$ [26, 27]. $n$ and $m$ are integer numbers defining the order of the polynomial, $\rho$ is the radial distance within the unit circle and $\varphi$ is the azimuthal angle. In our case, for simplicity, the two integer indices are replaced by a single index number in sequential order corresponding to [28] and the position in the field of view is specified by the pixel number $p$ instead of $(\rho, \varphi)$. The form deviations of the reference faces of the objectives can then be described by

$${}^j\mathcal{O}_{(p)} = \sum_{z=4}^{z_{\max}} {}^j o_z \;\; Z_z(p) \quad (3.6)$$

with $j \in \{1, 2\}$ for objectives 1 and 2 and the coefficients ${}^{s,j}o_z$ with the limit $z_{\max}$. The parameterisation starts at $z = 4$, because the offset, tilt and defocus cannot be separated unambiguously from the misalignment terms.

The latter are expressed by three spherical harmonics which correspond to translations along the three orthogonal spatial directions:

$$ {}^{j}\mathcal{C}_{(p,\,s)} = \sum_{m=-1}^{1} {}^{s,j}c_{1m} \;\; Y_{1m}(\Theta_{(p)}, \Phi_{(p)}) \tag{3.7} $$

with $j \in \{1, 2\}$ for directions 1 and 2 and the coefficients ${}^{s,j}c_{1m}$ describing the misalignments along the optical axis and perpendicular to it for each measured segment $s$. Here, the spherical coordinates $(\Theta_{(p)}, \Phi_{(p)})$ are not related to a position on the sphere's surface, but specify a position in the field of view referred to by the pixel $p$. The pure offset terms are omitted from ${}^{j}\mathcal{C}_{(p,\,s)}$ and treated as separate parameters

$$ {}^{j}\mathfrak{m}_{(s)} = {}^{s,j}c_{00} \;\; Y_{00}(\Theta_{(p)}, \Phi_{(p)}) \tag{3.8} $$

each of which is parameterised by the coefficient ${}^{s,j}c_{00}$ with $j \in \{\mathbb{S}1, \mathbb{S}2, \mathbb{E}\}$. Finally, the representation of the mean radius is

$$ \bar{r} = k_{00} \;\; Y_{00}(\vartheta, \varphi) \tag{3.9} $$

so that the shape of the sphere is given by the expression

$$ \mathcal{R}_{(\vartheta,\,\varphi)} = \bar{r} + \delta\mathcal{R}_{(\vartheta,\,\varphi)} \tag{3.10} $$

$$ = k_{00} \;\; Y_{00}(\vartheta, \varphi) + \sum_{\ell=2}^{\ell_{\max}} \sum_{m=-\ell}^{\ell} k_{\ell m} \;\; Y_{\ell m}(\vartheta, \varphi) \tag{3.11} $$

$$ = \sum_{\ell=0}^{\ell_{\max}} \sum_{m=-\ell}^{\ell} k_{\ell m} \;\; Y_{\ell m}(\vartheta, \varphi) \tag{3.12} $$

with $k_{1-1} = k_{10} = k_{11} = 0$.

The described model yields an over-determined system of linear equations (cf equation (3.3)) of around $10^5$ to $10^6$ rows and $10^3$ to $10^4$ columns depending on the number of measured values and the number of parameters, respectively. The values of the parameters which represent the topography of the sphere, the form deviations of the reference faces and the alignment corrections can be determined via a least squares optimisation. To be able to calculate the uncertainties of the model parameters based on the uncertainties of the input data, the solving algorithm implements a weighted least squares approach in which the covariance matrix of the measured values is taken into account (details can be found in [24]).

### 3.4.3 Examples of results

The evaluation of the measured data ends up in a topography of the form deviations of the sphere under test (as an example see figure 3.8). The topography is represented by a set of spherical harmonic functions from which the volume can then be obtained via an integration over the full solid angle. The achieved measurement uncertainty for a single radius is usually of the order of 1 nm. Typical values of a silicon sphere are a mean diameter of 93.6 mm for a sphere of natural isotopic consistence. The diameter values vary by about 17 nm between the minimum and the maximum diameter, which shows that the sphere is the result of an outstanding manufacturing process. Such a sphere, scaled up to the size of the Earth, would only have mountains of 1.20 m in height and the deepest sea would have a greatest depth of again not more than 1 m. Furthermore, the distances between these extremes would amount to a 6000 km distance on such an earth, so that estimating the inclination would not be easy. According to the analysis method described, the mean diameter is for 20 °C and 0 Pa and the correction for the oxide layer has not yet been taken into account. Corrections for special usage will be applied according to the requirements. While the form deviations of the sphere under test have no significant influence on the volume calculation based on the average radius [14], they directly affect the interference patterns from which the radius data are derived. To assess the significance of this effect, ray-tracing simulations have been applied [15]. Special ray-tracing software was set up to model the light path from the fibre end surfaces through the whole interferometer to the camera surfaces. Since there are different ray paths occurring only inside the Fizeau cavities due to multiple reflections, sequential ray tracing was implemented. Several ray bundles emerging from different positions and under different angles out of the fibre end surface are used. All light paths were calculated, especially regarding the multiple reflection orders in the Fizeau cavity. To determine the influence of the real sphere radius topography, the software was adopted to simulate measurements for arbitrary orientations of an input topography, e.g. the results of a real sphere. This allows the simulation of a complete sphere topography measurement with all the steps for the new orientation of the sphere, possible optical adjustments and a full evaluation including the reconstruction of the radii. Thus, by preparing a topography of a measured sphere and receiving a simulated topography result, it is possible to determine corrections of the measurement results. For the usual precision spheres which are measured in

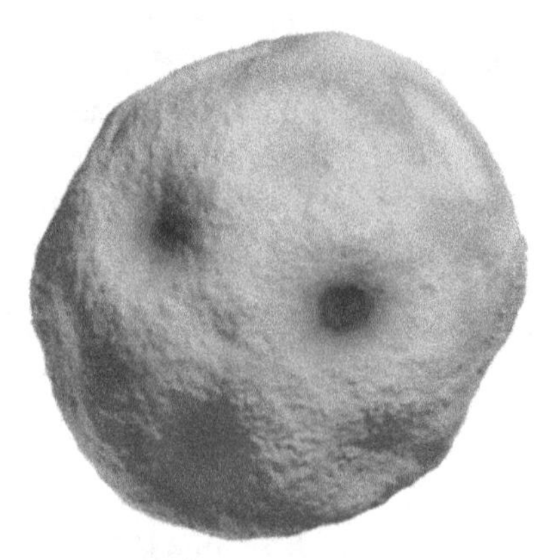

**Figure 3.8.** Rendered image of a reconstructed topography of a sphere with exaggeratedly illustrated form deviations. The colour scale spans 15 nm.

PTB's sphere interferometer, the uncertainty contribution due to form deviations is less than 0.2 nm, so that a relative uncertainty for the volume of $7 \times 10^{-9}$ could be reached.

## 3.5 Outlook

PTB's sphere interferometer was designed to determine the volume of spherical artefacts utilised as 1 kg mass standards. In this approach, uncertainties in the range of a few ppb could be reached. An advantage of the new definition of the mass unit is not being limited to 1 kg and therefore not being limited to spheres of 94 mm in diameter. In principle, any sphere which just fits into the two Fizeau objectives and any smaller sphere can be measured. However, the stability of an interferometer increases with its mechanical stability, so that a small-sized interferometer with smaller gaps would be of advantage for a very small sphere. For mechanical applications, smaller spheres are of greater interest. The knowledge of some absolute radii, or better still the radii of a complete sphere, would serve to trace back diverse contacting measuring devices, as used in 3D coordinate measuring systems for all kinds of industrial production. This requires the complete measurement of the sphere under test, and therefore puts high demands on the precise micro-positioning of such small spheres. Reducing the diameter by a factor of ten reduces the mass by a factor of 1000. Automated handling would therefore present great challenges to mechanical design and manufacturing processes. The optical design is not limited to spherical artefacts—e.g. a cylindrical geometry is conceivable, too. This requires cylindrical optics transforming the plane wave into an optical wave which fits the artefact to be investigated. Deviations of form and gauge would be available with interferometrical resolution and uncertainties in the nm and sub-nm region.

## References


[1] Richard P, Fang H and Davis R 2016 Foundation for the redefinition of the kilogram *Metrologia* **53** A6

[2] Leistner A and Zosi G 1987 Polishing a 1-kg silicon sphere for a density standard *Appl. Opt.* **26** 600–01

[3] Fujii K, Bettin H, Becker P, Massa E, Rienitz O, Pramann A, Nicolaus A, Kuramoto N, Busch I and Borys M 2016 Realization of the kilogram by the XRCD method *Metrologia* **53** A19

[4] Guillaume Ch Éd 1910 Détermination du volume du kilogramme d'eau (mesures par la mthode des contacts) *Trav. Mem. Bur. Int. Poids Mes.* **14** 1–276

[5] Chappuis P 1910 Détermination du volume du kilogramme d'eau (mesures par la premire mthode interferentielle) *Trav. Mem. Bur. Int. Poids Mes.* **14** 1–163

[6] Bowman H A, Schoonover R M and Carroll C L 1974 A density scale based on solid objects *J. Res. Natl. Bur. Stand.* **78A** 13–40

[7] Saunders J B Sr 1972 Ball and cylinder interferometer *J. Res. Natl. Bur. Stand.* **76C** 11–20

[8] Sacconi A, Peuto A, Panciera R, Pasin W, Pettorruso S and Rasetti M 1988 Density standards for a redetermination of the Avogadro constant *Measurement* **6** 41–5

[9] Kuramoto N and Fujii K 2009 Improvement in the volume determination for Si spheres with an optical interferometer *IEEE Trans. Instrum. Meas.* **58** 915–18

[10] Giardini W, Manson P, Wouters M, Warrington B, Ward B, Bignell N, Walsh C, Jaatinen E and Kenny M 2009 Density of a single-crystal natural silicon sphere *IEEE Trans. Instrum. Meas.* **58** 908–14

[11] Nicolaus R A and Bönsch G 1998 Doppelseitige Fizeau-Interferometer mit Phasenverschiebeauswertung für dimensionelle Messungen *Tech. Mess.* **65** 83–90

[12] Nicolaus R A and Geckeler R D 2007 Improving the measurement of the diameter of Si spheres *IEEE Trans. Instrum. Meas.* **56** 517–22

[13] Bönsch G and Böhme H 1989 Phase-determination of Fizeau interferences by phase-shifting interferometry *Optik* **82** 161–64

[14] Bartl G, Bettin H, Krystek M, Mai T, Nicolaus A and Peter A 2011 Volume determination of the Avogadro spheres of highly enriched $^{28}$Si with a spherical Fizeau interferometer *Metrologia* **48** S96–103

[15] Mai T and Nicolaus A 2017 Optical simulation of the new PTB sphere interferometer *Metrologia* **54** 487

[16] Birch K P and Downs M J 1988 The results of a comparison between calculated and measured values of the refractive index of air *J. Phys. E: Sci. Instrum* **21** 694

[17] Nicolaus R A and Bönsch G 2009 Aperture correction for a sphere interferometer *Metrologia* **46** 668–73

[18] Bönsch G, Schuster H-J and Schödel R 2001 Hochgenaue Temperaturmessung mit Thermoelementen (High-precision temperature measurements with thermo couples) *Tech. Mess.* **68** 550–57

[19] Imkenberg F, Nicolaus A and Abou-Zeid A 1999 Tunable 633 nm diode lasers and application for phase stepping interferometry, *Proc. 1st Int. EUSPEN Conf.: Precision Engineering-Nanotechnology* **vol 2** pp 243–46

[20] Rowley W R C 1977 A digital frequency-offset lock system designed for use with stabilized lasers.

[21] Nicolaus R A, Bönsch G and Kang C-S 2000 Interferometrische Kalibrierverfahren für die Schrittweitensteuerung eines Phasenverschiebeinterferometers *Tech. Mess.* **67** 328–33

[22] Johnson D P 1974 Geometrical considerations in the measurement of the volume of an approximate sphere *J. Res. Natl. Bur. Stand.* **78A** 41–8

[23] Mana G 1994 Volume of quasi-spherical solid density standards *Metrologia* **31** 289–300

[24] Bartl G, Krystek M and Nicolaus A 2014 PTB's enhanced stitching approach for the high-accuracy interferometric form error characterization of spheres *Meas. Sci. Technol.* **25** 064002

[25] Cohen-Tannoudji C, Diu B and Laloë F 1977 *Quantum Mechanics–Part 1* (New York: Wiley)

[26] Born M and Wolf E 1989 *Principles of Optics: Electromagnetic Theory of Propagation, Interference, and Diffraction of Light* 6th edn (New York: Pergamon)

[27] Wyant J C and Creath K 1992 Basic wavefront aberration theory for optical metrology *Applied Optics and Optical Engineering Series*vol XI (New York: Academic) ch 1

[28] Noll R J 1976 Zernike polynomials and atmospheric turbulence *J. Opt. Soc. Am.* **66** 207–11

**IOP** Publishing

Modern Interferometry for Length Metrology
Exploring limits and novel techniques
**René Schödel (Editor)**

# Chapter 4

# Laser interferometry for high resolution metrology in space

**Thilo Schuldt**

## 4.1 Introduction

Over recent decades, laser interferometry has become a standard tool for dimensional metrology in many Earth-based applications. A variety of different interferometer designs and implementations have been developed, increasing measurement sensitivity and accuracy and addressing specific requirements, e.g. placed in harsh environments within industrial production. At the same time, interferometry began to be investigated as a high performance tool for inter- and intra-spacecraft distance and angle metrology, enabling new mission concepts in science and Earth observation.

This chapter reviews the current status of space missions employing high resolution laser interferometry, a technology which just recently entered space operation. In December 2015, the LISA Pathfinder spacecraft was launched, successfully demonstrating picometer and nanoradian interferometry between two free flying proof masses on board the satellite [8]. It serves as a technology demonstrator for the Laser Interferometer Space Antenna (LISA), which is selected by the European Space Agency (ESA) for launch in 2034. Furthermore, the Gravity Recovery and Climate Experiment (GRACE) follow-on mission—launched in May 2018—will demonstrate the first nanometer sensitivity laser ranging instrument between two satellites flying at a distance of approximately 200 km in a low-Earth orbit [86]. These missions are dedicated to space-based gravitational wave detection and the mapping of the Earth's gravity field, respectively. This chapter focusses on these two application areas, showing the highest maturity within space missions using laser interferometry.

Gravitational waves have been directly detected for the first time in a simultaneous measurement of the two Earth-based gravitational wave detectors LIGO (Laser Interferometer Gravitational-Wave Observatory) in September 2015 [1]—an event that was recognized by the 2017 Nobel Prize in Physics. Gravitational waves cause a time-dependent strain in space, stretching and squeezing spacetime. This strain can be measured by monitoring distance variations between proof masses which are placed at different positions in spacetime. In the case of LIGO, the proof masses are realized as end mirrors of an interferometer which are 4 km apart. Distance variations are monitored using highest sensitivity laser interferometric techniques with a strain sensitivity $\Delta l/l$ below $10^{-23}/\sqrt{\mathrm{Hz}}$ for frequencies around 100 Hz [59]. LIGO is operating in the frequency band 10 Hz to 10 kHz where the low-frequency limit is mainly given by displacement noise from seismic motion. The proposed space-based gravitational wave detector LISA is complementary to the Earth-based detectors covering the low-frequency range 0.1 mHz to 0.1 Hz with a targeted strain sensitivity of $10^{-20}/\sqrt{\mathrm{Hz}}$. Distance variations between undisturbed proof masses placed within spacecraft which are separated by 2.5 million kilometers will be measured using heterodyne laser interferometry with the required 10 $\mathrm{pm}/\sqrt{\mathrm{Hz}}$ sensitivity in the LISA frequency band (cf figure 4.1, left). In 2017, LISA was selected by ESA as it's third large-class mission with a target launch in 2034. Within the LISA Pathfinder mission, key technologies for LISA were already successfully demonstrated in space [8, 9].

Spatial and time variations of the Earth's gravitational field can be measured using satellite-to-satellite tracking (SST) between two spacecraft in a low-Earth orbit (LEO) with altitudes of several hundred kilometers. Differential accelerations on the two satellites cause variations in the inter-satellite distance. This was successfully demonstrated within the NASA-DLR GRACE mission consisting of two identical satellites flown in a circular polar orbit [88]. The inter-satellite distance is maintained between 170 km and 270 km and variations in this distance are measured with μm-sensitivity using a two-way microwave link. While the GRACE mission was terminated in 2017 after 15 years of successful operation, the successor mission, GRACE follow-on, was launched in May 2018 in order to ensure continuation in data collection (cf figure 4.1, right). GRACE follow-on is mostly a rebuild of the GRACE satellites, but it includes an additional laser ranging instrument as technology demonstrator with 80 $\mathrm{nm}/\sqrt{\mathrm{Hz}}$ sensitivity in the measurement band between 2 mHz and 0.1 Hz. The distance variation measurement is about two orders

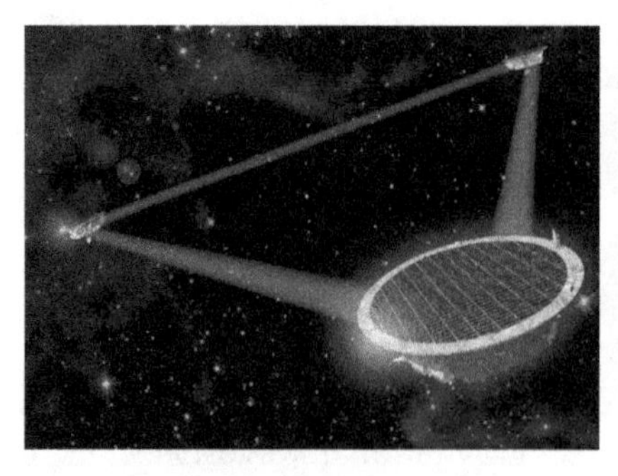
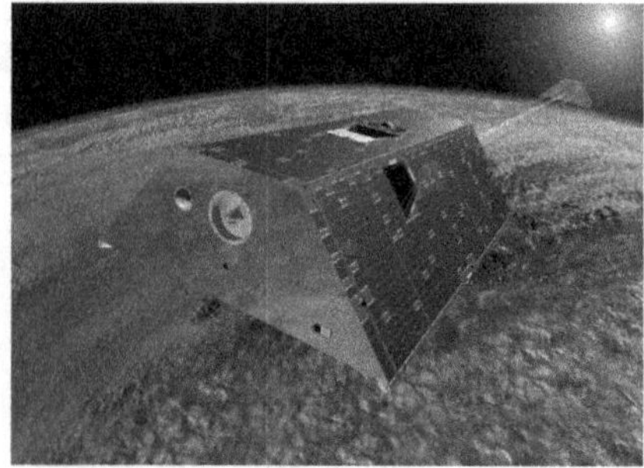

**Figure 4.1.** Artists' impressions of the space missions LISA (left, image credit: © Airbus GmbH) and GRACE follow-on (right, image credit: NASA). The laser links between the satellites are shown in red.

of magnitude better than the microwave instrument [86]. Within the Next Generation Gravity Mission (NGGM) program by ESA, several mission concepts for long-term monitoring of the time-variable Earth's gravity field with further improved temporal and spatial resolution are currently investigated including concepts similar to GRACE follow-on, also using laser ranging technology.

In addition, laser interferometry has also been investigated in the context of formation-flying missions where a satellite formation formed by several individual spacecraft needs to be controlled with high accuracy, requiring specific (absolute and relative) distance metrology systems in order to establish a rigid formation. This is needed, for example, for telescope missions using aperture synthesis and was investigated by ESA within the Darwin mission [22, 32] and by NASA within the Space Interferometry Mission (SIM) [58] and Terrestrial Planet Finder (TPF) [54] mission proposals. These mission concepts were put on hold in 2007, 2010 and 2011, respectively, due to budgetary reasons and only smaller mission concepts are currently under investigation with respect to feasibility, see e.g. the review given in [72].

Space operations yield to specific requirements for the optical systems including optimization with respect to power, mass and volume, autonomous operation and mechanical and thermal stability of the instrument structure. In the context of space qualification, components and systems need to undergo environmental testing including vibration testing, thermal cycling in vacuum, shock and radiation hardness tests. Pre-launch environment, launch phase and operational spacecraft environment needs to be considered, each with its own specific requirements. Especially electronic and optical components and systems need to be verified with respect to lifetime, often affected by space radiation environment. Within the instrument design, single-point failures need to be avoided and critical components/subsystems to be realized in redundancy in order not to jeopardize the mission in case of their failure. The so-called RAMS (reliability, availability, maintainability, safety) management process is typically applied which includes amongst others detailed risk analysis. Space operation is therefore a main driver for technology development, including innovations in new light-weight and stiff materials and specific assembly-integration technologies, e.g. for optical systems.

The realization of space instruments typically takes place in different development steps (on subsystem and system level). Starting from laboratory experiments not optimized for space, a dedicated breadboard model, called the elegant breadboard (EBB), is often developed, demonstrating the functional behavior using standard laboratory equipment. In a next step, an engineering model (EM) is realized showing the full functionality of the (sub)system. It takes into account space design requirements, e.g. with respect to compactness and robustness and is flight representative in form, fit and function. Critical functions are demonstrated in a relevant environment. The EM typically uses a combination of commercial and flight-grade components and does not include full redundancy. The qualification model (QM) is fully representative to the flight hardware and is subjected to complete functional and environmental testing where qualification test levels are applied. The flight model (FM) will go to space. It is subjected to functional and environmental testing, where acceptance test levels are applied. These are lower than the qualification test levels and may be reduced in time. The acceptance test will verify the manufacturing process.

Schedule, cost and development risks are crucial parameters for space missions and a sufficient technology maturity is needed to avoid failure in space where repair is not possible. Therefore, the technology maturity, from component level to system level, is assessed and quantified by the so-called technology readiness level (TRL) [34]. The highest level (TRL9) is reached for flight-proven components and systems which have demonstrated their performance in successful mission operation, the lowest level (TRL1) corresponds to the observation and reporting of basic principles. Functional verification of components and/or breadboards in a typical laboratory environment (including EBB) corresponds to TRL4, the full-scale EM to TRL6. The QM corresponds to TRL7 where the full system/subsystem is verified by successful environmental testing and TRL8 is assigned to the flight qualified FM.

In sections 4.3 and 4.4, the two application areas mentioned above—space-based gravitational wave detection and Earth gravity field measurement—are discussed. Beside presenting the mission concepts, the focus is placed on the laser interferometer concept for distance metrology and the corresponding technology development for space. A comparison of the key parameters of the inter-satellite laser ranging in both applications is given in table 4.1. They both rely on heterodyne interferometry using laser technology at a wavelength of 1064 nm. The fundamental principles of the interferometric distance measurement are detailed in section 4.2.

**Table 4.1.** Key parameters of LISA and GRACE follow-on laser ranging measurements, based on the current system designs.

| | LISA | GRACE follow-on |
|---|---|---|
| Inter-satellite distance | 2 500 000 km | 170 km to 270 km |
| Orbit | Heliocentric<br>Earth-trailing<br>with no atmospheric drag | Low-Earth orbit<br>(490 km altitude)<br>with high atmospheric drag |
| Satellite relative velocity | $\pm 15$ m s$^{-1}$ | $\pm 3$ m s$^{-1}$ |
| Measurement frequency band | 0.1 mHz to 0.1 Hz[a] | 2 mHz to 0.1 Hz |
| Ranging noise[b] | 10 pm/$\sqrt{\text{Hz}}$ × NSF($f$) | 80 nm/$\sqrt{\text{Hz}}$ × NSF($f$) |
| Laser power (at fiber output) | 2 W | 25 mW |
| Telescope aperture | 30 cm | 8 cm[c] |
| Received optical power[d] | ~100 pW | ~100 pW |

[a] With a goal sensitivity from 20 μHz to 1 Hz.
[b] A mission specific noise shaping function (NSF) applies with relaxations towards lower frequencies.
[c] There is no telescope, the beam is clipped at an aperture stop.
[d] Power received at the photodetector.

## 4.2 Interferometric distance and tilt metrology

Interferometers for distance measurement are realized in many variations, subdivided with respect to different properties such as e.g. optical path (e.g. Fabry–Perot, Michelson, Mach–Zehnder), homodyne versus heterodyne detection, and the use of polarizing optics versus non-polarizing optics. The choice of the interferometer design is mainly driven by the required performance but also needs to take into account application specific design drivers. Compared to homodyne detection, heterodyne interferometry offers higher noise immunity, and is straightforward in implementation for inter-satellite metrology due to Doppler shifts caused by relative velocities of the two spacecraft. It is used within LISA and GRACE follow-on, where details on the interferometer implementation are given in the corresponding sections 4.3 and 4.4, respectively.

In a heterodyne interferometer, two laser beams with frequencies $f_1$ and $f_2 = f_1 + \Delta f$ interfere. The frequency offset $\Delta f$ between the two beams can be realized, e.g. using an electro-optic or acousto-optic modulator (AOM) or a second offset phase locked laser. In an inter-satellite measurement using a laser with wavelength $\lambda$, a relative velocity $v_{\mathrm{rel}}$ between the two satellites causes a one-way Doppler shift $f_{\mathrm{Doppler}} = -v_{\mathrm{rel}}/\lambda$. Using a laser with a wavelength of 1064 nm, relative velocities in the order of a few m $\mathrm{s}^{-1}$ (as present in the case of GRACE follow-on and LISA, cf table 4.1) correspond to frequency shifts of the order of a few megahertz [86].

A simplified schematic of an interferometric inter-satellite distance measurement in a so-called transponder scheme is given in figure 4.2. It shows a heterodyne interferometer where the laser on spacecraft one (S/C1) is split into a local reference beam (often also referred to as 'local oscillator', LO) and a beam transmitted to the distant spacecraft (TX beam). The light received from the distant spacecraft (RX beam) is superimposed with the local oscillator and detected at a photodetector (PD).

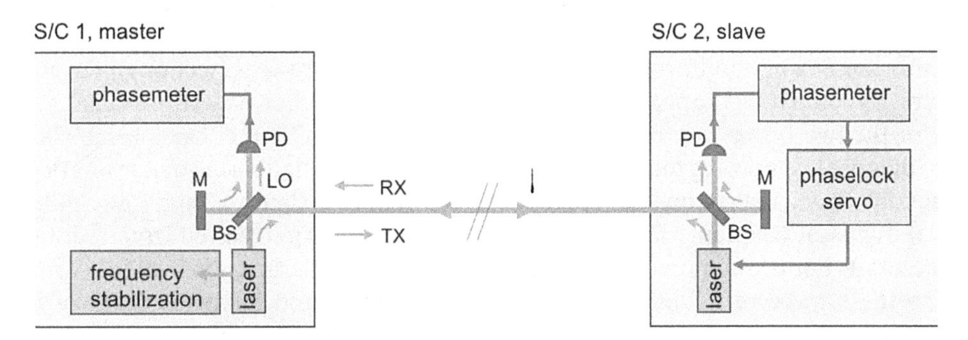


**Figure 4.2.** Schematic of the inter-satellite interferometry using a transponder scheme. The laser on spacecraft one (S/C1) acts as the master laser and is actively stabilized in frequency. It is split at a beamsplitter (BS) into a beam transmitted to the distant spacecraft (TX beam, sent to S/C2) and into a local oscillator (LO) within S/C1. On S/C1, the laser received from S/C2 (RX beam) is superimposed with the LO on a photodetector (PD) and the phase $\phi$ between the beams is measured with a phasemeter. On S/C2, the transmitted laser beam is superimposed with the local laser. A phase measurement is performed and the laser on S/C2 is (offset-) phase locked to the laser beam received from S/C (slave). In the configuration shown, the interferometer measures distance variations between the two mirrors (M) on distant spacecraft.

A dedicated measurement of the relative phase between the two beams (performed by the phasemeter) yields information on changes in inter-satellite distance. On spacecraft two (S/C2), a local laser is superimposed with the received light from S/C1 which experienced a Doppler shift on its way to S/C2. The local laser on S/C2 is phase locked to the received light with a frequency offset. The offset frequency is chosen such that the beat signal on S/C1 (measured between the local master laser and the received laser light, where the latter experienced an additional Doppler shift on its way back) is within the frequency range of the photodetector and phasemeter. Due to the large inter-satellite distances only a small fraction of the transmitted laser light is collected at the distant spacecraft with optical powers of the order of 100 pW at the photodetectors, cf table 4.1. Depending on the available laser output power and spacecraft separation distance, direct reflection of the laser light is only feasible for small inter-satellite distances (~100 km), as e.g. investigated within the Next Generation Gravity Mission (NGGM) program, cf section 4.4.2.

The interferometer includes an active beam steering (not shown in figure 4.2) which actively counteracts local spacecraft attitude jitter, keeping LO and RX wavefronts co-aligned and the interferometer contrast maximized. Furthermore, in the case of the LISA long armlengths, the angle of the TX laser beam needs to be controlled due to the finite travel time of the light in the LISA arms and the individual orbits of the satellites, causing a time-varying angular offset between optimal received and transmitted beams. Wavefront tilts between TX and RX beams need to be monitored and effects caused by tilt-to-length coupling (also referred to as 'piston effect')—i.e. a coupling of a tilt to an apparent longitudinal displacement, e.g. caused by a misalignment of the laser beam with respect to the center of rotation—need to be considered.

In an interferometer, the relative tilt between the two interfering wavefronts can easily be measured using differential wavefront sensing (DWS) technique [47, 61, 62, 65]. Therefore, a quadrant photodiode (QPD) is used and individual phase measurements for each quadrant are performed. A tilt between the two interfering laser beams causes a non-uniform interferometer pattern over the cross section of the laser beam. By comparing the phase at two positions over the beam cross section, a change in tilt of the two beams can be measured. In the case of a QPD, either the signals of two diagonally opposing quadrants or of the sums of two neighboring quadrants (left–right, up–down) are compared in phase. The schematics in figure 4.3 show an untilted local oscillator beam interfering with a tilted RX beam, e.g. received from a distant spacecraft. For translation measurement, the sum of all quadrants is used.

In these spacecraft configurations, the measurement and reference beam of the interferometer have highly unequal armlengths as the reference beam is located within the optical bench while the measurement beam represents the inter-satellite distance. When interfering both beams, laser frequency fluctuations do not cancel but result in phase variations in the measured interferometer signal. They influence the distance variation measurement, necessitating an active frequency stabilization of the laser. Several options are investigated, based on an optical resonator using the Pound–Drever–Hall locking technique [11, 64, 69], Doppler-free spectroscopy of

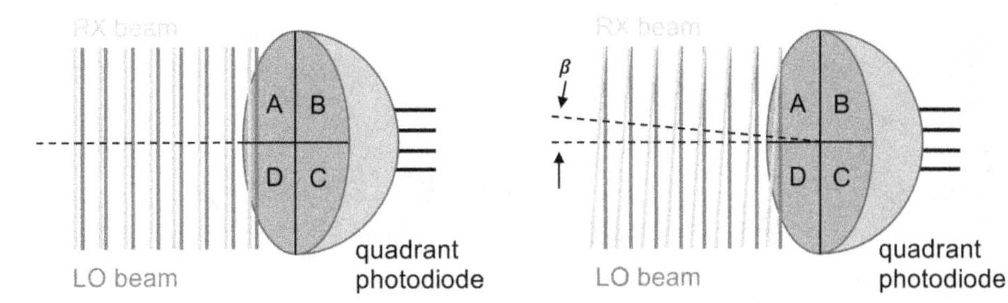


**Figure 4.3.** Schematic of differential wavefront sensing (DWS) in one tilt dimension (up–down). An untilted local oscillator (LO) laser beam is superimposed with an RX beam as e.g. received from the distant spacecraft (left: untilted RX beam, right: tilted RX beam). A phase measurement is individually processed for each segment of the quadrant photodiode. For small angles between LO and RX beams, the phase difference between the signals measured at the upper half (A+B) of the photodiode and the lower half (C+D) is proportional to the tilt angle $\beta$ between LO and RX beams. The second tilt dimension (left–right) is measured by comparing the signals (A+D) and (B+C).

molecular iodine [29, 44, 55, 82] or an unequal path length interferometer measuring the laser frequency fluctuations and correcting for in post-processing [40, 47].

Each application and interferometer implementation implies specific error contributions to the interferometric measurement. These include amongst others laser frequency noise, tilt-to-length coupling, thermally induced errors (e.g. temperature stability of the optical bench), laser shot noise, electronic noise in the phase measurement and stabilization loops (including digitizing noise), and laser intensity noise. Some of the error contributions can be handled within the interferometer design itself, others depend on the spacecraft design and mission scenario. For instance, thermal stability of the spacecraft and spacecraft jitter influence the interferometer performance. In the mission concepts discussed in this chapter, typically laser frequency noise and tilt-to-length coupling make the largest contribution to the error budget.

The currently lowest noise levels of 35 $\mathrm{fm}/\sqrt{\mathrm{Hz}}$ have been demonstrated by the LISA Technology Package onboard LISA Pathfinder, flying in a very benign and quiet environment. Using DWS, noise levels in tilt measurement at and below 1 $\mathrm{nrad}/\sqrt{\mathrm{Hz}}$ have been reported [19, 83].

## 4.3 Space-based gravitational wave detection

The principle of gravitational wave detection based on laser interferometry is schematically shown in figure 4.4, illustrating a Michelson interferometer for measuring changes in distance between widely separated proof masses. Due to its quadrupole nature, a gravitational wave propagating perpendicular to the plane of the interferometer at the same time increases the length of one interferometer arm and decreases the length of the other one. This (very simplified) measurement principle applies to Earth-based detectors such as LIGO [2, 9], VIRGO [3] and GEO600 [28, 42] and also to the space-based detector LISA which will be discussed in the following.

Only very large accelerated masses can produce significant and measurable gravitational wave signals. Gravitational waves have very small amplitudes, usually

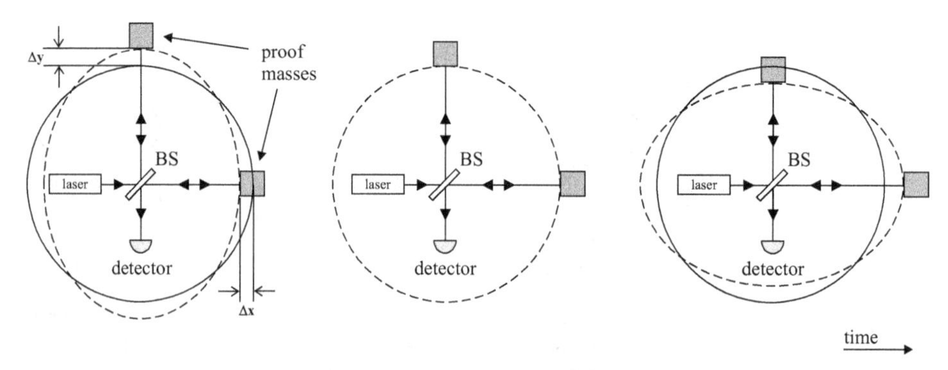


**Figure 4.4.** Effect of a gravitational wave on a ring of free proof masses where changes in distance are measured using a laser interferometer (BS: beamsplitter) [81].

denoted as strain $h$ with typical values of the order of magnitude of $\sim 10^{-20}/\sqrt{\text{Hz}}$ and below. To measure such strains, highest sensitivity distance metrology in combination with long armlengths of the interferometers is required.

The first ideas for space-based gravitational wave detectors using laser interferometry had already been presented in the 1980s by the Joint Laboratory for Astrophysics (JILA), USA [12, 35]. Since then, several mission concepts have been formulated and investigated, both in the US and in Europe. In 1993 the mission concept LISA was proposed to ESA by a team of European and US scientists. Subsequently, this concept was adapted several times, including different partners and matched to different budgetary constraints. A LISA mission concept study [37] and a pre-phase A study [56] were performed in 1998, followed by an industrial mission concept study [10, 45] and a mission formulation study (concluded in 2011) [33], both carried out by Astrium GmbH (now Airbus GmbH, Friedrichshafen, Germany). In 2017 ESA selected LISA as the third large-class mission in the ESA science program with a tentative launch in 2034. In order to verify key technologies for the LISA mission, a specific technology demonstration mission was proposed to ESA in 1998, called ELITE (European LISA Technology Experiment). In 2000 the mission was refined and renamed SMART-2 (Small Mission for Advanced Research in Technology) and finally LISA Pathfinder in 2004. It was launched in December 2015, successfully demonstrating the gravitational reference sensor, picometer interferometry, part of the drag-free system and the micro-Newton propulsion system.

As well as the LISA concept, which is the most evolved one, several other proposals for space-based gravitational wave detection—also using laser interferometry—are under investigation including ASTROD (Astrodynamical Space Test of Relativity using Optical Devices) [16, 67] and DECIGO (DECi-hertz Interferometer Gravitational wave Observatory) [53]. A thorough overview on space gravitational wave detectors is given in [66].

This section includes a description of the LISA mission based on the most recent document, i.e. the 2017 proposal in response to the ESA call for the L3 mission concepts [25]. However, specific designs presented in the following are also based on earlier LISA design concepts. The LISA Pathfinder mission is presented with a focus on interferometer design, implementation and test. A dedicated section is included on assembly-integration technologies for realization of robust space compatible optical benches as required, e.g. by LISA and LISA Pathfinder.

### 4.3.1 The Laser Interferometer Space Antenna

The aim of LISA is to detect gravitational waves in the low-frequency range between 0.1 mHz and 0.1 Hz (goal: 20 μHz to 1 Hz). Possible sources for these gravitational waves are inspirals of neutron star binaries, white dwarf binaries, super-massive black hole binaries as well as super-massive black hole formations and, possibly even (partly) the stochastic gravitational wave background from the early Universe.

The mission concept foresees three satellites which form a triangular configuration with a separation of 2.5 million kilometers. The current baseline is a heliocentric, Earth-trailing orbit where the LISA constellation is flying about 20° behind the Earth, corresponding to a distance of about 50 million kilometers to Earth, cf figure 4.5, left. The plane of the three satellites is inclined at 60° with respect to the ecliptic. Over the orbit, the triangular formation—with a nominal angle of 60°—is mainly maintained with a slow variation in the order of $\pm 1°$ over a one year's orbit (called the 'breathing angle') and Doppler shifts between the spacecraft up to 20 MHz. The armlengths change by $\sim$ 50 000 km over the orbit. The individual orbits of the three satellites cause the spacecraft constellation to rotate about its center once per year, yielding to a full sky coverage in gravitational wave detection (cf figure 4.5, right).

*Measurement principle*

A schematic of the LISA configuration is shown in figure 4.6. Any combination of two arms of the LISA triangle forms a Michelson interferometer with an arm length of about 2.5 million kilometers. Gravitational waves passing the LISA formation will be detected as differential length variations of the LISA arms using laser interferometry where distance variations caused by spacecraft orbital dynamics are outside the LISA measurement band. Each satellite in the baseline comprises two free flying proof masses which represent the end mirrors of the interferometers. Each proof mass is dedicated to one interferometer arm. Placed within the spacecraft, they

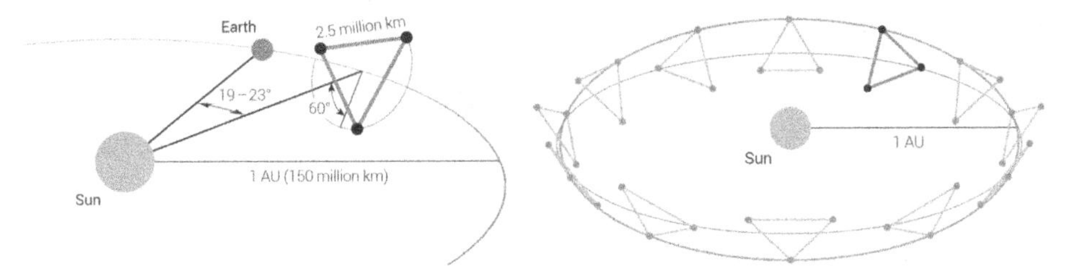


**Figure 4.5.** LISA orbit: The three satellites form a triangle constellation, flying in an Earth orbit around the Sun. The plane of the satellites has an inclination of 60° with respect to the ecliptic, allowing a full sky coverage in detection [25]. Image credit: LISA Consortium/AEI/UF/S. Barke (CC BY).

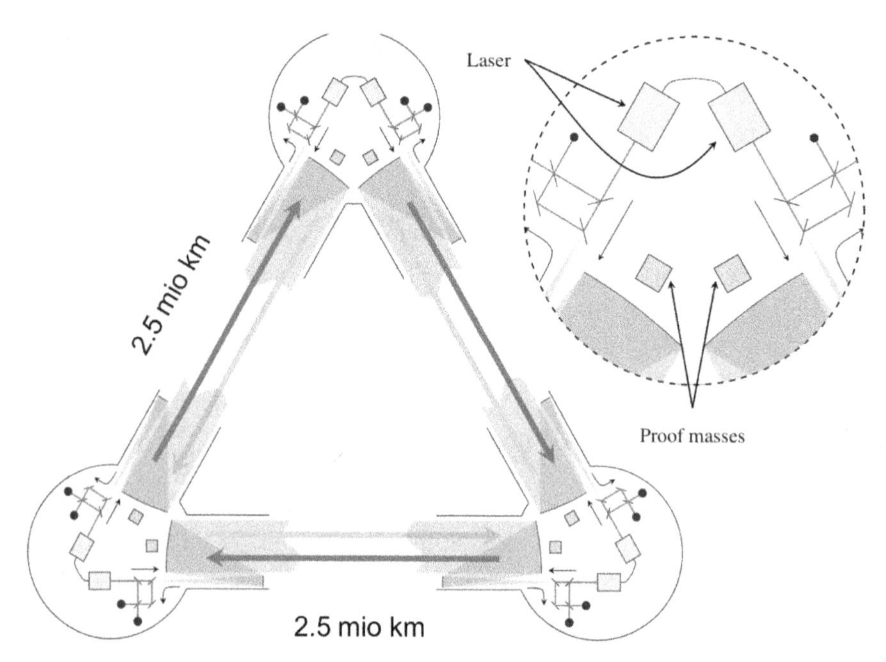


**Figure 4.6.** Schematic of the LISA configuration using heterodyne interferometry for measuring distance variations in each arm of the triangle. It implements a transponder scheme where the local laser is phase locked to the received laser light. Interferometric measurements are taken at the locations marked with blue dots (also showing redundant ports). The proof mass position interferometer readout is not included for clarity [52]. 

are shielded against external disturbances e.g. caused by solar radiation pressure. In order to avoid large displacements of the proof masses within their housings—which are rigidly connected to the spacecraft structure—the satellites are controlled in so-called drag-free mode where the spacecraft is forced to follow the two proof masses along the interferometer axes they define. Along the other degrees of freedom, the proof masses are electrostatically actuated. This drag-free attitude control system (DFACS) was successfully demonstrated onboard LISA Pathfinder, cf section 4.3.2 and figure 4.13.

Attitude and position of the satellites are controlled using micro-Newton thrusters in combination with dedicated feedback loops. While external disturbances are repelled, interacting forces between the satellite and the proof mass are still present. Such forces are e.g. the gravitational force caused by the satellite acting on the proof mass, forces due to electrostatic charging of the proof mass and gradients in magnetic field, setting stringent requirements on the design of the spacecraft. LISA foresees an active proof mass discharge using UV illumination and photo-electric emission from the proof mass and the electrode housing [77, 85, 93]. It was successfully demonstrated onboard LISA Pathfinder, together with a cold gas micro-propulsion system with heritage from the Gaia mission [7], where technology development is still ongoing, see e.g. [50] and references therein.

The requirements on residual acceleration of the proof mass $S_a^{1/2}$ and total displacement noise $S_{\rm ifo}^{1/2}$ of the interferometric measurement between two proof masses on distant spacecraft, both for frequencies between 0.1 mHz and 0.1 Hz, are given by [25]:

$$S_a^{1/2} \leqslant 3 \times 10^{-15} \frac{\mathrm{ms}^{-2}}{\sqrt{\mathrm{Hz}}} \cdot \sqrt{1 + \left(\frac{0.4\,\mathrm{mHz}}{f}\right)^2} \cdot \sqrt{1 + \left(\frac{f}{8\,\mathrm{mHz}}\right)^4} \tag{4.1}$$

$$S_{\rm ifo}^{1/2} \leqslant 10 \frac{\mathrm{pm}}{\sqrt{\mathrm{Hz}}} \cdot \sqrt{1 + \left(\frac{2\,\mathrm{mHz}}{f}\right)^4}. \tag{4.2}$$

The requirements for LISA are based on the performance demonstrated by LISA Pathfinder [9]. They are plotted in figure 4.7 where the shaded area corresponds to the extended frequency range which is the aimed for goal (between 20 μHz and 0.1 mHz and between 0.1 Hz and 1 Hz).

While in a classical Michelson interferometer, the incoming beam is reflected at the measurement mirror, LISA uses a transponder scheme due to the low optical power received at the distant spacecraft, cf section 4.2 and figure 4.2. An optical power of about 2 W is sent from one spacecraft where about 700 pW are received at the entrance pupil of the telescope and ~100 pW at the detector on the distant spacecraft. Within the transponder scheme, the laser on the distant spacecraft is (offset) phase locked to the incoming light and again about 2 W of optical power is transmitted back [60]. On the initial spacecraft, this light is superimposed with part of the original transmitted laser beam, resulting in a heterodyne signal at frequencies between 2 MHz and 20 MHz due to Doppler shifts caused by the individual orbits of the satellites. The phase measurement of this signal—referenced to the center of mass of the two proof masses—gives the information about changes in length of the interferometer arm.

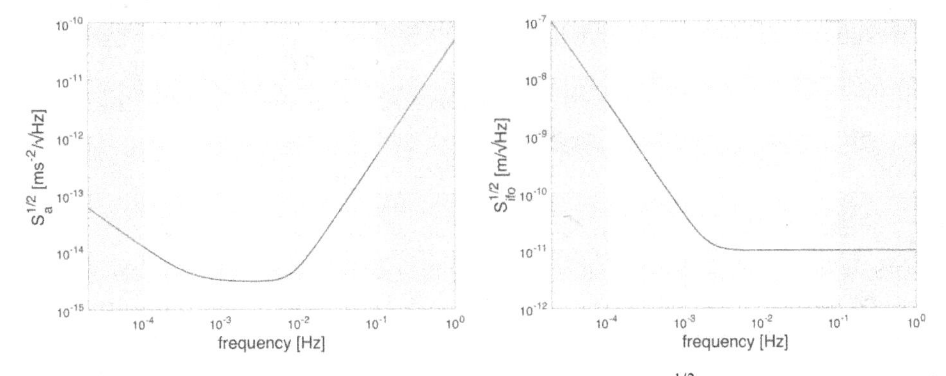


**Figure 4.7.** LISA requirements on residual acceleration of the proof mass $S_a^{1/2}$ (left) and displacement noise $S_{ifo}^{1/2}$ of the interferometric ranging measurement (right). The shaded areas below $10^{-4}$ Hz and above 0.1 Hz correspond to the extended goal frequency ranges.

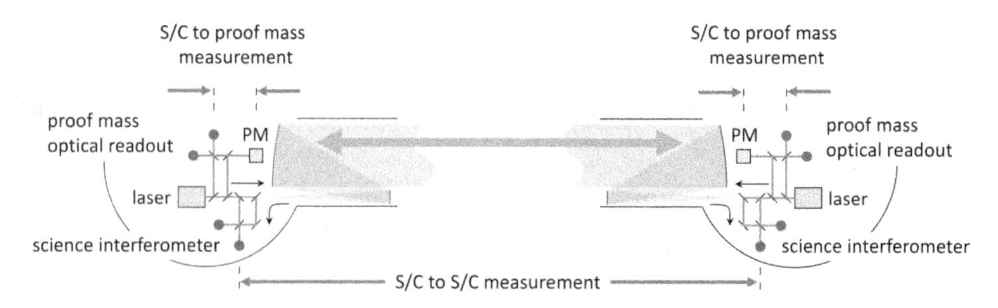


**Figure 4.8.** Schematic of the interferometric measurement within one LISA arm (also referred to as 'strap-down architecture'). The measurement between the two proof masses (PMs) on distant spacecraft (S/C) is split into three functionally decoupled interferometric measurements: two between proof mass and the corresponding optical bench (optical readout) and one between the two optical benches on distant spacecraft (science interferometer). Interferometric measurements are taken at the locations marked with blue dots (also showing redundant ports), adapted from [52]. © IOP Publishing. Reproduced with permission. All rights reserved.

The schematic shown in figure 4.8 shows the interferometric measurement within one LISA arm. The measurement between the two proof masses on distant spacecraft is subdivided into three technically and functionally decoupled measurements [75], easing assembly, integration, verification and test of the subsystems. Therefore, the laser light coming from the distant spacecraft is not directly reflected by the proof mass, but the (heterodyne) beat signal with the local beam is measured on the optical bench (science interferometer). Additionally, the distance between the optical bench and the associated proof mass is measured with the same sensitivity as in the distant spacecraft interferometry with a dedicated interferometer (proof mass optical readout).

All interferometric measurements are post-processed on the ground. The three measurements within one LISA arm are combined to a distance variation measurement between two proof masses on distant spacecraft. Furthermore, the measurements of the individual LISA arms are post-processed with respect to time-delay interferometry (TDI) [89] where configurations of Michelson and Sagnac interferometers are generated. By synthesizing virtual equal arm length Michelson interferometers, laser frequency noise is reduced by several orders of magnitude. Further suppression of the laser frequency noise to the required $10^{-20}/\sqrt{\mathrm{Hz}}$ level uses a laser pre-stabilization with a frequency stability $\leqslant 300\ \mathrm{Hz}/\sqrt{\mathrm{Hz}}$ in the LISA frequency band 20 μHz to 1 Hz. The absolute distances between the spacecraft need to be determined with ~10 cm accuracy for TDI and are measured using a pseudo-random noise (PRN) phase modulation on each of the laser beams [48].

*Optical payload*

Each satellite contains two identical assemblies consisting of an optical bench, a telescope and a gravitational reference sensor (GRS, including the proof mass with its corresponding electrode housing, launch lock mechanism and vacuum enclosure), cf figure 4.9. Two lasers in cold redundancy are attributed to each assembly, generating the laser beam transmitted to the distant spacecraft (TX). Their light is sent via single-mode polarization maintaining optical fibers to the optical bench

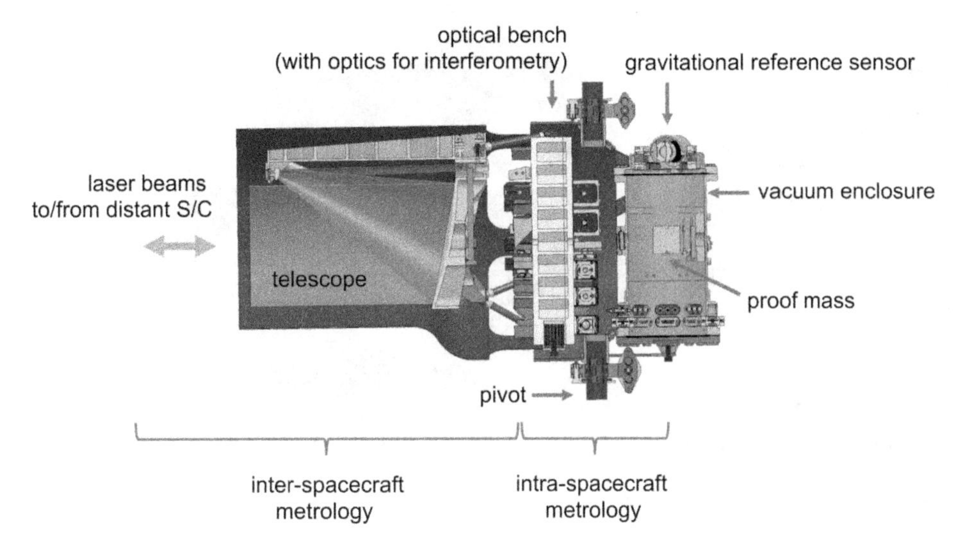


**Figure 4.9.** Rigidly mounted assembly consisting of an optical bench, telescope and gravitational reference sensor with a total mass of about 50 kg. The telescope has an aperture of 30 cm. Not shown is the electrode housing surrounding the proof mass within the vacuum enclosure and the launch lock mechanism [25]. Image credit: LISA Consortium/AEI/UF/S.Barke (CC BY).

which is mounted vertically behind the telescope, parallel to the primary mirror. The telescope processes both, transmitted (TX) and received (RX) laser beams in opposite directions. Current design parameters foresee a telescope diameter of 30 cm and a laser power of 2 W out of the fiber on the optical bench. The gold-coated proof mass is a cube with an edge length of 46 mm, made of a gold–platinum alloy with low magnetic susceptibility and a mass of about 2 kg. It is surrounded by an electrode housing with a gap of 3–4 mm between proof mass and housing. The electrode housing contains electrodes which are used for six degrees of freedom translational and rotational capacitive sensing of the proof mass position and tilt and also for electrostatic force and torque actuation of the proof mass. In the current baseline design, both assemblies are mounted 60° to each other in a frame which allows rotation of each assembly about the vertical axis by about 2° in order to compensate for the breathing angle, cf figure 4.10. The laser wavelength is 1064 nm, corresponding to the emission wavelength of solid-state Nd:YAG lasers.

The optical bench includes the (polarizing) optics for interferometry, a schematic of a possible implementation is shown in figure 4.11. Each optical bench includes three interferometers: (i) a science interferometer measuring the phase between the light received from the distant spacecraft and the local TX beam; (ii) a local interferometer (optical readout) monitoring position and orientation of the proof mass, and (iii) a reference interferometer. The optical bench includes two fiber switching units (FSU) which enable the switching between two redundant fiber outputs. Both fiber output ports are combined at a polarizing beamsplitter, followed by a rotatable waveplate which corrects the output polarization in case of switching of the input [26]. Power monitor photodiodes are placed behind the FSUs. The main

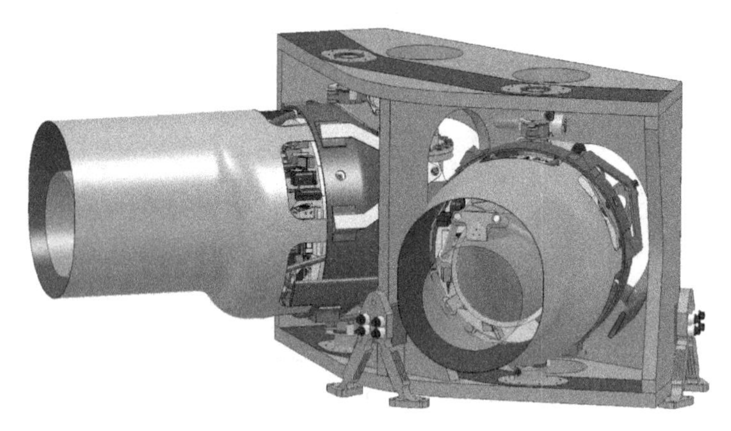

**Figure 4.10.** Two assemblies (as detailed in figure 4.9) mounted in one frame. Each assembly can be rotated within the frame by ±1° in order to compensate for the breathing angle [25]. Image credit: LISA Consortium/AEI/UF/S.Barke (CC BY).

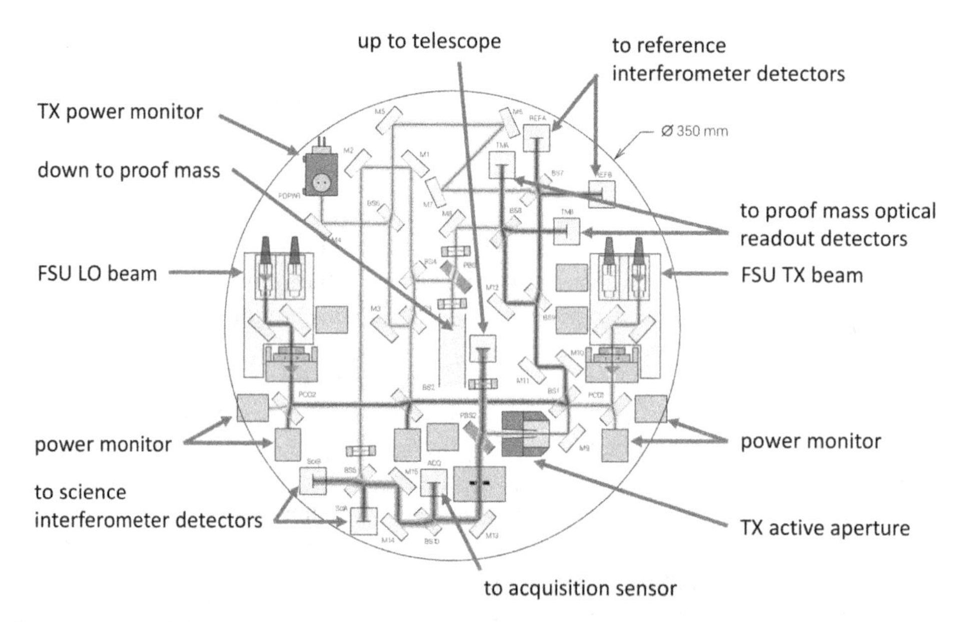


**Figure 4.11.** OptoCAD [78] model of a possible layout of the LISA optical bench with a diameter of 35 cm. Shown is the top surface of the optical bench with the main optical paths for interferometry. The bottom side of the optical bench includes photodetectors with dedicated optics imaging the entrance pupil of the telescope to the photodiodes and the camera for initial link acquisition [25]. Image credit: LISA Consortium/AEI/UF/S. Barke (CC BY).

part of the local transmitted laser beam (TX), shown in red, is reflected at a polarizing beamsplitter towards the telescope, part of it is further routed towards the proof mass and as reference beam to the science interferometer. The RX beam from the distant spacecraft is shown in green. Part of it is guided to a CCD for initial spacecraft acquisition [21]. It is then superimposed with part of the TX beam for the science interferometer. Part of the TX laser beam is fiber coupled at the second FSU and sent to the other optical bench (on the same spacecraft). Vice versa, the TX laser beam from the other optical bench is sent to this optical bench (marked in blue in figure 4.11) where it is used as local oscillator (LO). Knowledge of the phase between the two TX laser beams is required for TDI, measured at the detectors for the reference interferometer. This so-called 'backlink' is currently experimentally under investigation and can either be realized as a fiber connection (as shown in figure 4.11) or free-beam [51]. The phase between the light reflected at the proof mass and the LO is measured at the optical readout detectors, yielding the position of the proof mass.

All interferometric phase measurements use quadrant photo diodes enabling differential wavefront sensing (DWS) [47, 61, 62], cf section 4.2. The sum of all quadrants is used for translation and the sum of opposite halves for tilt measurements (pitch and yaw). The signals of the proof mass position and tilt are input to the DFACS control loop of the satellite attitude. All interferometer signals are processed onboard in a field-programmable gate array (FPGA) based phasemeter [18].

### 4.3.2 The LISA Pathfinder mission

The ESA mission LISA Pathfinder was launched on December 3, 2015, and successfully operated in space until June 2017. Its main purpose was to test LISA technologies which cannot be tested on the ground including the gravitational reference sensors, micro-Newton thrusters and ultra-stable laser interferometry in space environment. The spacecraft carried two payloads, the European built LISA Technology Package (LTP) consisting of two gravitational reference sensors (with proof masses and electrode housing) together with a laser metrology system and cold gas micro-Newton thrusters [4] and the US built disturbance reduction system (DRS) including colloidal micro-Newton thrusters and software for drag-free control on a dedicated processor [57]. LISA Pathfinder mimiced one arm of the LISA interferometer which was reduced from several million kilometers to 38.6 cm and placed within one spacecraft. In flight operation, a parasitic differential acceleration of the two proof masses of $5 \times 10^{-15}$ m $s^{-2}/\sqrt{Hz}$ for frequencies between 1 mHz and 30 mHz was demonstrated, which is more than one order of magnitude better than the LISA Pathfinder requirements and even below the LISA requirements within its frequency band from 20 μHz to 1 Hz [5, 9]. For frequencies between 1 mHz and 30 mHz, the amplitude spectral density of the differential acceleration is limited by Brownian noise, for frequencies above 30 mHz by interferometer measurement noise which is $< 35$ fm/$\sqrt{Hz}$.

A schematic at system level of the LTP interferometer is given in figure 4.12. It consists of several functional subsystems: (i) laser head, (ii) modulation bench, (iii) LTP core assembly with optical bench and gravitational reference sensors (GRS), (iv) phasemeter and (v) data management unit (DMU). The GRS contains proof mass, electrode housing, launch lock ('caging mechanism', rigidly clamping the proof mass during launch and cruise phase) and charge management of the proof mass within a vacuum tank. The position of the proof mass is measured using a

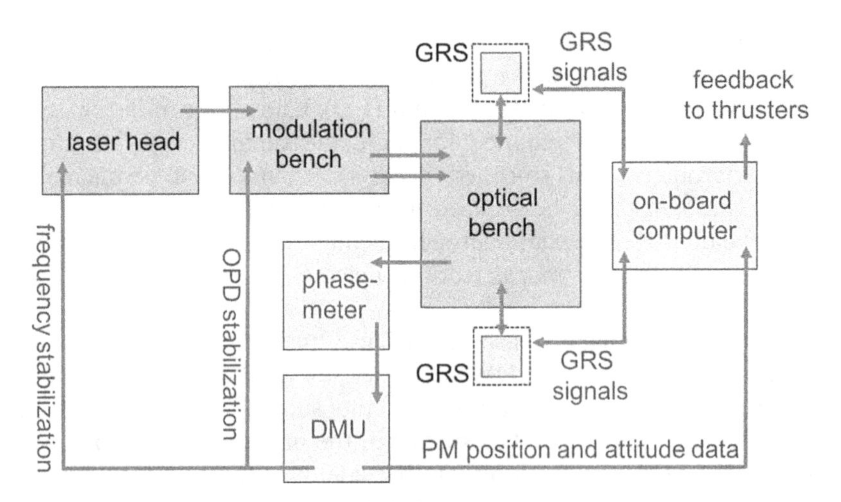


**Figure 4.12.** System level schematic of the LTP interferometer. It consists of several optical and electronic subsystems and includes laser head, modulation bench, optical bench, gravitational reference sensors (GRS), phasemeter and data management unit (DMU). The DFACS software and the LTP experiment control are located on the on-board computer. The GRS signals include the signals for readout and actuation of proof mass positions, actuation of the launch lock mechanisms (Piezo) and charge control of the proof masses. Not shown are the control electronics for laser head and modulation bench and corresponding power supplies. Red arrows mark optical interfaces, gray arrows electrical interfaces.

heterodyne laser interferometer along some degrees of freedom (the line-of-sight of the two proof masses) and along all degrees of freedom using a capacitive sensing system within the electrode housing. In-flight measurements show a capacitive sensor noise of a few nm/$\sqrt{\text{Hz}}$ and about 100 nrad/$\sqrt{\text{Hz}}$, respectively [5]. These signals can be used for drag-free attitude control of the spacecraft. Furthermore, the electrodes can also be used for actuation of the proof mass, i.e. centering the proof mass in its housing. The principle of drag-free control of the spacecraft with the LTP is shown in figure 4.13, also showing the layout of the optical bench.

The laser head has an output power of 25 mW and is fiber coupled to the modulation bench where it is split and both beams are frequency shifted using acousto-optic modulators (AOM). The AOM driving frequencies differ by a few kHz which corresponds to the heterodyne frequency of the interferometer measurements. Both beams are then fiber coupled to the optical bench where four separate interferometers are implemented for: (i) a distance measurement between the two proof masses and their differential alignment, (ii) a distance measurement between one proof mass and the optical bench and alignment of the proof mass, (iii) providing a reference phase, and (iv) measuring laser frequency fluctuations by interfering laser beams with intentionally unequal pathlengths. The reference phase as superposition of both laser beams is taken for a digital control loop for stabilization of the optical path length difference (OPD) with feedback to two piezo actuators within the modulation bench [15, 90]. More details on the optical bench layout can be found e.g. in [46]. The Mach–Zehnder type interferometer uses heterodyne signal detection with quadrant photodetectors enabling a tilt measurement of the proof masses using differential wavefront sensing, cf section 4.2. The interferometer is realized using non-polarizing optics in order to avoid any (potential) problems related to polarizing optics including polarization mixing e.g. caused by thermally induced changes in properties of the optical components. Therefore, the incident angle of the laser beam on the proof mass is slightly larger than 0°, as shown in figure 4.13. In the context of LISA, the use of polarization optics was investigated in detail showing its feasibility also at the $\mathrm{pm}/\sqrt{\mathrm{Hz}}$ sensitivity level [26, 83].

A CAD illustration of the LTP core assembly is depicted in figure 4.14. The two vacuum enclosures are shown with gold–platinum proof masses centered within their electrode housings and the optical bench in-between. The optical bench is realized on a 20 cm by 20 cm baseplate made of Zerodur glass ceramic where the optical components are integrated using hydroxide-catalysis bonding, cf section 4.3.2.

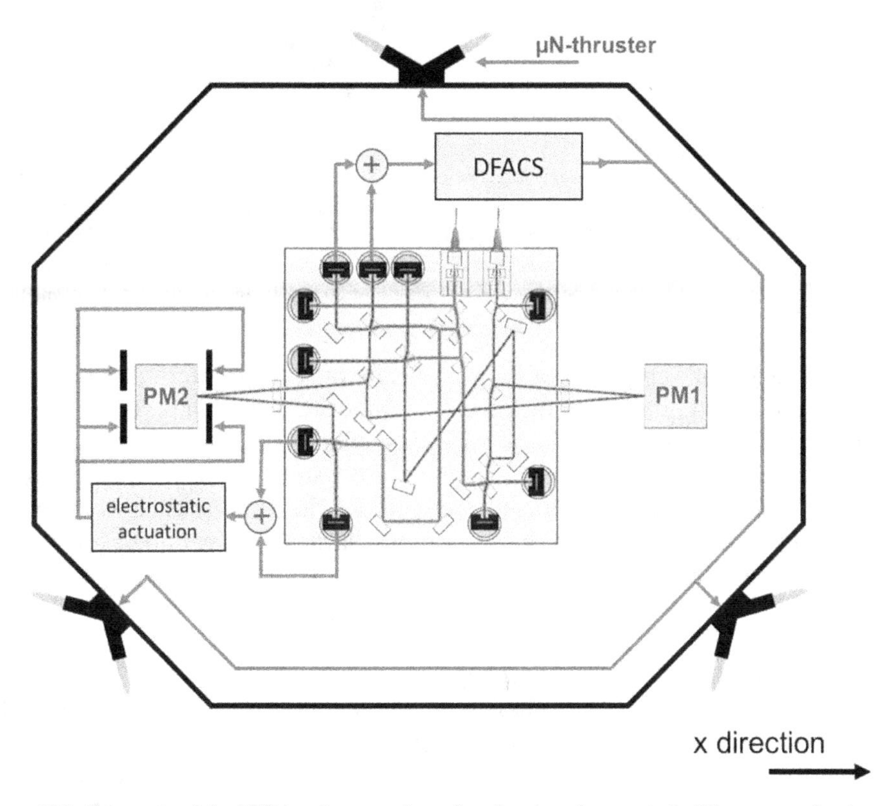


**Figure 4.13.** Schematic of the LTP in science mode performing drag-free control of the spacecraft using proof mass 1 (PM1) as reference, i.e. the spacecraft is forced to follow PM1 which is kept centered in the electrode housing by actuation of the spacecraft's relative position and attitude. PM2 is centered within its housing by using an electrostatic actuation system. The position of the proof masses (in the $x$-direction) is measured using the laser interferometer signals implemented on the optical bench where the differential displacement between the two proof masses is the science signal. The schematic of the optical bench is adapted from [73]. 

**Figure 4.14.** The LISA Technology Package core assembly aboard LISA Pathfinder. The two gold–platinum proof masses are shown within the electrode housings and vacuum enclosures (gravitational reference sensor, GRS). Between the two proof masses, the optical bench containing a high sensitivity laser metrology system is placed. Not shown are the Zerodur side walls which rigidly connect GRS and optical bench and act as interface to the spacecraft. The optical bench has a side length of 20 cm. Image credit: ESA/ATG medialab.

In order to get an impression of the complexity of such an interferometer system to be accommodated on a spacecraft, the flight model CAD of all parts belonging to the LTP experiment is shown in figure 4.15 where all parts belonging to the DRS and the spacecraft bus and structure are omitted. Also shown is a photograph of the flight model of the LTP core assembly integration into the science module. The different LTP subsystems and the corresponding harness (optical fibers and electric cables) need to be distributed in a way, that self-gravity effects within the spacecraft are compensated down to $5 \times 10^{-11}g$ at the proof mass location [6].

*Assembly-integration technologies for space optical systems*

The LISA and LISA Pathfinder missions set very challenging requirements on the thermal and mechanical stability of the integrated optical bench, the mechanical connections to the gravitational reference sensors and on the telescope structure in the case of LISA. The requirement in displacement noise in the interferometric measurements is as low as a few $\text{pm}/\sqrt{\text{Hz}}$ for frequencies in the mHz-regime, cf equation (4.2). Furthermore, the optical assembly needs to withstand launch loads and environmental conditions in orbit. The LTP engineering model was successfully subjected to thermo-vacuum tests with several cycles between 0 °C and 40 °C and to vibrational tests with sine-wave and random excitations in three axes up to 25 g [17].

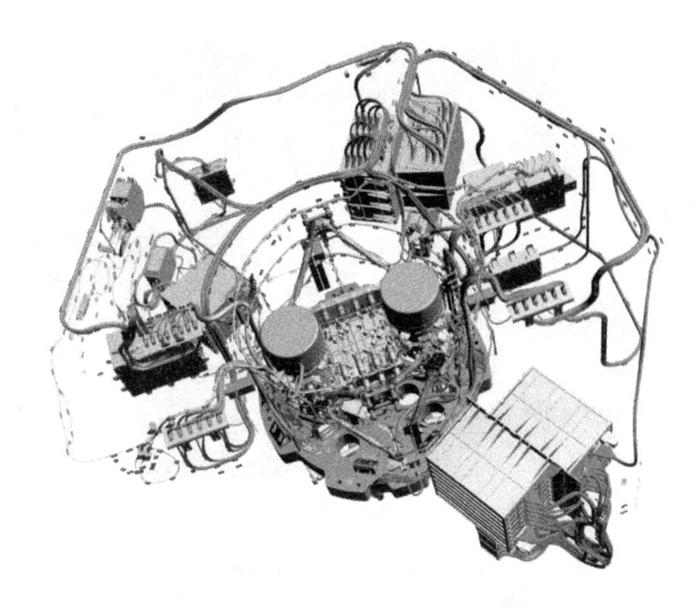

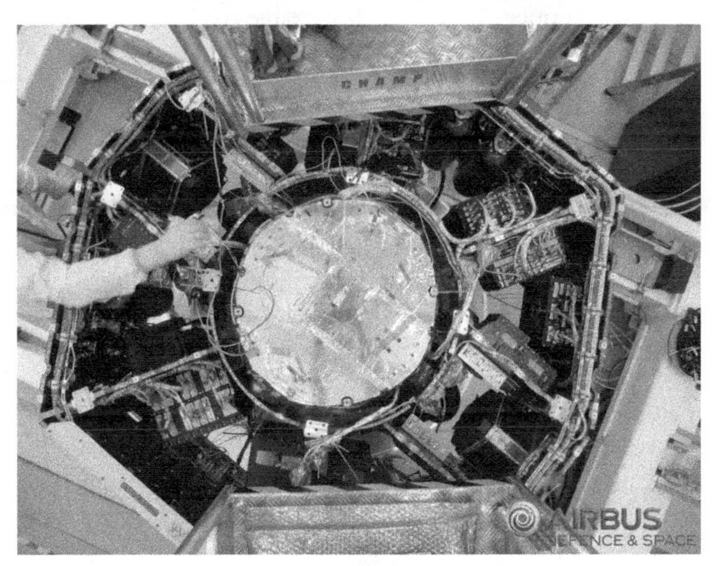

**Figure 4.15.** Top: Flight model CAD of all parts belonging to the LTP experiment, distributed over the spacecraft, including optical and electrical harness. Image credit: Airbus GmbH. Bottom: Integration of the LTP core assembly flight model into the science module at IABG Ottobrunn. The science module has an outer diameter of 2.1 m and a height of 1 m. Image credit: Airbus GmbH.

Glass ceramics such as Zerodur or ultra-low expansion (ULE) glass show very low coefficients of thermal expansion of $2 \times 10^{-8}$/K near room temperature and are suitable candidates for realizing high thermal and mechanical stable optical benches. For integration of the Gravity Probe B star-tracking telescope assembly, the technique of hydroxide-catalysis bonding was developed [43] and has been further applied in the GEO600 Earth-based gravitational wave detector [74]. It was used for

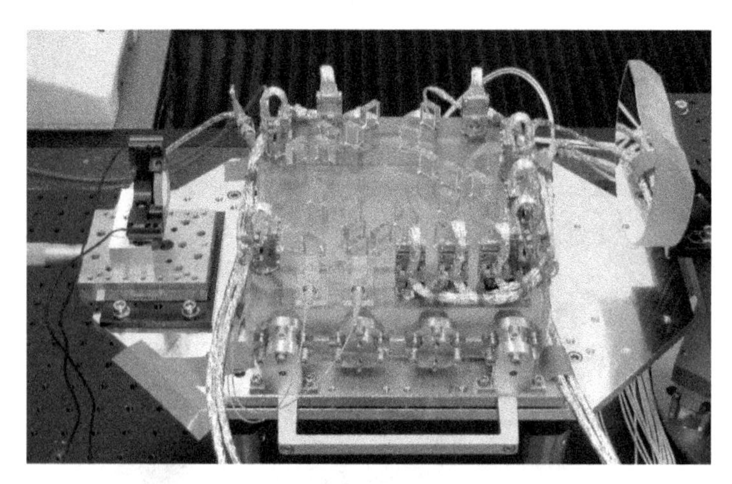

**Figure 4.16.** The LISA Pathfinder flight optical bench. The optical components are integrated on an optical bench made of Zerodur using hydroxide-catalysis bonding. The test masses are replaced by two mirrors, seen to the left and to the right. Image credit: University of Glasgow and University of Birmingham.

integration of the LTP optical bench and is baseline for LISA [31]. Between two flat surfaces (with global flatness $\leqslant\lambda/10$) of a silica containing material, a bond can be established by creating silicate-like networks, resulting in a quasi-monolithic setup.

A photograph of the LTP flight model optical bench is shown in figure 4.16. The optical components are made of fused silica and are joined to the Zerodur baseplate using hydroxide-catalysis bonding. The optical bench is mounted within the spacecraft using Titanium inserts glued into the side walls of the optical bench. With the successful launch and operation of the LISA Pathfinder spacecraft, these assembly-integration technologies have been space proven and are baseline for LISA.

An alternative integration technology using a space-qualified two-component epoxy has been investigated for integration of optical systems for space [71]. Compared to the above-mentioned hydroxide-catalysis bonding technology, adhesive bonding offers longer curing and hence adjustment times and less stringent requirements on the integration environment. A picometer sensitivity heterodyne interferometer was realized in the context of the LISA mission [80] and several optical frequency references based on Doppler-free spectroscopy of molecular iodine [29, 82], see figure 4.17. This technology is of special interest for optical setups which do not require interferometric, i.e. picometer dimensional stability, as long-term stability at this level has not yet been demonstrated.

## 4.4 Mapping Earth's gravitational field using satellite-to-satellite tracking

The Gravity Recovery and Climate Experiment (GRACE) was operated in space from 2002 until 2017 using a two-way microwave link for distance variation measurements between two identical satellites with an inter-satellite distance between 170 km and 270 km. Data of variations of the Earth's gravitational field are recovered with high spatial (~500 km) and temporal (~1 month) resolution,

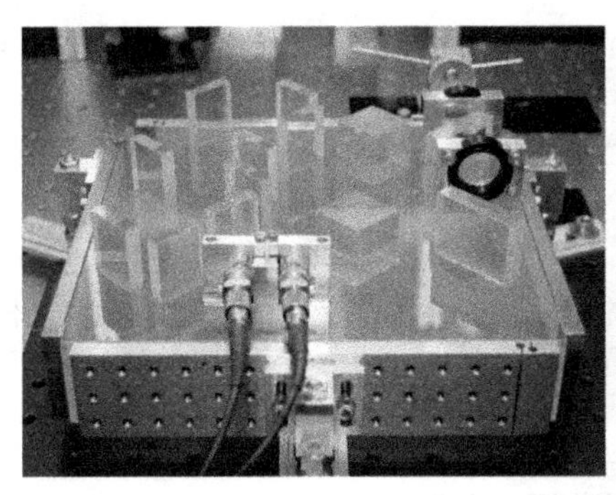 

**Figure 4.17.** Optical systems developed for space, using adhesive bonding technology (developed in a cooperation of University of Applied Sciences Konstanz (HTWG), University of Bremen, Humboldt-University Berlin, Airbus GmbH Friedrichshafen and DLR). Left: interferometer showing picometer and nanoradian sensitivity, developed in the LISA context [41, 80]; Right: Doppler-free iodine spectroscopy, used for laser frequency stabilization at the $10^{-15}$ level in frequency stability [29] (reprinted by permission from Springer Nature).

which gives information on geophysical phenomena where large mass redistributions take place, e.g. relevant for climate evolution assessment. These include ice mass unbalance, glacier melting, sea level changes, ocean circulations and solid Earth related changes e.g. caused by post-glacial rebound and tectonics [88, 92]. The GRACE follow-on mission was launched in May 2018 and provides data continuation. It is a close rebuild of the original GRACE satellites with an additional laser ranging instrument (LRI) for relative distance metrology. The LRI is a technology demonstrator, backing up the microwave instrument in the case of demonstrated functionality. Its technology relies on the LISA heritage with less stringent requirements concerning displacement noise. Nevertheless, future missions beyond GRACE follow-on are already being discussed and under investigation, e.g. by ESA within the Next Generation Gravity Mission (NGGM) program. One focus here is satellite-to-satellite tracking (SST) in low-Earth orbit with an LRI as the primary instrument, similar to GRACE follow-on. However, further mission architectures using other spacecraft constellations and/or orbits are investigated, including gradiometer concepts using one single satellite (similar to the Gravity field and steady-state Ocean Circulation Explorer, GOCE), cf section 4.4.2.

The measurement principle of SST is shown in figure 4.18. Two satellites are flying in the same orbit with an inter-satellite distance of $d$ (measured at their centers of mass, CoM). Both satellites experience their own gravitational acceleration, resulting in a change of the inter-satellite distance $\Delta d$ over time. Each satellite carries an accelerometer, measuring the non-gravitational accelerations acting on the spacecraft, mainly produced by atmospheric drag, i.e. residual atmosphere. The accelerometer's proof mass is placed at the CoM of the satellites where gravitational forces by the spacecraft acting on the proof mass are compensated. The orbit of the satellite is monitored using a GNSS (global navigation satellite system). These measurements are combined and post-processed resulting in the measurement of the Earth gravity potential.

The requirement for the GRACE follow-on laser distance measurement is given by

$$S_{\text{ifo}}^{1/2} \leqslant 80\frac{\text{nm}}{\sqrt{\text{Hz}}} \cdot \sqrt{1 + \left(\frac{f}{3\text{ mHz}}\right)^{-2}} \cdot \sqrt{1 + \left(\frac{f}{10\text{ mHz}}\right)^{-2}} \tag{4.3}$$

within the relevant frequency band between 2 mHz and 0.1 Hz, cf figure 4.19. The frequency range is determined by the orbit frequency of the satellite [68]. NGGM

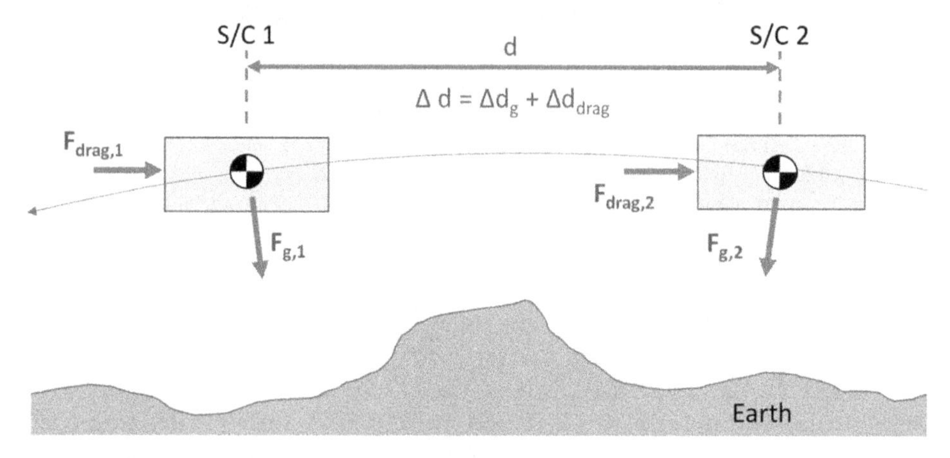


**Figure 4.18.** Principle of an Earth gravity measurement using satellite-to-satellite tracking. Two satellites are flying at a distance $d$. Variations in distance ($\Delta d$) are caused by differential gravitational acceleration ($\Delta d_g$) of the two satellites and by drag forces acting on the spacecraft ($\Delta d_{\text{drag}}$). The latter is measured by accelerometers and corrected for in post-processing. The distance variation $\Delta d$ is measured with a dedicated metrology system.

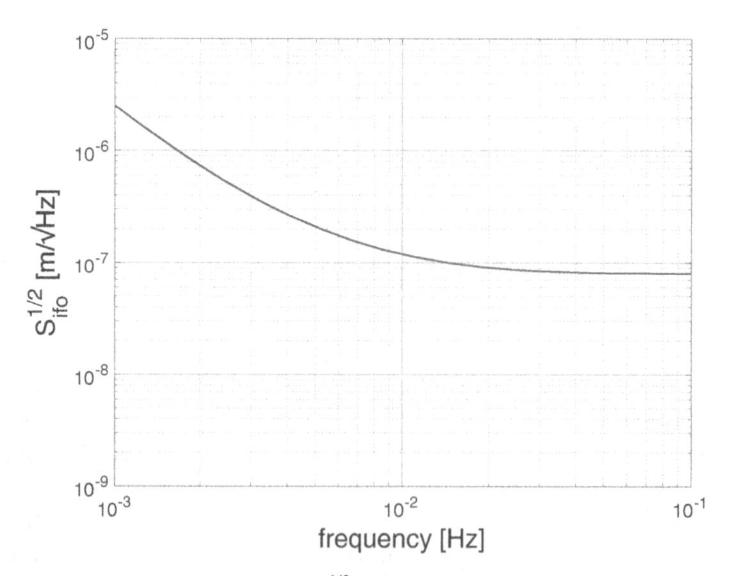


**Figure 4.19.** Requirement on displacement noise $S_{\text{ifo}}^{1/2}$ of the GRACE follow-on laser ranging instrument. The relevant frequency band is 2 mHz to 0.1 Hz.

has similar requirements, depending on the science objectives and the mission concept. For an inter-satellite distance of 100 km, the current design attributes an overall relative distance measurement error of 20 nm/$\sqrt{\text{Hz}}$ [20]. The requirement for the distance metrology is derived from the top-level science requirements stating required accuracies for spatial and temporal resolutions for the geophysical signals of interest, see e.g. [87].

The Attitude and Orbit Control System (AOCS) of the GRACE and GRACE follow-on satellites guarantees a ±4 mrad pointing accuracy using a star camera for attitude information [49]. While pointing variations at this level can be managed by the microwave ranging instrument due to the wide beam and wide receive field of view, the LRI requires higher pointing accuracies of about ±100 μrad. Therefore, an acquisition scan with active beam steering needs to be performed in order to initially establish the laser link between the two distant spacecraft. Active beam steering is furthermore needed for continuous compensation of spacecraft jitter.

### 4.4.1 The GRACE follow-on laser ranging instrument

As the GRACE follow-on satellites were mainly a rebuild of the GRACE satellites, the additional laser ranging instrument needed to be accommodated within the given satellite design with minimum impact on it. This posed many restrictions to the LRI realization, especially as the microwave K/Ka-band instrument is on the direct line-of sight of the two satellites. A specific configuration has been worked out where the laser beams are guided around the microwave instrument and the cold gas tank. The laser beam routing results in the so-called 'racetrack' configuration which is shown in figure 4.20. A specific structure—called the triple mirror assembly (TMA)—is

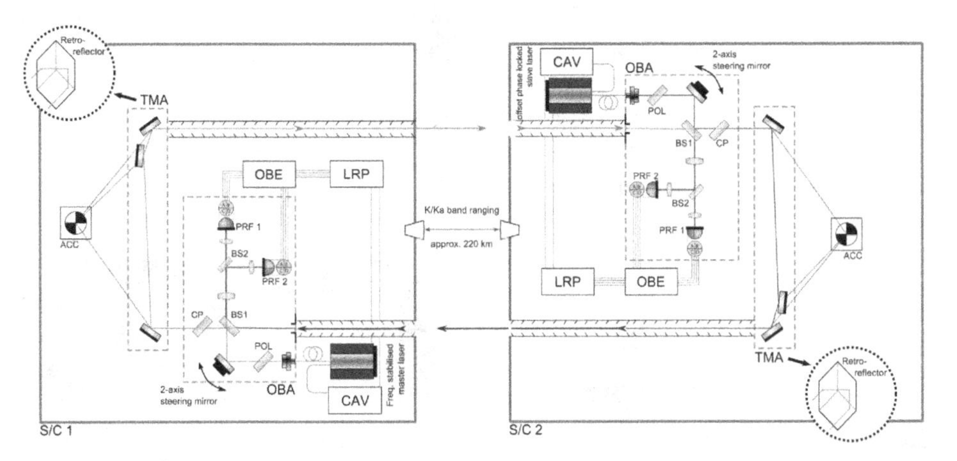


**Figure 4.20.** Optical layout of the GRACE follow-on laser ranging instrument with an inter-satellite distance of ~ 220 km. Each satellite carries a laser, a cavity-based frequency stabilization (CAV), an optical bench assembly (OBA), dedicated optical bench electronics (OBE), a laser ranging processor (LRP) and a triple mirror assembly (TMA). The accelerometer (ACC) is placed at the vertex of the TMA. The optical bench includes the optics for interferometry, including polarizer (POL), beamsplitter (BS), compensation plate (CP) and (redundant) quadrant photodetectors with photodiode receiver frontends (PRF). Image credit: © SpaceTech GmbH.

used for routing the laser beam around the microwave instrument and the cold gas tanks. It includes three mirrors, perpendicular to each other, representing a corner cube configuration where the vertex, i.e. the intersection point of the three mirror planes, of this retroreflector is placed within the accelerometer at the CoM of the satellite. The TMA resembles a retroreflector where the mirrors are only realized in the parts hit by the laser beam, together with a supporting structure. In an ideal case without misalignments and no temperature influences, this design ensures that the round-trip path length stays invariant under rotation around the vertex and the reflected beam is always anti-parallel to the incident beam.

Each satellite carries a laser system, an optical bench and a TMA together with corresponding phasemeter and control electronics. The LRI uses a transponder scheme similar to LISA where the laser on one satellite acts as the master laser, frequency stabilized to an optical cavity, cf section 4.2. The light is sent via the TMA to the distant spacecraft where the local laser (slave laser) is offset phase locked to the weak received laser beam (RX) and then sent as a transmitted beam (TX) via the TMA to the initial spacecraft. The offset frequency can be chosen within the range from 4 MHz to 20 MHz. Using a laser with an output power of 25 mW, less than 1 nW is detected at the distant spacecraft.

As the satellite attitude control is not sufficient for LRI operation, the instrument itself needs to internally compensate for spacecraft jitter. This is handled by a tip-tilt steering mirror which uses DWS signals from quadrant photodetectors as input for a corresponding control loop. The relative tilt signals (yaw and pitch) between RX and TX laser beams are controlled to zero, i.e. both beams are kept co-aligned and the heterodyne efficiencies of the interferometer signals at the photodetectors—also on the distant spacecraft—are maximized.

Figure 4.20 includes a schematic of the optical bench, an illustration of the flight model implementation is shown in figure 4.21. The light source is a solid-state Nd: YAG non-planar ring oscillator (NPRO) with an output wavelength of 1064 nm. It is fiber coupled to the optical bench where it is collimated within the fiber injector and reflected at a tip-tilt steering mirror. At a beamsplitter, 90% of the light is reflected towards the distant spacecraft and 10% is transmitted and sent to a

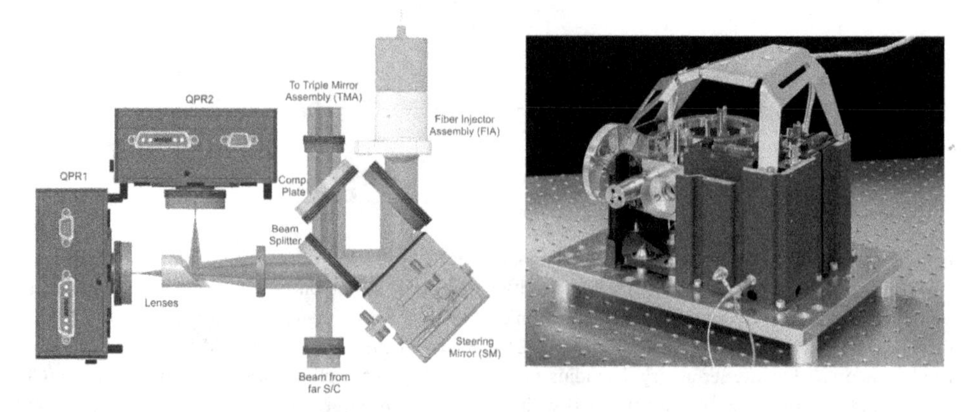


**Figure 4.21.** LRI optical bench assembly showing the optical path (top view, left) and a photograph of the flight model (right). It uses a titanium baseplate with optics made of BK7. Image credit: © SpaceTech GmbH.

quadrant photodetector. This beam acts as a reference beam for the phase measurement yielding translation and tilt informations, cf section 4.2. The RX beam from the distant spacecraft is clipped at an aperture and superimposed with the reference beam at the beamsplitter. The imaging optics (consisting of two lenses) in front of the photodetectors images both the RX beam aperture and the TX beam at the steering mirror onto the surface of the photodetector. Therefore, a tilt of the laser beam around the aperture or the steering mirror will cause no beam walk at the photodetector and (ideally) no piston effect (i.e. tilt-to-length coupling) occurs. The optical path length of the laser beam within the beamsplitter is dependent on its incident angle and spacecraft jitter would couple into the distance measurement. In order to reduce this coupling factor to a value of a few nm/mrad, an additional compensation plate, placed at 90° to the beamsplitter, is included in the RX beam.

Due to the TMA geometry, the interferometric measurement is mostly immune to spacecraft jitter. As the TMA is within the measurement path of the interferometer, it has to fulfill stringent requirements with regard to mechanical and thermal stability. Co-alignment between incoming and outgoing laser beams needs to be below 40 μrad with 5 μrad $K^{-1}$ thermal stability and optical path length stability needs to be below 400 nm $K^{-1}$ [23]. Non-operational temperatures between −20° and 45° as well as operational temperatures between 10° and 30° need to be handled by the instrument. An engineering model of the TMA is shown in figure 4.22. It consists of a tube made of carbon-fiber reinforced polymer (CFRP) with a very low coefficient of thermal expansion (CTE). Two glass inserts are bonded into the CFRP tube using a two-component epoxy with high shear strength. The mirrors are attached to mounting prisms which are joined to the glass inserts. For these glass to glass bonds, hydroxide-catalysis bonding is used as detailed in section 4.3.2.

The dominant noise sources in the laser distance measurement are laser frequency noise and pointing induced noise. Laser frequency noise is coupling to the interferometric measurement due to the unequal arm length configuration where laser frequency stability is required to be $\leqslant 30\ \mathrm{Hz}/\sqrt{\mathrm{Hz}}$ in the frequency band 0.1 mHz to 1 Hz. This necessitates an active laser frequency stabilization where

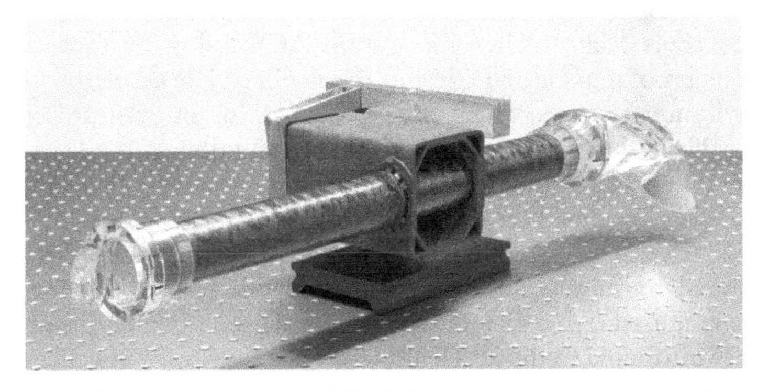

**Figure 4.22.** Photograph of the triple mirror assembly (TMA) engineering model. It uses a CFRP spacer where the three mirrors are attached to glass mounting brackets bonded into the ends of the CFRP tube. Image credit: © SpaceTech GmbH.

GRACE follow-on employs stabilization to an optical resonator using the standard Pound–Drever–Hall technique [11, 38].

In order to verify the performance of the LRI on the ground, a specific measurement setup—as part of the so-called Optical Ground Support Equipment (OGSE)—is required. It simulates the inter-satellite link taking into account the far-field intensity profile, Doppler shifts and spacecraft jitter as present in real flight operation [76]. Furthermore, the performance of the TMA needs to be assessed, especially with respect to co-alignment, i.e. parallelism, of the incoming and outcoming laser beams. Using a specific interferometer setup, the TMA flight units have been characterized showing that they will enable the LRI ranging measurements with the required sensitivity [23].

### 4.4.2 Next Generation Gravity Mission

The LRI on board GRACE follow-on has a ranging noise up to two orders of magnitude lower than the microwave ranging instrument. However, it will not directly improve the geoid determination as other noise sources are limiting, namely accelerometer noise followed by ocean tide and non-tidal mass variation errors [36]. Future mission concepts for measuring the Earth's gravity field are currently under investigation, both by ESA and NASA. They aim for a higher precision in the determination of the Earth's gravitational field together with a higher spatial and temporal resolution. Different architectures are evaluated where the GRACE-like two-satellite collinear pair is one option [20, 91]. Two of these pairs can be combined for a dual-GRACE mission ('Bender configuration'), flying in polar and medium inclinations [13]. More satellite pairs can be added, further improving temporal and spatial resolution. More complex satellite configurations are also investigated such as the 'pendulum' and 'cartwheel' formations, see e.g. [20], and also combinations of satellites in low- and medium-Earth orbits. The mission concepts rely on inter-satellite laser distance metrology with nanometer sensitivity and LEO–LEO inter-satellite distances up to a few hundred kilometers. They foresee improved accelerometers or drag-free operation of the satellites similar to the LISA and LISA Pathfinder missions.

Current concepts such as GRACE and GRACE follow-on foresee an accelerometer with its proof mass at the CoM of the satellite. The accelerometer measures the non-gravitational forces which are accounted for in post-processing of the ranging data. Using laser interferometry, direct reflection of the laser light at the proof mass can also be considered, see e.g. [70]. By operating the satellite in drag-free mode where the satellite is centered around the proof mass by feedback to dedicated (μN-) thrusters, external disturbances on the ranging measurement as well as effects caused by misalignments of the LRI with respect to the CoM of the satellite are minimized *a priori*. While promising higher ranging sensitivities due to avoiding noise attributed to the acceleration measurement, it yields higher complexity and non-modularity of the payload components.

As detailed above, different mission configurations are investigated within NGGM. The in-line formation of one or more satellite pairs in LEO, similar to

GRACE and GRACE follow-on, is currently the most evolved configuration, also investigated in detail within the preparatory NGGM study by ESA [14, 20]. Here, two different implementations of the interferometer for laser ranging are investigated, either using a transponder or a retroreflector scheme. The transponder scheme is similar to the GRACE follow-on LRI where one spacecraft acts as master and the laser on the second satellite is offset phase locked to the incoming laser light (cf section 4.2 and figure 4.2). This requires relatively low optical powers and enables larger inter-satellite distances. However, two laser systems need to be operated simultaneously on the two satellites. In the retroreflector scheme, the laser light from the master spacecraft is retro reflected at the distant spacecraft, cf the schematics shown in figure 4.23. Here, higher optical powers are needed and only limited inter-satellite distances can be realized. On the other side, the complexity is lower compared to the transponder scheme as the slave satellite only consists of the retroreflector and only one laser needs to be operated at a time. Both implementations foresee heterodyne interferometry in order to distinguish the sign of the inter-satellite distance variation.

The choice for realization is mainly dependent on the inter-satellite distance and the available laser output power. Trade-offs on system level need to take into account the mission requirements and reflect these different options.

Several projects are ongoing with respect to technology development for NGGM. A mission concept based on a retroreflector scheme with an inter-satellite distance of 100 km is currently investigated [14]. A corresponding interferometer breadboard has been realized using a separate angle metrology unit [63]. Within the High Stability Laser (HSL) project, a laser source delivering 500 mW laser output power at a wavelength of 1064 nm using a fiber amplifier is being developed which includes a compact setup for laser frequency stabilization on an optical resonator [24]. While the GRACE follow-on cavity stabilization (with a laser output power of 25 mW) is being developed by the Jet Propulsion Laboratory (JPL, USA), the HSL project is carried out within a European consortium.

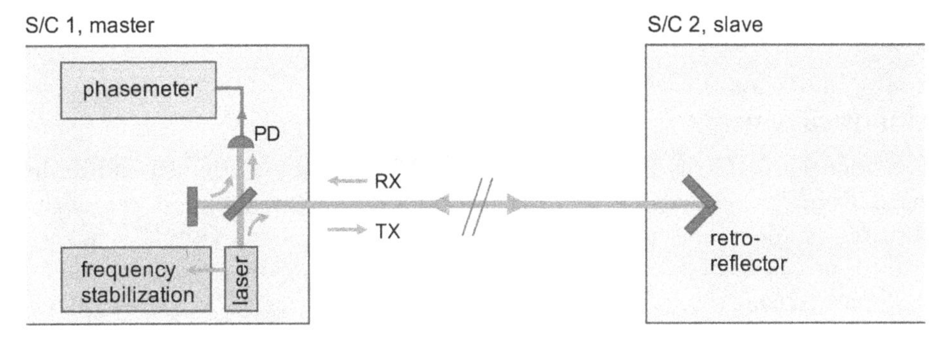


**Figure 4.23.** Possible implementations of the NGGM interferometer using a retroreflector on the distant spacecraft. Not shown are the accelerometers measuring the non-gravitational forces acting on the two spacecraft.

## 4.5 Summary and outlook

Several mission concepts foresee laser interferometry for inter- and intra-spacecraft distance variation measurements with high sensitivity. It is employed by missions dedicated to measuring the Earth's gravity field with high spatial and temporal resolution and to space-based gravitational wave detection. This chapter focussed on these two application areas and presented the corresponding missions.

Within the LISA Pathfinder mission, the technique of laser interferometry was very successfully demonstrated in space. Distance variations between two free flying proof masses were measured with noise levels below 35 $\mathrm{fm}/\sqrt{\mathrm{Hz}}$ using a heterodyne interferometer and a dedicated, quasi-monolithic, assembly-integration technology for the realization of the optical bench. GRACE follow-on was launched in May 2018 and will for the first time demonstrate $\mathrm{nm}/\sqrt{\mathrm{Hz}}$ sensitivity laser ranging between two distant satellites with an inter-satellite separation of ~200 km. The space-based gravitational wave detector LISA is scheduled for launch in 2034 employing $\mathrm{pm}/\sqrt{\mathrm{Hz}}$ interferometry between satellites which have a separation of about 2.5 million kilometers.

While these missions employ laser interferometry, mission concepts using cold atom interferometry are also under investigation. Due to smaller (de Broglie) wavelengths, atom interferometry promises higher sensitivities, e.g. in the measurement of distance variations and accelerations. This technology is already being investigated in the context of space-based gravitational wave detection [27, 39] as well as missions for mapping Earth's gravity field [30], mainly on study level and first technology demonstration activities. The first steps towards the engineering of a space atom interferometer are ongoing, see, e.g. [84]. Compared to laser interferometry, atom interferometry implies a much higher complexity of the instrument and hence a large effort to realize space compatibility. Nevertheless, atom interferometry is seen as a very promising technology for the next but one generations of space-based gravitational wave detectors and Earth gravity missions.

This development of the so-called second generation quantum technologies, where individual quantum systems are influenced (see e.g. [79]), is currently the focus of several national and international funding programs such as the European Quantum Technology Flagship Program, also addressing space applications of quantum technologies.

## Acknowledgements

The author thanks Claus Braxmaier from DLR/University of Bremen and Rüdiger Gerndt from Airbus GmbH (Friedrichshafen) for discussions and very helpful comments on the manuscript. Kolja Nicklaus from Spacetech GmbH is acknowledged for provision of schematics and photographs of the GRACE follow-on laser ranging instrument.

## References


[1] Abbott B P *et al* 2016 Observation of gravitational waves from a binary black hole merger *Phys. Rev. Lett.* **116** 061102

[2] Abbott B P *et al* 2017 Gw170817: Observation of gravitational waves from a binary neutron star inspiral *Phys. Rev. Lett.* **119** 161101

[3] Acernese F *et al* 2017 Status of the Advanced Virgo gravitational wave detector *Int. J. Mod. Phys.* A **32** 1744003

[4] Anza S *et al* 2005 The LTP experiment on the LISA Pathfinder mission *Class. Quantum Grav.* **22** S125–38

[5] Armano M *et al* 2017 Capacitive sensing of test mass motion with nanometer precision over millimeter-wide sensing gaps for space-borne gravitational reference sensors *Phys. Rev.* D **96** 062004

[6] Armano M *et al* 2016 Constraints on lisa pathfinder's self-gravity: design requirements, estimates and testing procedures *Class. Quant. Grav.* **33** 235015

[7] Armano M *et al* 2015 A strategy to characterize the lisa-pathfinder cold gas thruster system *J. Phys.: Conf. Ser.* **610** 012026

[8] Armano M *et al* 2016 Sub-femto-*g* free fall for space-based gravitational wave observatories: Lisa pathfinder results *Phys. Rev. Lett.* **116** 231101

[9] Armano M *et al* 2018 Beyond the required LISA free-fall performance: new LISA Pathfinder results down to 20 μHz *Phys. Rev. Lett.* **120** 061101

[10] Astrium 2000 LISA–study of the Laser Interferometer Space Antenna; final technical report *Technical Report* LI-RP-DS-009

[11] Bachman B *et al* 2017 Flight phasemeter on the Laser Ranging Interferometer on the GRACE follow-on mission *J. Phys.: Conf. Ser.* **840** 012011

[12] Bender P L, Faller J E, Hall J L, Hils D, Stebbins R T and Vincent M A 1990 Optical interferometer in space *NASA, Relativistic Gravitational Experiments in Space* ; Hellings R W pp 80–8

[13] Bender P L, Wiese D N and Nerem R S 2008 A possible dual-grace mission with 90 degree and 63 degree inclination orbits *Proc., 3rd Int. Symp. on Formation Flying, Missions and Technologies. European Space Agency Symp. Proc., ESA SP-654* (Noordwijk, The Netherlands: ESA) JILA Pub. 8161

[14] Bonino L, Cesare S, Massotti L, Mottini S, Nicklaus K, Pisani M and Silvestrin P 2017 Laser metrology for Next Generation Gravity Mission *IEEE Instrum. Meas. Mag.* **20** 16–21

[15] Born Mand the LPF collaboration 2017 LISA Pathfinder: OPD loop characterisation *J. Phys.: Conf. Ser.* **840** 012036

[16] Braxmaier C *et al* 2012 Astrodynamical space test of relativity using optical devices i (astrod i)–a class-m fundamental physics mission proposal for cosmic vision 2015–2025: 2010 update *Exp. Astron.* **34** 181–201

[17] Braxmaier C *et al* 2004 LISA pathfinder optical interferometry *Proc. SPIE* **5500** 5500-1–10

[18] Bykov I, Delgado J J E, Marín A F G, Heinzel G and Danzmann K 2009 Lisa phasemeter development: Advanced prototyping *J. Phys.: Conf. Ser.* **154** 012017

[19] Cervantes F G *et al* 2013 LISA Technology Package Flight Hardware Test Campaign *9th LISA Symp., Astronomical Society of the Pacific Conf. Series* **vol 467**; Auger G, Binétruy P and Plagnol E 141 p

[20] Cesare S, Allasio A, Anselmi A, Dionisio S, Mottini S, Parisch M, Massotti L and Silvestrin P 2016 The European way to gravimetry: From GOCE to NGGM *Adv. Space Res.* **57** 1047–64

[21] Cirillo F and Gath P F 2009 Control system design for the constellation acquisition phase of the lisa mission *J. Phys.: Conf. Ser.* **154** 012014

[22] Cockell C S *et al* 2009 Darwin–an experimental astronomy mission to search for extrasolar planets *Exp. Astron.* **23** 435–61

[23] Dahl C *et al* 2017 Laser ranging interferometer on GRACE follow-on *Proc. SPIE* **10562** 10562-1–9

[24] Dahl K *et al* 2017 High stability laser for interferometric earth gravity measurements *Proc. SPIE* **10562** 10562-1–6

[25] Danzmann K *et al* 2017 LISA: A proposal in response to the ESA call for L3 mission concepts

[26] d'Arcio L *et al* 2017 Optical bench development for LISA *Proc. SPIE* **10565** 10565-1–7

[27] Dimopoulos S, Graham P W, Hogan J M, Kasevich M A and Rajendran S 2009 Gravitational wave detection with atom interferometry *Phys. Lett.* B **678** 37–40

[28] Dooley K Land the LIGO Scientific Collaboration 2015 Status of GEO 600 *J. Phys.: Conf. Ser.* **610** 012015

[29] Döringshoff K, Schuldt T, Kovalchuk E V, Stühler J, Braxmaier C and Peters A 2017 A flight-like absolute optical frequency reference based on iodine for laser systems at 1064 nm *Appl. Phys.* B **123** 183

[30] Douch K, Wu H, Schubert C, Müller J and dos Santos F P 2018 Simulation-based evaluation of a cold atom interferometry gradiometer concept for gravity field recovery *Adv. Space Res.* **61** 1307–23

[31] Elliffe E J, Bogenstahl J, Deshpande A, Hough J, Killow C, Reid S, Robertson D, Rowan S, Ward H and Cagnoli G 2005 Hydroxide-catalysis bonding for stable optical systems for space *Class. Quant. Grav.* **22** S257

[32] Ergenzinger K, Schuldt T, Berlioz P, Braxmaier C and Johann U 2017 Dual absolute and relative high precision laser metrology *Proc. SPIE* **10565** 10565-1–7

[33] ESA 2011 LISA assessment study report (Yellow Book) Technical Report *ESA/SRE(2011)3*

[34] ESA-ESTEC Requirements & Standards Division 2017 ECSS-E-HB-11A–Technology readiness level (TRL) guidelines *Technical report* European Cooperation for Space Standardization

[35] Faller J E, Bender P L, Hall J L, Hils D and Vincent M A 1985 Space antenna for gravitational wave astronomy *Proc. Colloquium Kilometric Optical Arrays in Space, ESA SP-226* pp 157–63

[36] Flechtner F, Neumayer K-H, Dahle C, Dobslaw H, Fagiolini E, Raimondo J-C and Güntner A 2016 What can be expected from the grace-fo laser ranging interferometer for earth science applications? *Surv. Geophys.* **37** 453–70

[37] Folkner W M, Bender P L and Stebbins R T 1998 LISA mission concept study *Technical Report* JPL-Publ-97-16

[38] Folkner W M *et al* 2010 Laser frequency stabilization for GRACE-2 *Proc. of the Earth Science Technology Forum (2010)*

[39] Gao D-F, Wang J and Zhan M-S 2018 Atomic interferometric gravitational-wave space observatory (aigso) *Commun. Theor. Phys.* **69** 37

[40] Gerberding O, Isleif K-S, Mehmet M, Danzmann K and Heinzel G 2017 Laser-frequency stabilization via a quasimonolithic Mach-Zehnder interferometer with arms of unequal length and balanced dc readout *Phys. Rev. Appl.* **7** 024027

[41] Gohlke M, Schuldt T, Döringshoff K, Peters A, Johann U, Weise D and Braxmaier C 2015 Adhesive bonding for optical metrology systems in space applications *J. Phys.: Conf. Ser.* **610** 012039

[42] Grote Hand the LIGO Scientific Collaboration 2010 The geo 600 status *Class. Quant. Grav.* **27** 084003

[43] Gwo D-H 1998 Ultraprecision bonding for cryogenic fused-silica optics *Proc. SPIE* **3435** 3435-1–7

[44] Halloin H *et al* 2017 Molecular laser stabilization for LISA *Proc. SPIE* **10566** 10566-1–8

[45] Hammesfahr A 2001 Lisa mission study overview *Class. Quant. Grav.* **18** 4045

[46] Heinzel G, Braxmaier C, Schilling R, Rüdiger A, Robertson D, te Plate M, Wand V, Arai K, Johann U and Danzmann K 2003 Interferometry for the lisa technology package (ltp) aboard smart-2 *Class. Quant. Grav.* **20** S153

[47] Heinzel G *et al* 2004 The ltp interferometer and phasemeter *Class. Quant. Grav.* **21** S581

[48] Heinzel G, Esteban J J, Barke S, Otto M, Wang Y, Garcia A F and Danzmann K 2011 Auxiliary functions of the LISA laser link: ranging, clock noise transfer and data communication *Class. Quant. Grav.* **28** 094008

[49] Herman J, Presti D, Codazzi A and Belle C 2004 Attitude Control for GRACE *18th Int. Symp. on Space Flight Dynamics ESA Special Publication* **vol 548** 27 p

[50] Hey F G 2018 *Micro Newton Thruster Development* (Wiesbaden: Springer) ISBN 978-3-658-21209-4

[51] Isleif K-S *et al* 2018 Towards the LISA backlink: experiment design for comparing optical phase reference distribution systems *Class. Quant. Grav.* **35** 085009

[52] Jennrich O 2009 LISA technology and instrumentation *Class. Quant. Grav.* **26** 153001

[53] Kawamura S *et al* 2008 The japanese space gravitational wave antenna - decigo *J. Phys.: Conf. Ser.* **122** 012006

[54] Lawson P R *et al* 2008 Terrestrial Planet Finder Interferometer: 2007-2008 progress and plans *Proc. SPIE* **7013** 7013-1–15

[55] Leonhardt V and Camp J B 2006 Space interferometry application of laser frequency stabilization with molecular iodine *Appl. Opt.* **45** 4142–46

[56] LISA study team 1998 LISA–Laser Interferometer Space Antenna; Pre-Phase A Report *Technical Report* MPQ233

[57] Maghami P G Jr, O'Donnell J R, Hsu O H, Zeimer J K and Dunn C E 2017 Drag-free performance of the ST7 disturbance reduction system flight experiment on the LISA Pathfinder *Proc. of the ESA Conf. on Guidance, Navigation and Control Systems*

[58] Marr J C 2003 Space interferometry mission (SIM): overview and current status *Proc. SPIE* **4852** 4852-1–15

[59] Martynov D V *et al* 2016 Sensitivity of the advanced ligo detectors at the beginning of gravitational wave astronomy *Phys. Rev.* D **93** 112004

[60] McNamara P W 2005 Weak-light phase locking for lisa *Class. Quant. Grav.* **22** S243

[61] Morrison E, Meers B J, Robertson D I and Ward H 1994 Automatic alignment of optical interferometers *Appl. Opt.* **33** 5041–49

[62] Morrison E, Meers B J, Robertson D I and Ward H 1994 Experimental demonstration of an automatic alignment system for optical interferometers *Appl. Opt.* **33** 5037–40

[63] Mottini S, Biondetti G, Cesare S, Castorina G, Musso F, Pisani M and Leone B 2017 Laser metrology for a next generation gravimetric mission *Proc. SPIE* **10566** 10566-1–9

[64] Mueller G, McNamara P, Thorpe I and Camp J 2005 Laser frequency stabilization for LISA *Technical Report* NASA/TM-2005-212794, NASA

[65] Müller H, Chiow S-W, Long Q, Vo C and Chu S 2005 Active sub-rayleigh alignment of parallel or antiparallel laser beams *Opt. Lett.* **30** 3323–25

[66] Ni W-T 2016 Gravitational wave detection in space *Int. J. Mod. Phys.* D **25** 1630001

[67] Ni W-T, Sandford M C W, Veillet C, Wu A-M, Fridelance P, Samain E, Spalding G and Xu X 2003 Astrodynamical space test of relativity using optical devices *Adv. Space Res.* **32** 1437–41 Fundamental Physics in Space

[68] Nicklaus K *et al* 2017 Optical bench of the laser ranging interferometer on GRACE follow-on *Proc. SPIE* **10563** 10563-1–9

[69] Nicklaus K *et al* 2017 High stability laser for Next Generation Gravity Missions *Proc. SPIE* **10563** 10563-1–8

[70] Pierce R, Leitch J, Stephens M, Bender P and Nerem R 2008 Intersatellite range monitoring using optical interferometry *Appl. Opt.* **47** 5007–19

[71] Ressel S, Gohlke M, Rauen D, Schuldt T, Kronast W, Mescheder U, Johann U, Weise D and Braxmaier C 2010 Ultrastable assembly and integration technology for ground- and space-based optical systems *Appl. Opt.* **49** 4296–303

[72] Rinehart S A, Savini G, Holland W, Absil O, Defrere D, Spencer L, Leisawitz D, Rizzo M, Juanola-Paramon R and Mozurkewich D 2016 The path to interferometry in space *Proc. SPIE* **9907** 9907-1–15

[73] Robertson D I *et al* 2013 Construction and testing of the optical bench for lisa pathfinder *Class. Quant. Grav.* **30** 085006

[74] Rowan S, Twyford S M, Hough J, Gwo D-H and Route R 1998 Mechanical losses associated with the technique of hydroxide-catalysis bonding of fused silica *Phys. Lett.* A **246** 471–78

[75] Sallusti M, Gath P, Weise D, Berger M and Schulte H R 2009 Lisa system design highlights *Class. Quant. Grav.* **26** 094015

[76] Sanjuan J, Gohlke M, Rasch S, Abich K, Görth A, Heinzel G and Braxmaier C 2015 Interspacecraft link simulator for the laser ranging interferometer onboard grace follow-on *Appl. Opt.* **54** 6682–89

[77] Saraf S *et al* 2016 Ground testing and flight demonstration of charge management of insulated test masses using uv-led electron photoemission *Class. Quant. Grav.* **33** 245004

[78] Schilling R OptoCAD: Tracing Gaussian $TEM_{00}$ beams through an optical set-up http://home.mpcdf.mpg.de/~ros/optocad.html.

[79] Schleich W P *et al* 2016 Quantum technology: from research to application *Appl. Phys.* B **122** 130

[80] Schuldt T, Gohlke M, Kögel H, Spannagel R, Peters A, Johann U, Weise D and Braxmaier C 2012 Picometre and nanoradian heterodyne interferometry and its application in dilatometry and surface metrology *Meas. Sci. Technol.* **23** 054008

[81] Schuldt T 2010 An optical readout for the LISA gravitational reference sensor *PhD Thesis* (Humboldt-Universität zu Berlin: Mathematisch-Naturwissenschaftliche Fakultät I) 16241

[82] Schuldt T, Döringshoff K, Kovalchuk E V, Keetman A, Pahl J, Peters A and Braxmaier C 2017 Development of a compact optical absolute frequency reference for space with $10^{-15}$ instability *Appl. Opt.* **56** 1101–06

[83] Schuldt T, Gohlke M, Weise D, Johann U, Peters A and Braxmaier C 2009 Picometer and nanoradian optical heterodyne interferometry for translation and tilt metrology of the LISA gravitational reference sensor *Class. Quant. Grav.* **26** 085008

[84] Schuldt T *et al* 2015 Design of a dual species atom interferometer for space *Exp. Astron.* **39** 167–206

[85] Shaul D N A, Aaraujo H M, Rochester G K, Schulte M, Sumner T J, Trenkel C and Wass P 2008 Charge management for LISA and LISA Pathfinder *Int. J. Mod. Phys.* D **17** 993–1003
[86] Sheard B S, Heinzel G, Danzmann K, Shaddock D A, Klipstein W M and Folkner W M 2012 Intersatellite laser ranging instrument for the GRACE follow-on mission *J. Geod.* **86** 1083–95
[87] Sneeuw N, Flury J and Rummel R 2004 Science requirements on future missions and simulated mission scenarios *Earth Moon Planets* **94** 113–42
[88] Tapley B D, Bettadpur S, Watkins M and Reigber C 2004 The gravity recovery and climate experiment: Mission overview and early results *Geophys. Res. Lett.* **31** L09607
[89] Tinto M and Dhurandhar S V 2014 Time-delay interferometry *Liv. Rev. Relat.* **17** 6
[90] Wand V *et al* 2006 Noise sources in the ltp heterodyne interferometer *Class. Quant. Grav.* **23** S159
[91] Wiese D N, Folkner W M and Nerem R S 2009 Alternative mission architectures for a gravity recovery satellite mission *J. Geod.* **83** 569–81
[92] Wouters B, Bonin J A, Chambers D P, Riva R E M, Sasgen I and Wahr J 2014 Grace, time-varying gravity, earth system dynamics and Climate Change *Rep. Prog. Phys.* **77** 116801
[93] Ziegler T, Fichter W, Schulte M and Vitale S 2009 Principles, operations, and expected performance of the lisa pathfinder charge management system *J. Phys.: Conf. Ser.* **154** 012009

**IOP** Publishing

Modern Interferometry for Length Metrology
Exploring limits and novel techniques
**René Schödel (Editor)**

# Chapter 5

# Interferometry in air with refractive index compensation

**Florian Pollinger**

The SI definition of the metre

> The metre is the length of the path travelled by light in vacuum during a time interval of 1/299 792 458 s.

is of indisputable beauty in terms of universality and conceptual rigour. In application, however, many length measurements cannot be performed in vacuum, but have to be performed in air. In a medium, the light propagation velocity changes and thus, the natural scale of a length measurement as defined by the SI definition. In practice, measurement technology and interferometry in particular are so well advanced that the uncertainty of the correction of this change in speed of light often determines the overall uncertainty of the length measurement. In this chapter, the concept of the index of refraction is discussed, major quantitative models described and various approaches for the compensation presented.

## 5.1 The index of refraction

When traversing a medium, an electromagnetic wave interacts with the atoms of the medium. Snell's law shows that the speed of light $c$ in a medium deviates from the vacuum speed $c_0$ by

$$c = c_0/n \tag{5.1}$$

defining the refractive index $n$. White light separation in prisms indicates that the index of refraction depends on the light frequency $\nu$. This is called 'dispersion'. In air, the index of refraction also depends on the thermodynamic properties. Heat can induce shimmering of the surface of a road. For metrological applications, these phenomenological observations need to be quantified. Today, sophisticated semi-

doi:10.1088/2053-2563/aadddcch5  

empirical formulae are available to describe dispersion and dependence on intrinsic thermodynamic parameters like temperature $\vartheta$, air pressure $p$, humidity $f$, and $CO_2$ content $x_{CO_2}$. Their numerical value is based on adjustment to sophisticated experiments. The functional dependencies can be 'heuristically derived' using classic electrodynamics and thermodynamics. The following discussion of the dispersion factor is largely based on the treatments by Barrel and Sears [1] and Born and Wolf [2], while the discussion of the density term follows mainly Edlén [3] and Ciddor [4].

### 5.1.1 The dispersion factor

Assume a gaseous dielectric medium consisting of nonpolar particles of number density $N$. An external electromagnetic field $E$ interacts with the medium and induces polarisation, modifying the local electric field. In consequence, an effective field, $E'$, acts on each particle. The electric dipole moment $p_{\mathrm{dp}}$ of this particle is proportional to the effective field $E'$ by the polarizability $\alpha$,

$$p_{\mathrm{dp}} = \alpha E'. \tag{5.2}$$

The total electric dipole moment $P_{\mathrm{dp}}$ per unit volume is then given by

$$P_{\mathrm{dp}} = N\alpha E', \tag{5.3}$$

taking $\alpha$ as the mean polarizability. The polarizability $\alpha$ can be understood as the microscopic origin of the relative permittivity $\varepsilon_r$ and thus, by Maxwell's relation, to the index of refraction. In a homogeneous dielectric medium, the Lorentz–Lorenz formula quantifies the relation between mean polarizability $\alpha$ and the index of refraction $n$

$$\alpha = \frac{3\varepsilon_0}{N}\frac{n^2 - 1}{n^2 + 2}. \tag{5.4}$$

In such a medium, it is also possible to derive a simple model for the mean polarizability. For this, an electromagnetic wave is assumed by an $E$-field oscillating in the $x$ direction that traverses this medium. Each particle contains one electron of mass $m_{\mathrm{e}}$ and charge $e$ bound to the molecule with a resonance frequency $\omega_0$. The effective field $E'$ induces a force $F = -eE_0'e^{-i\omega t}$ on a given electron. This leads to a classical expression for an externally driven oscillator

$$\frac{\mathrm{d}^2x}{\mathrm{d}t^2} + \omega_0^2 x = -\frac{e}{m_{\mathrm{e}}}E_0' e^{-i\omega t} \tag{5.5}$$

with the solution

$$x(t) = \frac{e^2}{m_{\mathrm{e}}}\frac{E_0'}{(\omega_0^2 - \omega^2)}e^{-i\omega t}, \tag{5.6}$$

describing a displacement of the charge $e$. The resulting dipole moment $p_{\mathrm{dp}}$ of a single particle is given by $p_{\mathrm{dp}} = ex(t, z)$, and the polarisation density $P_{\mathrm{dp}}$ of the medium by

$$P_{\mathrm{dp}}(t) = Np_{\mathrm{dp}} = N\frac{e^2}{m_{\mathrm{e}}}\frac{1}{(\omega_0^2 - \omega^2)}E'. \tag{5.7}$$

Substituting equations (5.3) and (5.4) into equation (5.7), a simple functional dependence for the index of refraction $n$ on the optical frequency $\omega$ can be derived

$$\frac{n^2 - 1}{n^2 + 2} = \frac{1}{3\varepsilon_0}N\frac{e^2}{m_{\mathrm{e}}}\frac{1}{(\omega_0^2 - \omega^2)}. \tag{5.8}$$

The wavelength-dependence of the index of refraction is hence, in practice, determined by damped absorption resonances. In general, there will be more than one resonance. They can be assumed to superpose linearly, weighted by the fractions $f_i$ of electrons of resonance frequencies $\omega_i$ ($i = 0, 1, 2, 3,\ldots$):

$$\frac{n^2 - 1}{n^2 + 2} = \frac{1}{3\varepsilon_0}N\frac{e^2}{m_{\mathrm{e}}}\sum_i\frac{f_i}{(\omega_{0i}^2 - \omega^2)}. \tag{5.9}$$

Figure 5.1 shows experimental data sensitive to the dispersion in the spectral range of a specific oxygen transition ($b^1\Sigma_g^+(\nu = 0) \leftarrow X^3\Sigma_g^+(\nu = 0)$). Each transition induces the functional footprint of a damped resonance on the dispersion curve. The simple model is capable of describing this functional dependence.

Much of the formalism was developed in the beginning of the twentieth century. Then, the driving interest for the knowledge of the index of refraction was the need for traceability of spectroscopic observations. As the SI definition of the metre was then based on an artefact, wavelengths in air had to be determined rather than frequencies. Therefore, the optical frequencies $\omega$ are still commonly expressed in terms of vacuum wavenumbers $\sigma \equiv 1/\lambda_0$ in $\mu\mathrm{m}^{-1}$ in this context.

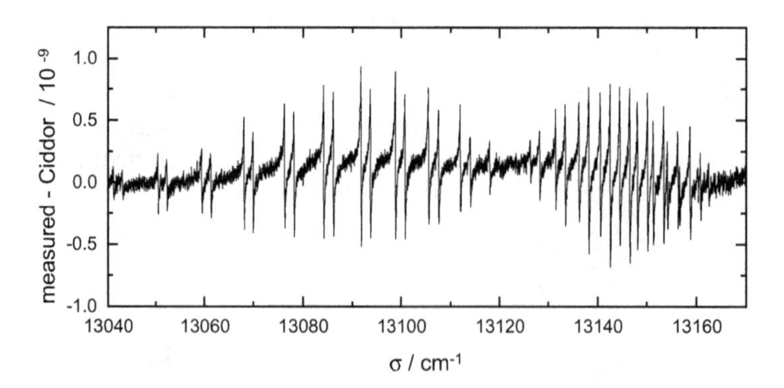


**Figure 5.1.** Dispersion at the oxygen 'A-band' $b^1\Sigma_g^+(\nu = 0) \leftarrow X^3\Sigma_g^+(\nu = 0)$ as observed by Balling and co-workers in Fourier-transform interferometry [5]. Depicted is the measured index of dispersion in deviation from the Ciddor equation, normalised to the refractive index of the second harmonic of ethyn at $\sigma_{C_2H_2} = 12966.94\ \mathrm{cm}^{-1}$: $(n(\sigma)/n(\sigma_{C_2H_2}))_{\mathrm{meas}} - (n(\sigma)/n(\sigma_{C_2H_2}))_{\mathrm{Ciddor}}$. The oxygen resonances, which are not explicitly included in the Ciddor model, are clearly visible. They show the functional dependence of damped resonances as predicted by the simple model leading to equation (5.9).

$$\frac{n^2 - 1}{n^2 + 2} = N \times \sum_i \frac{k_i}{(\sigma_i^2 - \sigma^2)}. \tag{5.10}$$

The sum is referred to as dispersion factor $K(\sigma)$ in the following:

$$K(\sigma) \equiv \sum_i \frac{k_i}{(\sigma_i^2 - \sigma^2)}. \tag{5.11}$$

Approximations of different orders to the general expression (5.10) can be used to explain various empiric approximation formulae for the index of refraction, like Cauchy's formula or the Sellmeier equation [1]. In air, for wavelengths in the visible and the infra-red, Edlén [6] proposed to approximate the dispersion factor by two resonances and a constant term

$$K(\sigma) \approx k_0 + \frac{k_1}{(\sigma_1^2 - \sigma^2)} + \frac{k_2}{(\sigma_2^2 - \sigma^2)}. \tag{5.12}$$

It should be noted that the dispersion factor $K(\lambda)$ does not depend on the environmental parameters, but only on the wavelength.

### 5.1.2 The density term

So far, only the wavelength-dependence has been developed. Generally, the Lorentz–Lorenz term $L \equiv (n^2 - 1)/(n^2 + 2)$ is not the quantity of interest, but the refractivity $(n - 1)$. Expression (5.10) can be developed further towards a practical tool for length metrology. For the typical optical range of interest in air, the magnitude of the index of refraction can be estimated to be $n \approx 1 + O(10^{-4})$. Assuming $n$ within the interval [0.999, 1.001], the Lorentz–Lorenz term $L$ can then be approximated by [3]

$$L = \frac{n^2 - 1}{n^2 + 2} \approx (n - 1)\frac{2}{3} \times \left(1 - \frac{n - 1}{6}\right) \tag{5.13}$$

with an absolute deviation below $1.5 \times 10^{-10}$. Applying the approximation $(1 - x)^{-1} \approx (1 + x)$ one can transform equation (5.12) to

$$(n - 1) \approx K(\lambda) \times N \times \frac{3}{2} \times \left(1 + \frac{n - 1}{6}\right). \tag{5.14}$$

It is known from experiment that the term $(1 + (n - 1)/6)$ changes in the visible, i.e. between wavelengths of 400 and 800 nm, in the order of $1.3 \times 10^{-6}$ and in the near-infra-red, i.e. between 800 and 1600 nm, in the order of $3.0 \times 10^{-7}$. In all current approximation formulae for the index of refraction, this term is approximated by a constant term $\kappa$, e.g. based on the mean value $\bar{n}_{\mathrm{exp}}$ of the index of refraction of the experimental data for the largest and the smallest wavelength available for standard conditions [7].

$$\frac{3}{2}\left(1+\frac{n(\lambda, \vartheta, p, h, x_{CO_2})-1}{6}\right) \approx \frac{3}{2}\left(1+\frac{\bar{n}_{exp}-1}{6}\right) \equiv \kappa. \tag{5.15}$$

This is a relatively hard approximation. Experimental data shows, however, that this approximation seems legitimate (see also results discussed in section 5.3). The wavelength-dependence is in practice very well absorbed by the adjustment of the dispersion factor polynomial in equation (5.12) to the experimental data.

The number density $N$ in equation (5.12) is unfortunately not straightforwardly accessible. By classical thermodynamics, it depends, however, on the measurable absolute temperature $T$ and the composition of the gas. This can be derived starting from the equation of state for an ideal gas of point-like particles. It is given, dependent on the pressure $p$, the gas volume $V$, the absolute temperature $T$, the amount of substance $\nu_{mol}$ and the molar gas constant $R$, by

$$pV = \nu_{mol}RT. \tag{5.16}$$

For a real gas like air, attractive and repulsive particle–particle interactions must be taken into account. Formally, these can be introduced by the compressibility $Z$ [8], leading to

$$pV = \nu_{mol}RTZ. \tag{5.17}$$

Using $\nu_{mol} = m_\Sigma / M_{mol}$ with the molar mass $M_{mol}$ and the total gas mass $m_\Sigma$, one can deduce

$$\frac{m_\Sigma}{V} = \frac{p}{RTZ} M_{mol}. \tag{5.18}$$

This equation describes the dependence of the volumetric mass density $\rho \equiv m_\Sigma V^{-1}$ on the thermodynamic intrinsic constants. Number density $N$ and volumetric mass density $\rho$ are connected by the average single gas particle mass $m_{sp}$ to $\rho = Nm_{sp}$. Equation (5.18) hence provides a link to deduce the number density in equation (5.12) from measurable quantities. The parameters $T$, $Z$, and $M_{mol}$ on the right-hand side of equation (5.18), however, still need to be developed to a more application-convenient form.

For practical reasons, the temperature $\vartheta$ in °C is usually used in this context instead of the absolute temperature $T$. Using the conversion formula $T = T_0 + \vartheta$ from the Kelvin to the Celsius scale with $T_0 = 273.15$ K, the temperature dependence $1/T$ in equation (5.18) can be rewritten by

$$\frac{1}{T} = \frac{1}{T_0} \times \frac{1}{1+\alpha\vartheta}. \tag{5.19}$$

Due to its functional dependence, the factor $\alpha = T_0^{-1}$ is sometimes referred to as the 'theoretical expansion coefficient of a perfect gas' [1].

As argued before, the compressibility $Z$ describes the deviation of the behaviour of a real gas from the ideal gas law. In the gas phase, it can be attributed to Van der

Waals interactions. It is known from statistical mechanics that the real gas law can be expressed in the so-called virial expansion in powers of pressure $p$ [9]:

$$Z = (1 + B'(T, x_w)p + C'(T, x_w)p^2 + \cdots). \tag{5.20}$$

The second virial coefficient $B'(T, x_w)$ is a measure of the interaction between two molecules, the third virial coefficient $C'(T, x_w)$ between three, and so on. In principle, the virial coefficients depend only on temperature $T$. In the case of mixtures like air/water vapour, the coefficients can be expressed in inner-species and inter-species interactions, weighted by the mole fractions $x_{da}$ and $x_w$ of the components. Giacomo [8] and Davis [10] recommend using the expansion to the third virial coefficient, leading to an equation of the form

$$\begin{aligned} Z(p, \vartheta, x_w) = 1 - \frac{p}{\vartheta + T_0}[a_0 + a_1\vartheta + a_2\vartheta^2 \\ + (a_2 + a_3\vartheta)x_w + (a_4 + a_5\vartheta)x_w^2] \\ + \frac{p^2}{(\vartheta + T_0)^2}(a_6 + a_7 x_w^2), \end{aligned} \tag{5.21}$$

with $a_i$, $i = 0, 1, \ldots, 7$, representing experimentally determined constants. Ciddor incorporates this formula completely into his expression for the index of refraction [4]. Barrel and Sears proposed an (effective) first order approximation of the type

$$Z(p, \vartheta) = 1 - \varepsilon(\vartheta)p. \tag{5.22}$$

This has since been used by many other approximations, for example the formulae of Edlén [3], Birch and Downs [11], or Bönsch and Potulski [7].

Summarising the previous discussion in this chapter, it becomes clear that for standard dry air, the quantity $(n - 1)_{sda}/K(\lambda)$ can be approximated in the following algebraic structure

$$D_{sda}(p, \vartheta) = \frac{\kappa}{m_{sp}T_0 R} \times \frac{Z(p, \vartheta, x_w = 0)^{-1}}{1 + \alpha\vartheta} M_{mol}(x_{CO_2}, x_w), \tag{5.23}$$

the so-called density factor. It depends only on the intrinsic thermodynamic parameters and, by the molar mass, on the composition. For standard dry air, the refractivity hence factorises to an expression of the form

$$(n - 1)_{sda} = D_{sda}(p, \vartheta) \times K(\lambda). \tag{5.24}$$

### 5.1.3 Carbon dioxide and water contents

Dry air needs to be specially prepared for laboratory applications. In length metrology, however, one usually has to deal with moist air in the composition as given. The water vapour contents (in mole fractions $x_w$) depends, e.g. on general weather conditions. Furthermore, when working in closed rooms, the carbon dioxide contents $x_{CO_2}$ can deviate significantly from the standard air reference value of $x_{CO_2}^{sda} = 400$ ppm (or 450 ppm in the case of Ciddor [4]).

In the case of carbon dioxide, however, the dispersion term does not change considerably for deviations of $x_{CO_2}$ from $x_{CO_2}^{sda}$. Experimental results discussed by Edlén [3] and Ciddor [12] show that the refractive index of standard air and $CO_2$ differ wavelength-dependently by a factor of 3.6 nm/$\lambda$ only. In a wavelength range between 300 and 1600 nm, this corresponds to a total relative variation of the order of 1%. This can be approximated by its mean value to a constant scaling factor for the refractivity. The magnitude of this scaling factor is dominated by the density factor. The molar mass of standard dry air $M_{sda}$ is given by 28.9635 kg mol$^{-1}$ [8]. For a deviating $CO_2$ fraction, Giacomo proposes it is assumed that each additional carbon dioxide molecule $CO_2$ corresponds to a replaced oxygen molecule $O_2$ [8]. Today, the Bureau International des Poids et Mesures (BIPM) proposes for the molar mass of standard dry air $M_a$ [10]

$$M_a = [28.9635 + 12.011 \times (x_{CO_2} - 400 \times 10^{-6})] \times 10^{-3} \frac{\text{kg}}{\text{mol}}. \tag{5.25}$$

Changes in the molar mass scale the density term (5.12) linearly,

$$D(p, \vartheta, x_{CO_2}) = D_{sda}(p, \vartheta)\, g_{CO_2}(x_{CO_2}), \tag{5.26}$$

and thus, the index refraction $n$ as well:

$$(n - 1)_x = (n - 1)_{sda}\, g_{CO_2}(x_{CO_2}) \tag{5.27}$$

with

$$g_{CO_2}(x_{CO_2}) = 1 + \chi_0 \times \left(x_{CO_2} - x_{CO_2}^{sda}\right) \tag{5.28}$$

Bönsch and Potulski choose the reference level $x_{CO_2}^{sda}$ to 400 ppm, and a scaling factor of $\chi_0$ of 0.5327. A change by 100 ppm rescales the refractivity by a factor $5.3 \times 10^{-5}$. This magnitude corresponds well to the associated change of the molar mass $M_{mol}$ (and thus the density term $D$ of approximately $4.1 \times 10^{-5}$).

The water vapour dispersion factor $K(\lambda)$ varies substantially from the dispersion factor of standard air. Experimental data for the refractivity of water vapour is compiled in [13]. Therefore, the contribution of the water refractivity to moist air needs to be treated in a more sophisticated way. One can assume that the polarizability of individual molecules are unchanged in a mixture of various substances. This seems legitimate for moist air, as the number density of water molecules is relatively small in comparison to nitrogen molecules. Under this approximation, the Lorentz–Lorenz term $L$ of the mixture can be approximated by the weighted sum of the contributions due to each substance $i$ [2, 14]

$$L = \frac{n^2 - 1}{n^2 + 1} = \sum_i \frac{\rho_i}{\rho_{ip}} L_i = \sum_i \frac{\rho_i}{\rho_{ip}} \times \frac{n_i^2 - 1}{n_i^2 + 1} \tag{5.29}$$

with $\rho_i$ representing the density of the substance, $\rho_{ip}$ the density of the pure substance for the given external conditions and $n_i$ the refractive indices of the pure substance. Ciddor argued that this formula can be simplified further in air to

$$(n-1)_{\mathrm{xw}} = \frac{\rho_{\mathrm{x}}}{\rho_{\mathrm{xp}}} \times (n-1)_{\mathrm{x}} + \frac{\rho_{\mathrm{w}}}{\rho_{\mathrm{wp}}} \times (n-1)_{\mathrm{w}} \tag{5.30}$$

by an accuracy of a few parts in $10^{-10}$ because of the very low concentration of water vapour [4].

Edlén [3], Birch and Downs [11], and Bönsch and Potulski [7] assume one resonance for the refractivity of water in the visible. They approximate equation (5.30) by

$$n - 1 = (n-1)_{\mathrm{x}} - p_{\mathrm{w}} g_{\mathrm{w}}(\sigma) \tag{5.31}$$

with the correction term $g_{\mathrm{w}}(\sigma)$

$$g_{\mathrm{w}}(\sigma) = \gamma_0 + \gamma_1 \sigma^2, \tag{5.32}$$

$\gamma_{0,1}$ being constants. As a weighting factor, the partial pressure of water $p_{\mathrm{w}}$ is used here.

Ciddor [4], on the other hand, works with the polynomial equation introduced by Owens [14]

$$(n-1)_{\mathrm{w}} = \alpha_{\mathrm{corr}} \sum_{k=0}^{4} w_i \sigma^{2i}, \tag{5.33}$$

with $\alpha_{\mathrm{corr}}$ and $w_i$ representing constants. He uses equation (5.30) to describe the refractivity of moist air.

### 5.1.4 Summary on the refractivity model and numerical values

The (relatively tedious) derivation in the previous sections justifies the following generalised expression for phase refractivity in air:

$$(n-1) = \frac{\rho_{\mathrm{x}}}{\rho_{\mathrm{xp}}} \times D(p, \vartheta, x_{\mathrm{CO_2}}) K(\lambda) + \frac{\rho_{\mathrm{w}}}{\rho_{\mathrm{wp}}} \times g_{\mathrm{w}}(\sigma). \tag{5.34}$$

In the course of the last 120 years, several explicit numerical expressions have been derived. A survey can be found in reference [15]. In this work, we restrict the discussion to the expression by Bönsch and Potulski which was developed based on the latest experimental data in 1998 [7]. Its results vary from the Ciddor formula on the order of $1 \times 10^{-8}$ in the visible, well within the estimated uncertainty of both expressions. Bönsch and Potulski effectively simplify expression (5.34) to

$$\begin{aligned} n(\sigma, p, \vartheta, x_{\mathrm{CO_2}}, p_{\mathrm{w}}) - 1 = {} & K(\sigma) \times D_{\mathrm{sda}}(p, \vartheta) \\ & \times g_{\mathrm{CO_2}}(x_{\mathrm{CO_2}}) - p_{\mathrm{w}}/\mathrm{Pa} \times g_{\mathrm{w}}(\sigma). \end{aligned} \tag{5.35}$$

They derive the following explicit expressions [7]:

$$
\begin{aligned}
K(\sigma) &= 10^{-8}\left[8091.37 + \frac{2\,333\,983}{130 - (\sigma/\mu\mathrm{m}^{-1})^2} + \frac{15\,518}{38.9 - (\sigma/\mu\mathrm{m}^{-1})^2}\right] \\
g_{\mathrm{CO_2}}(x_{\mathrm{CO_2}}) &= 1 + 0.5327 \times (x_{\mathrm{CO_2}} - 0.0004) \\
D_{\mathrm{sda}}(p, \vartheta) &= \frac{p/\mathrm{Pa}}{93\,214.60} \\
&\quad \times \frac{1 + 10^{-8} \times (0.5953 - 0.009\,876 \times \vartheta/^{\circ}\mathrm{C}) \times p/\mathrm{Pa}}{1 + 0.003\,661\,0 \times \vartheta/^{\circ}\mathrm{C}} \\
g_{\mathrm{w}} &= 10^{-10}\,[3.8020 - 0.0384 \times (\sigma/\mu\mathrm{m}^{-1})^2].
\end{aligned}
\tag{5.36}
$$

In practice, the partial pressure of water is often derived from a measurement of the relative humidity and the air temperature. Conversion formulae are given by Hardy [16].

The uncertainty of the empiric formulae by Ciddor and by Bönsch and Potulski is difficult to quantify. Both rely on multiple inputs from other measurements, performed in part decades before their analysis. Ciddor provides a conservative estimate of the order of 2 to $5 \times 10^{-8}$ [4], Bönsch–Potulski claim an uncertainty of approximately $1 \times 10^{-8}$ for their experimental conditions (environmental conditions as well as spectral range). Both formulae agree to $10^{-8}$ level. There is some experimental indication that below $10^{-8}$, the model needs substantial refinement (see figure 5.1). Hence, a combined uncertainty of the order of $3 \times 10^{-8}$ seems a reasonable assumption for the uncertainty of the compensation formula itself. As mentioned by Ciddor and Bönsch and Potulski themselves, the uncertainty is usually dominated by the determination of the environmental parameters. This is covered in sections 5.2 and 5.3 of this chapter.

### 5.1.5 Experimental refractometry

The experimental determination of the index of refraction is highly demanding on sensor systems and experimental design. In this section, the two main experimental approaches pursued in the last two decades for the investigation of the phase refractive index are briefly introduced. For a deeper discussion, the cited original work should be consulted.

The most accurate refractometers are based on highly stable Fabry–Perot interferometers. The change of laser frequency upon stabilisation on a Fabry–Perot interferometer is used to measure the index of refraction [17–23]. As an example, the high-performance refractometer designed by Egan *et al* [17] as depicted in figure 5.2 is briefly discussed. Two HeNe-lasers are stabilised on a Fabry–Perot pair made out of ultra-low expansion (ULE) glass (see figure 5.2(b)). When locked to a resonance of a cavity of length $l_{\mathrm{cav}}$, the laser frequency $\nu$ is given by [23]

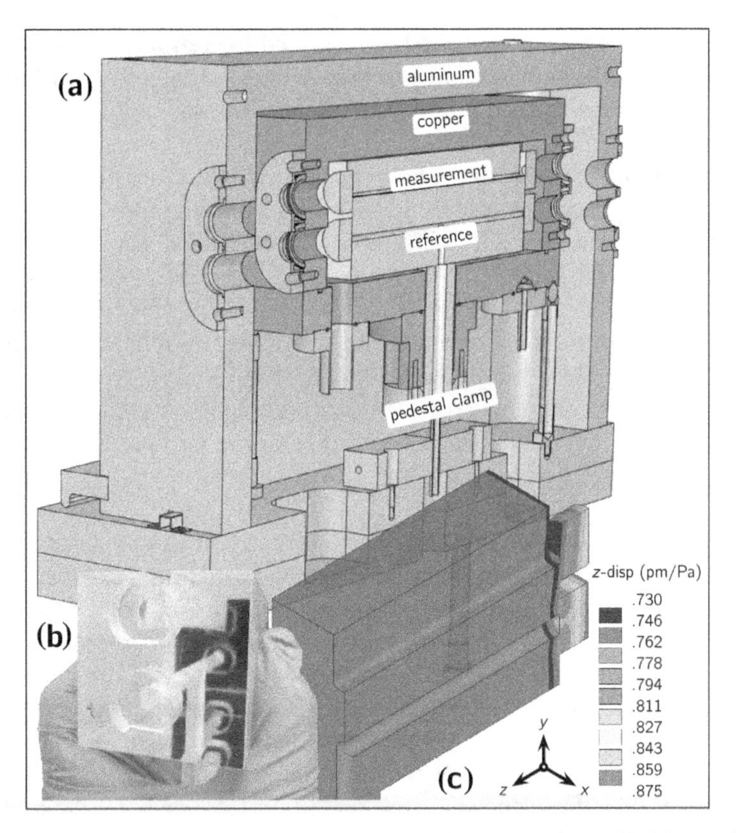


**Figure 5.2.** Dual Fabry–Perot cavity refractometer developed by Egan *et al* [17]. (a) Scheme of the thermal isolation and evacuation system. (b) Photograph of the cavity made entirely of ultra-low expansion (ULE) glass. (c) Finite element analysis of distortions of the cavity after evacuation. Reprinted with permission from reference [17], © (2015) Optical Society of America.

$$\nu = \frac{c}{2nl_{\mathrm{cav}}}\left[m + \frac{\phi(l_{\mathrm{cav}})}{\pi} - \frac{\phi_{\mathrm{R}}(\nu)}{2\pi}\right] \tag{5.37}$$

where $m$ represents number of modes within the cavity length, $\phi(l_{\mathrm{cav}})$ the diffraction phase shift of the propagating beam and $\phi_{\mathrm{R}}(\nu)$ reflection phase shifts due to the mirrors. When evacuating one of the two cavities, the resonance frequency of this cavity and thus the beat note of the two lasers changes. The absolute refractivity can then be derived from [17]

$$n - 1 = \frac{\Delta f_{\mathrm{eff}}}{\nu} + nd_{\mathrm{m}}p + d_{\mathrm{r}}p, \tag{5.38}$$

with $\frac{\Delta f_{\mathrm{eff}}}{\nu}$ representing the measured effective fractional change in beat note frequency and two correction terms for pressure-induced longitudinal distortions. A detailed discussion can be found in reference [17]. In general, beat note frequencies can be determined with low uncertainty. Measurement uncertainties as low as

$3 \times 10^{-9}$ for the refractive index have been claimed for refractometers based on this design [17]. One disadvantage of this approach is the required high finesse of the resonators. As a consequence, the refractometer works only for the dedicated wavelength.

A broader spectral window can be achieved by phase-sensitive refractometers based on spatial resolution. A vacuum cell is located in the measurement path. For the most sensitive designs [7, 13, 24], the beam is expanded, the centre of the beam passing the vacuum cell, the outer part traversing the windows, but the rest of the beam path being constantly in the ambient medium. By observation of the phase change under evacuation of the vacuum cell, the change in optical path length can be measured and thus, the change in refractive index. This symmetric alignment compensates for the remaining geometric misalignments. Variations in the optical thickness of the windows as well as the influence of the vacuum on the centre of the windows [7] are critical. This method is discussed in greater depth in chapter 2 of this book.

Broadband high-coherence sources like frequency combs might pave the way for effective broadband investigations of the index of refraction. An efficient measurement scheme was proposed by Balling and co-workers in 2012 [5]. If individual comb modes of a frequency comb pulse can be resolved in a Fourier-transform interferometer, there is information on the air wavelength $\lambda_{\text{air}}$ as a result of the Fourier transform in reference to the displacement of the moving reference mirror, e.g. by a reference interferometer. Complementary, by the mode order $i$ of the comb mode, the absolute optical frequency $\nu_i$ of the comb mode is known to be dependent on carrier envelope offset frequency $f_{\text{ceo}}$ and repetition frequency $f_{\text{rep}}$ by $\nu_i = f_{\text{ceo}} + i \times f_{\text{rep}}$. Taking one arbitrary comb line as reference, one can derive the relative dispersion $n(\nu)/n(\nu_{\text{ref}})$ right from the data by

$$\frac{n(\nu)}{n(\nu_{\text{ref}})} = \frac{\lambda_{\text{ref, air}} \times \nu_{\text{ref}}}{\lambda_{i,\text{ air}} \times \nu_i}. \tag{5.39}$$

According to the generalised refractivity equation (5.34), this ratio is dominated by the dispersion term. Balling *et al* claim an agreement of this ratio with the predictions by the Ciddor equation of the order of $(-0.8 \pm 0.3) \times 10^{-9}$ per 100 nm in the wavelength range between 720 and 880 nm [5].

In principle, refractometers can be used to compensate the index of refraction for an optical distance measurement directly. The concept is to place the refractometer in the direct vicinity of the length measurement beam and to determine the index of refraction in parallel. Recently, Kruger and Chetty [25] proposed a relatively simple design for such a refractometric weather station. The achievable uncertainty is of the order of a few parts in $10^{-8}$ [23, 25]. The intricacies of high accuracy refractometry, however, like the interaction of water with the window surfaces, limit the achievable uncertainties [23]. Furthermore, the probing space of such a refractometer is limited. Gradients in the index of refraction, e.g. due to temperature variations as discussed in section 5.2.1 are hence very difficult to measure accurately this way.

### 5.1.6 A note on group velocity

So far, only the phase refractive index $n_p$ has been discussed. As mentioned already in chapter 1 of this book, in the case of pulsed light, the group refractive $n_g$ is of interest:

$$n_g(\nu, p, \vartheta, x_{CO_2}, p_w) = n_p(\nu, p, \vartheta, x_{CO_2}, p_w) + \nu \frac{dn_p(\nu, p, \vartheta, x_{CO_2}, p_w)}{d\nu}, \quad (5.40)$$

or

$$n_g(\lambda, p, \vartheta, x_{CO_2}, p_w) = n_p(\lambda, p, \vartheta, x_{CO_2}, p_w) - \lambda \frac{dn_p(\nu, p, \vartheta, x_{CO_2}, p_w)}{d\lambda}. \quad (5.41)$$

The group refractive index can be derived by explicit derivation of the formulae discussed before. Ciddor and Hill [27] have indeed published an analytical expression for the group refractive index which can be implemented[1]. In general, however, provided a careful numerical implementation, a numerical derivation of the phase refractive index based on the various expressions for the $n_p(\lambda, p, \vartheta, x_{CO_2}, p_w)$ provides stable results.

It is interesting to realise, however, that the group index of refraction can also be directly measured. Yang and co-workers [26, 28] have demonstrated this for refractometers based on comb-based Fourier-transform interferometry [28] and on frequency-sweeping-interferometry (cf figure 5.3). The basic equation of frequency-sweeping-interferometry (FSI) as discussed in chapter 8 of this book, for example, offers direct access to the group refractive index by its basic equation [29]

$$\frac{d\Phi(\nu)}{d\nu} = \frac{4\pi l}{c_0} \times n_g(\nu). \quad (5.42)$$

Yang *et al* use a classical refractometer with an evacuated reference beam path and a measurement path in air of equal path lengths [26]. They run the interferometer in FSI mode, sweeping the frequency of an external cavity diode laser (ECDL) in their case with a centre frequency of 194 061.02 GHz or 1544.8360 nm vacuum wavelength. The frequency hub $\Delta\nu$ is referenced to a frequency comb, and recorded together with the phase change $\Phi(\nu(t)) - \Phi(\nu(0))$. The derived group refractivity agrees within $1 \times 10^{-7}$ with Ciddor's prediction. This relatively large deviation can be expected given the relatively poor knowledge of the environmental parameters of the investigated air in this proof-of-principle experiment. A better control of the environmental parameters could improve the uncertainty of the respective measurements. In general, a better secured validity of the group index compensation formulae particularly in the infra-red around 1560 nm would

[1] It should be noted, however, that equation (B2) in their publication [27] contains a sign error. By proper application of the derivation chain rule, one can easily show that it should read $10^8 dn_s/d\sigma = +2\sigma \sum_p a_p/(b_p - \sigma^2)^2$.

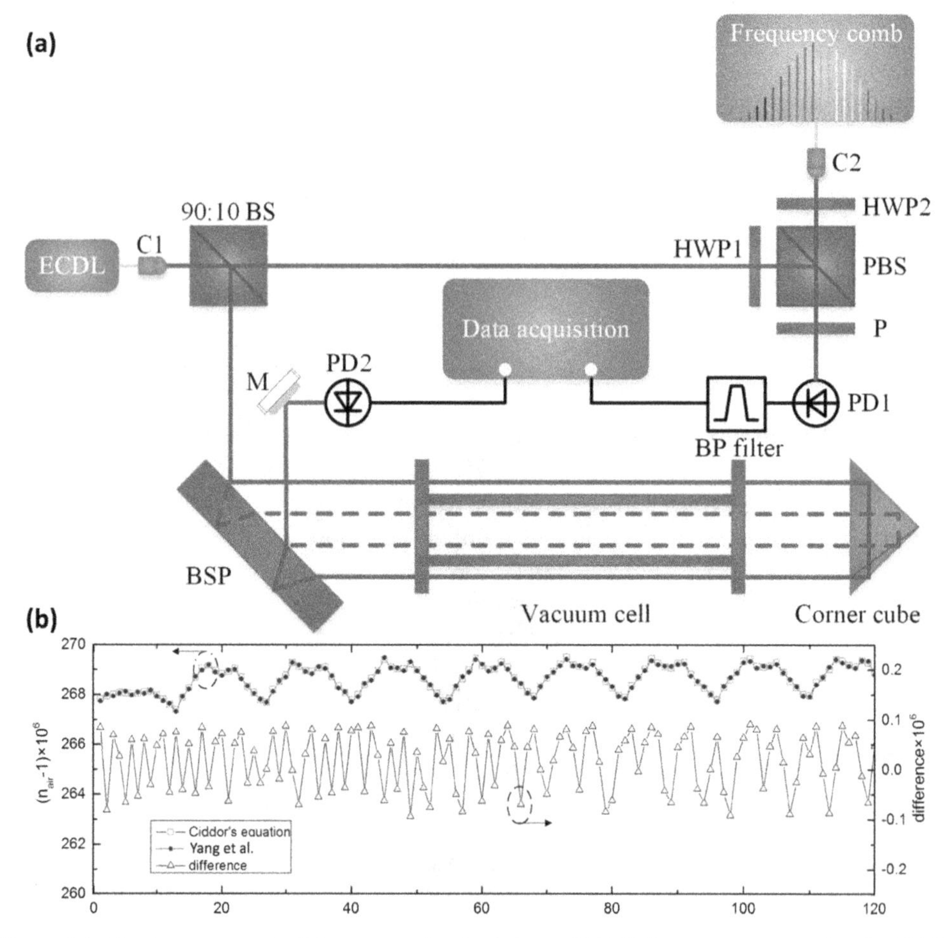


**Figure 5.3.** Direct measurement of the group refractive index at 1544.8360 nm vacuum wavelength. (a) FSI refractometry experiment based on a frequency comb referenced external cavity diode laser ECDL. (b) Absolute value for the group refractivity. Comparison against $n_g$ values derived from the Ciddor equation [4]. Adapted with permission from reference [26], © (2017) Optical Society of America.

be of great importance. It is the typical spectral band for length measurement applications based on fibre-doped frequency combs or economically appealing telecommunication lasers.

## 5.2 Determining the intrinsic parameters

In the previous section, empiric compensation formulae for the refractive index have been derived. As input, the effective temperature, the ambient pressure, and the water vapour as well as the carbon dioxide content is needed. An efficient implementation of sufficient accuracy is hence the challenge for the experiment designer. In this section, different approaches are presented.

### 5.2.1 Sensor networks

The most straightforward and most popular approach to refractive index compensation in air measurement is the measurement of the environmental parameters with dedicated sensors. Every realisation is different and needs to be optimised for the measurement task in question. A fundamental discussion of the various sensor types and intricacies goes beyond the scope of this section. It is intended, nevertheless, to provide some guidance for designing a suitable sensor network. The starting point for all considerations are the sensitivity coefficients $c_X$ of the index of refraction with respect to the environmental parameter temperature $\vartheta$, ambient pressure $p$, humidity $h$, and $CO_2$ contents $p_{CO_2}$. They are given in table 5.1 for a wavelength of 532 nm.

The sensitivity towards changes in the carbon dioxide content is relatively small. However, human presence in closed rooms can increase the $CO_2$ conten considerably. Low-cost, compact spectroscopic sensors with uncertainties of the order of $u(x_{CO_2}) = 50$ ppm are commercially available and should be used in these cases. They can be calibrated against a traceable reference gas.

For most reasonable measuring conditions, temporal and local variation of the air pressure can be expected to be more moderate than in the case of temperature sensors. High quality absolute pressure sensors should be used with measurement uncertainties of the order of 0.1 hPa. By calibration, expanded uncertainties of $u(p)_{k=2,\,\mathrm{cal}} = 0.054$ hPa can be achieved. For extended distances, at least a measurement at starting and end point is recommended. Height differences need to be corrected with the help of the barometric formula.

There are three different classical sensor types for air humidity. The first sensor type is the dew point hygrometer. It observes the dew point temperature of the

**Table 5.1.** Refractivity compensation contribution to the measurement uncertainty budget according to GUM [30] for a calibration performed at the PTB 600 m baseline under optimum ambient conditions as published in reference [31]. Sensitivity coefficients $c_x(l)$ are given for a wavelength of 532 nm. Adapted with permission from reference [31]. 

| Quantity $X_i$ | Uncertainty $\delta X_i$ | Probability distribution function | Factor | Standard uncertainty $u(x)$ | Sensitivity $c_x(n)$ | Uncertainty contribution $u(n)$ |
|---|---|---|---|---|---|---|
| $\vartheta_{cal}$ | 80 mK | Normal | 2.0 | 0.040 K | $9.3 \times 10^{-7}$/K | $3.7 \times 10^{-8}$ |
| $\vartheta_{trans}$ | 0.10 K | Rectangular | $\sqrt{3}$ | 0.058 K | $9.3 \times 10^{-7}$/K | $5.4 \times 10^{-8}$ |
| $\langle\vartheta_i\rangle_l$ | 0.15 K | Normal | 1.0 | 0.15 K | $9.3 \times 10^{-7}$/K | $1.4 \times 10^{-7}$ |
| $p_{cal}$ | 5.1 Pa | Normal | 2.0 | 2.45 Pa | $2.7 \times 10^{-9}$ /Pa | $6.6 \times 10^{-9}$ |
| $\langle p_i\rangle_l$ | 20 Pa | Normal | 1.0 | 20.0 Pa | $2.7 \times 10^{-9}$ /Pa | $5.4 \times 10^{-8}$ |
| $h_{cal}$ | 0.35 %RH | Normal | 2.0 | 0.175 %RH | $1.0 \times 10^{-8}$ /%RH | $1.8 \times 10^{-9}$ |
| $\langle h_i\rangle_l$ | 2.0 %RH | Normal | 1.0 | 2.0 %RH | $1.0 \times 10^{-8}$ /%RH | $2.0 \times 10^{-8}$ |
| $\langle x_{CO_2}\rangle_l$ | 440 ppm | Rectangular | $\sqrt{3}$ | 254 ppm | $1.5 \times 10^{-10}$ /ppm | $3.8 \times 10^{-8}$ |
| Bönsch | $3.5 \times 10^{-8}\ l$ | Normal | 1.0 | $3.5 \times 10^{-8}$ | 1.0 | $3.5 \times 10^{-8}$ |
| Combined standard uncertainty $u(n)_{k=1} = 1.7 \times 10^{-7}$ | | | | | | |

ambient air. The basic unit is a temperature controlled mirror on which a condensate is formed. The dew point temperature is reached for zero net mass transfer between mirror and gas. This process is usually optically monitored by measuring the reflectance. The relative humidity is then derived from the measured dew point temperature and the air temperature. Dew point mirrors have a wide range of operation and achieve the most reliable measurement accuracies. Their response time, however, is limited. A second measurement principle which is of historic interest, particularly in geodetic measurements, is psychrometers. They measure the temperature difference of two thermometers, one surrounded by ambient air, one embedded in a wet piece of cloth. The temperature difference is a measure for heat flux due to water evaporisation from the wet cloth to ambient air. The relative humidity $h$ can be derived using tables and proper calibration. More suitable for continuous remote data acquisition and for flexible use, however, are impedance hygrometers. The core sensor consists of electrodes separated by a polymer layer. Absorption of water molecules alters the impedance, and can hence be used to measure relative humidity. The process is not always reversible. Long-term stability of impedance hygrometers is critical in uncontrolled environments. After thorough calibration, a measurement uncertainty of $u(h)_{k=2,\,\mathrm{cal}} = 0.35$ %RH can be achieved at 50 %RH. The various humidity sensor principles are reviewed in more detail in reference [32].

Often, the temperature measurement is considered most critical for the refractivity compensation. Two sensor types are predominantly used: platinum resistance thermometers and semiconductor-based thermistors. Platinum resistance thermometers exploit the almost linear temperature dependence of the resistance of metals. The noble metal platinum is frequently used. Its temperature coefficients are highly stable. Drifts over a year are typically of the order of a tens of Millikelvin maximum. Calibrated platinum temperature sensors achieve the highest accuracies in temperature measurements. The disadvantage of metal resistance thermometers is their relatively slow response time. Typically, platinum resistance thermometers are designed as wire coils or thin films protected by a glass housing and an additional cover. This material decreases the thermal response velocity considerably. If fast temperature changes are to be expected and relevant for the measurement principle, semiconductor thermistors can be an alternative. Temperature coefficients are typically much larger, increasing sensitivities. However, self-heating can play a role even at lower probing currents. Furthermore, they are known to age easily and need therefore tight monitoring. Temperature sensors are discussed in much more detail e.g. by Bonfig [33] or Childs [34].

A Pt-100 temperature sensor itself can be calibrated to uncertainties $u(\vartheta)_{k=2,\,\mathrm{cal}}$ below 10 mK. In air, however, this needs to be qualified. In temperature chambers, deviations of the order of 100 mK of calibrated sensors are observed. For the temperature sensor design used at the PTB 600 m base, e.g. an 'effective calibration uncertainty' of $u(\vartheta)_{k=2,\,\mathrm{cal}}$ of 80 mK is assumed (cf table 5.1). Outdoors, additional measures have to be taken to ensure that the sensor is actually sensitive to the air temperature. It is standard to shield the sensor against direct radiation heating with standard housings (see figure 5.4(b)). In these housings, however, a micro-climate

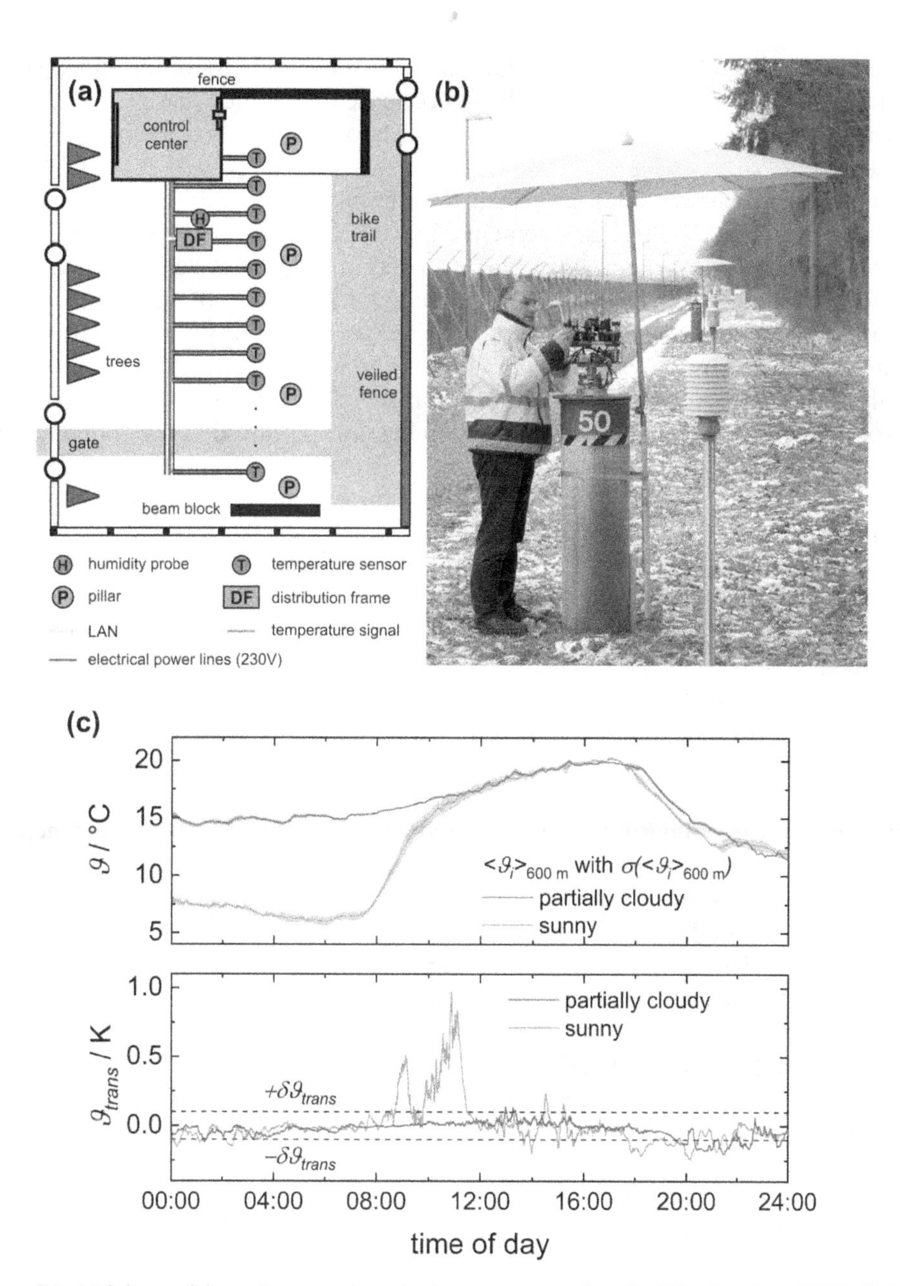


**Figure 5.4.** (a) Scheme of the environmental monitoring sensor network at the 600 m baseline of PTB. (b) PTB scientist modifying a frequency comb-based distance metre on the 50 m pillar of the 600 m baseline. The temperature sensors are protected against direct Sun exposure. (c) Temporal and spatial gradients along (long) and across (trans) the baseline depending on the Sun exposure. (a), (c) adapted with permission from reference [31]. 

will evolve which does not have to represent the ambient air conditions adequately. To resolve this problem, Eschelbach has proposed ventilated sensor shields [35]. However, the additional heat load of the ventilator also needs to be shielded from the sensor.

All sensors discussed so far are sensitive to the environment in their near vicinity. For interferometric length measurements, however, the effective parameters along the whole beam path are relevant. For extended measurement volumes, it is hence important to use multiple probes, in particular in the case of the more volatile parameter temperatures, and to a certain degree, also humidity. In general, sensors should be located in the vicinity of the beam and at beam height. They should be evenly distributed along the whole beam path. An example for such an extended sensor network is the refurbished PTB 600 m baseline discussed in depth in reference [31]. A network of 60 temperature sensors, six humidity and at least two pressure sensors is used to monitor the index of refraction with uncertainty well below $10^{-6}$ (see table 5.1). Sensor information needs to be collected and logged efficiently. The update times must be chosen suitable for the sensor time constants. There are multiple solutions to this problem. Network architectures based, e.g. on wifi, are commercially available. Wired connections, e.g. by USB, RS232, or for longer distances, by optical fibre are more robust against obstruction. But they require a larger financial investment and are less flexible in use. An example for an efficient implementation based on a Raspberry Pi processor has recently been described by Lewis and co-workers [36].

Important information derived from such data sets aside from the (weighted) mean value is the standard deviation, providing information on the homogeneity of the environmental parameters. Depending on air flow, local (or, in the case of the Sun, global) heat sources, substantial gradients can occur. Examples of longitudinal and transversal gradients as observed at the PTB 600 m baseline are presented in figure 5.4(c). The standard deviation observed during the measurement must be included in the uncertainty budget. In a moderately controlled environment, this standard deviation dominates the combined measurement uncertainty of the refractive compensation (see table 5.1). Besides spatial, temporal variations of the index of refraction also need to be resolved in agreement with the acquisition time of the length measurement. In uncontrolled environments, substantial changes of the index of refraction can be extremely fast. In figure 5.5, a relative change in a magnitude of approximately $0.5 \times 10^{-7}$ is observed within 600 ms for an outdoor measurement at moderate weather conditions (see section 5.3.3). In many applications, this effect averages out. If not resolved completely, these fluctuations lead to an increased standard deviation of the compensated length result.

In general, sensor networks are the standard method for the compensation of the index of refractions. Designed intelligently for the specific measurement task, high relative accuracies below $10^{-6}$ can be achieved. Local and temporal gradients in the measurement, however, do affect the achievable measurement uncertainty and need to be taken into account. Especially for larger scales, the effort for the maintenance,

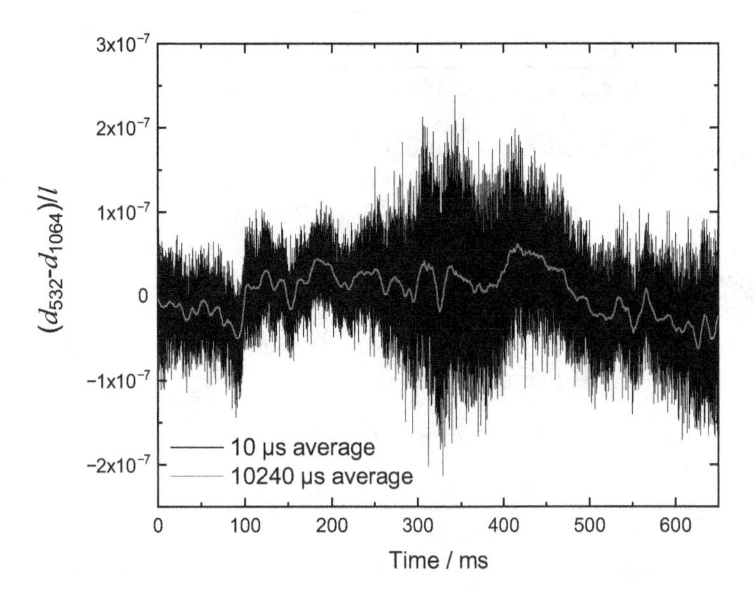


**Figure 5.5.** Optical path difference between 532 nm ($d_{532}$) and 1064 nm ($d_{1064}$) optical paths as observed by the two-colour TeleYAG interferometer (see section 5.3.3). The difference has been normalised to the absolute distance of $l = 864$ m for different signal averaging times. The fluctuations indicate a magnitude of refractive index changes on a short time scale. (It should be noted, however, that remanent system noise can also explain parts of the observed noise.)

e.g. due to calibration, grows significantly. The design of the sensor network defines also the measurement volume, limiting the flexibility of large-scale measurements in practice. In the following sections, alternative compensation approaches will be discussed which overcome many of these limitations, though often at the expense of measurement complexity.

### 5.2.2 Acoustic thermometry

Sound waves propagate as pressure waves in air. The speed $u_a$ depends on the same intrinsic parameters as electromagnetic waves, i.e. temperature, pressure and gas composition. Cramer derived a semi-empirical equation for the zero speed of sound $u_a^0$ of the form [39]

$$\begin{aligned} u_a^0(\vartheta, p, x_w, x_{CO_2}) = \; & a_0 + a_1\vartheta + a_2\vartheta^2 + (a_3 + a_4\vartheta + a_5\vartheta^2)x_w \\ & + (a_6 + a_7\vartheta + a_8\vartheta^2)p + (a_9 + a_{10}\vartheta + a_{11}\vartheta^2)x_{CO_2} \\ & + a_{12}x_w^2 + a_{13}p^2 + a_{14}x_{CO_2}^2 + a_{15}x_w p x_{CO_2}, \end{aligned} \tag{5.43}$$

with $a_i$ ($i = 0, 1, 2,\ldots, 15$) representing constants [38, 39] and $f_s$ the sound frequency. Morfey and Howell [40, 41] developed the dispersion relation

$$\frac{1}{u_{\mathrm{a}}^{0}(\vartheta, p, x_{\mathrm{w}}, x_{\mathrm{CO_2}})} - \frac{1}{u_{\mathrm{a}}(f_s, \vartheta, p, x_{\mathrm{w}}, x_{\mathrm{CO_2}})} = \sum_{\mathrm{r}} \frac{\alpha_r}{2\pi f_{\mathrm{r}}}, \tag{5.44}$$

where $f_s$ denotes the frequency of sound, $\alpha_{\mathrm{r}}$ the attenuation coefficient, and $f_{\mathrm{r}}$ the relaxation frequency of gas $r$. Korpelainen and Lassila realised that 'the relative effect of a change in air temperature is about 2000 times greater on the speed of sound than on the refractive index of air' [38]. They developed the idea to use a measurement of the speed of sound to derive the effective index of air. By combining the Cramer and Morfey–Howell equations (5.43) and (5.44) with the Bönsch–Potulski equation (5.36), they derived an explicit empiric polynomial for the temperature or the refractive index based on [38]:

$$\begin{aligned} g_{n/\vartheta}(u_{\mathrm{a}}, p, x_{\mathrm{w}}, x_{\mathrm{CO_2}}) = {} & b_0 + b_1 u_{\mathrm{a}} + b_2 u_{\mathrm{a}}^2 + b_3 x_{\mathrm{w}} + b_4 x_{\mathrm{w}}^2 + b_5 x_{\mathrm{w}}^3 \\ & b_6 p + b_7 p^2 + b_8 x_{\mathrm{c}} + b_9 x_{\mathrm{w}} u_{\mathrm{a}} + b_{10} p u_{\mathrm{a}} + b_{11} x_{\mathrm{w}} p. \end{aligned} \tag{5.45}$$

The constants $b_i$ with $i = 0, 1, \ldots, 11$ depend on the targeted quantity (refractive index $n$ or effective temperature $\vartheta$). It is clear from this equation that pressure $p$, and water and $CO_2$ contents $x_{\mathrm{w}}$ and $x_{\mathrm{CO_2}}$ need to be measured by auxiliary sensors for such a thermometer. For the determination of the speed of the sound, the length needs to be determined with sufficient accuracy, e.g. by a rough estimate from the imperfectly compensated interference measurement [38], a time-of-flight measurement of a modulated laser source [37], or an ultrasonic interference measurement [37, 42, 43]. Furthermore, the clock needs to be carefully synchronised. The frequency of the sound source must fit to the measurement length since attenuation is frequency dependent. The basic set-up is summarised in figure 5.6. Acoustic and optical sender and receiver units should be located in close proximity. To reduce sensitivity to possible air flow directions, the sound velocity measurement should be performed in both directions along the beam.

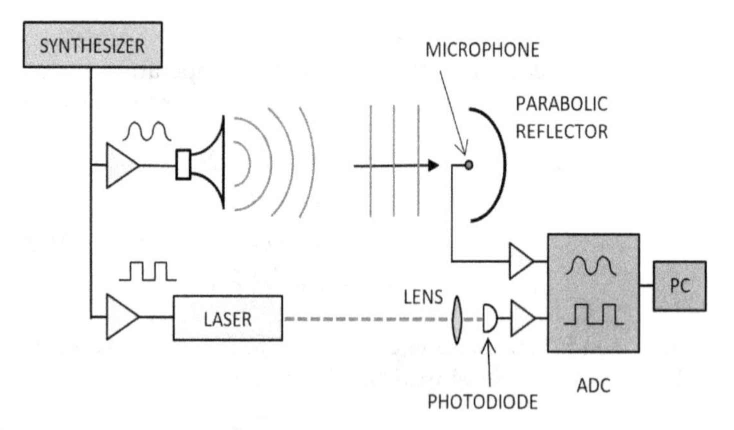


**Figure 5.6.** Schematic set-up of an acoustic thermometer. Reprinted with permission from [37]. 

Korpelainen and Lassila investigated the speed of sound for the frequency $f$ of 50 kHz thoroughly and adjusted the Cramer coefficients accordingly [38]. In a proof-of-principle experiment, they monitor a fixed mechanical length of 4.7 m over 510 s, while introducing local heat by a radiator. As can be seen in figure 5.7, apparent length changes in the raw interferometer signal vanish when applying the acoustically determined index of refraction compensation. The acoustic thermometer indicates a change of approximately 0.2 K in effective temperature. For the given set-up, Korpelainen and Lassila claim a combined measurement uncertainty of 25 mK for the effective temperature $\vartheta$, and an uncertainty of $2.6 \times 10^{-8}$ for the derived index of refraction $n$.

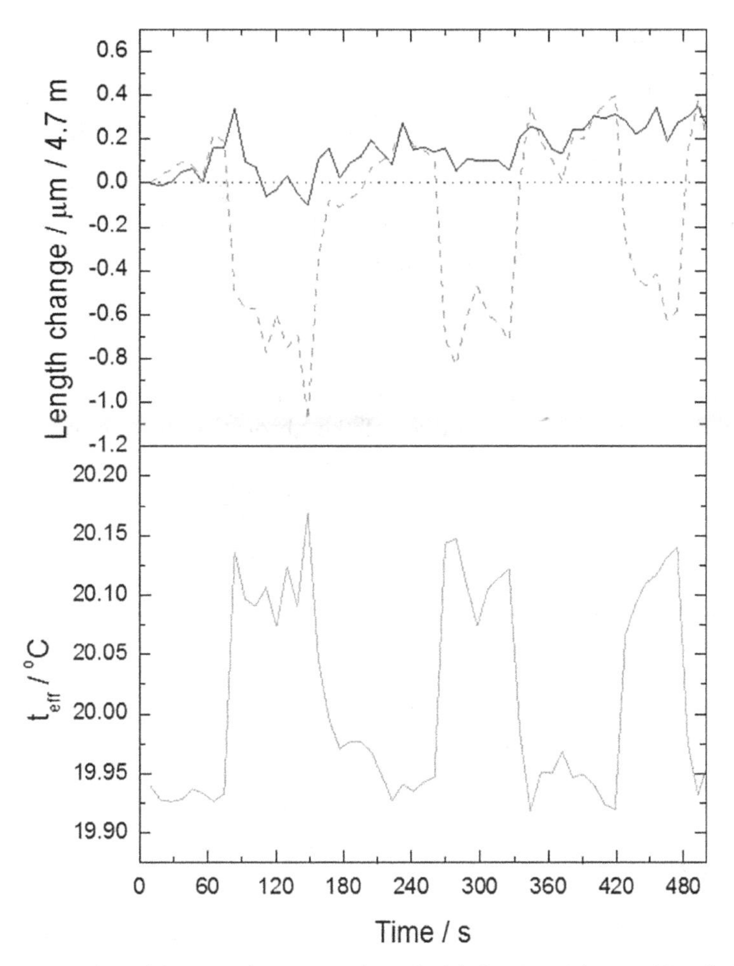


**Figure 5.7.** Demonstration of the acoustic compensation principle by Korpelainen and Lassila [38]. A fixed mechanical length of 4.7 m is monitored by an interferometer while a radiator is switched on and off for 510 s. The dashed pink line indicates the uncompensated length measurement, the solid blue line is derived from the acoustically determined index of refraction. The lower diagram depicts the acoustically measured change in temperature. Adapted with permission from reference [38]. Copyright (2011) Society of Photo Optical Instrumentation Engineers.

For more extended measurement volumes and more flexible use, the realisation gets more and more complicated. Echoes from sound reflections need to be identified in the data processing. Attenuation due to absorption and beam spreading limits the achievable distance. Furthermore, local gradients and in water and $CO_2$ contents need to captured by the auxiliary sensor system accurately. Pisani *et al* [37] report on measurements up to 180 metres. In this case, agreement with the sensor temperature is only within 2 K. In well-defined measurement volumes, however, acoustic thermometry or refractometry obviously has great potential. Last but not least, the implementation is relatively straightforward after a thorough device calibration, and the application does not require expert skills from the operator.

### 5.2.3 Spectroscopic parameter determination

The measurement of the effective intrinsic parameters as seen by the light seems an appealing concept. Light as a probe instead of pressure waves might suggest itself when compensating an optical interference measurement. One way to realise such an optical probe is based on quantitative molecular absorption spectroscopy and will be briefly discussed in this section. The starting point is the Beer–Lambert law which relates the transmitted optical intensity $I$ in a medium to the optical depth $\tau$ at the optical frequency $\nu$ by

$$I(\nu) = I_0 e^{-\tau(\nu,\, l)}. \tag{5.46}$$

For a specific transition between the molecular states $\eta$ and $\eta'$ at frequency $\nu_{\eta\eta'}$, the optical depth depends on the product of the absorption line strength $S_{\eta\eta'}(T)$, the molecular absorption line shape function $g(\nu,\, \nu_{\eta\eta'},\, T,\, p)$ (normalised to area = 1), the number density $N$ and the absorption path length $l$ by

$$\tau(\nu,\, l) = NS(T)\, g\left(\nu,\, \nu_{\eta\eta'},\, T,\, p\right) l. \tag{5.47}$$

The line shape function $g(\nu,\, \nu_{\eta\eta'},\, T,\, p)$ depends on the broadening mechanism. Both Doppler and pressure broadening are relevant for oxygen thermometry under ambient conditions [44]. The convolution of these two effects leads to a Voigt profile. Recent results indicate that collision-induced velocity changes also need to be taken into account, narrowing the linewidth ('Dicke-narrowing') [45]. This can be done by a Galatry profile or a pressure-dependent-narrowing of the Doppler contribution of the Voigt profile [45]. The line strength $S(T)$ can be expressed dependent on the line strength $S(T_{\text{ref}})$ for the reference temperature $T_{\text{ref}}$ for any temperature $T$ by

$$S(T) = S(T_{\text{ref}}) \frac{Q(T_{\text{ref}})}{Q(T)} \exp\left[\frac{-hcE_\eta}{k_{\text{B}}}\left(\frac{1}{T} - \frac{1}{T_{\text{ref}}}\right)\right]\left(\frac{1 - \exp\left(-hc\nu_{\eta\eta'}/k_{\text{B}}T\right)}{1 - \exp\left(-hc\nu_{\eta\eta'}/k_{\text{B}}T_{\text{ref}}\right)}\right) \tag{5.48}$$

where $Q(T)$ represents the total internal partition function, $E_\eta$ the lower state energy involved in the transition, and $h$ the Planck and $k_{\text{B}}$ the Boltzmann constant. The last term accounts for stimulated emission and can be neglected in the visible and typical environmental conditions in question here [46].

There are different approaches to derive the temperature from spectroscopic measurements. As absorption species, oxygen transitions seem a smart choice: the concentration of oxygen is almost constant, and there are many well-isolated transitions in the near-infra-red (760–770 nm). This spectral region is well-accessible for tunable diode laser sources [45]. In 'two-line oxygen thermometry', two absorption lines of different lower state energy ($E_{\eta_1} \neq E_{\eta_2}$) are experimentally observed and their transmitted intensity measured [47]. Either the complete absorption intensity $A_i = -\int_\nu \ln (I(\nu)/I(\nu)_0)\mathrm{d}\nu$ or, for faster data acquisition, the peak height normalised according to $A_i = -\ln (I(\nu_i)/I(\nu_i)_0)/g(\nu_i, \nu_{\eta_i\eta'}, T, p)$ is monitored for each peak $i = 1, 2$ [44, 45, 48]. According to equations (5.46)–(5.48), the intensity ratio $R = A_1/A_2$ depends on

$$R = \frac{S_1(T_{\mathrm{ref}})}{S_2(T_{\mathrm{ref}})} \exp\left[\frac{-hc\left(E_{\eta_1} - E_{\eta_2}\right)}{k_{\mathrm{B}}}\left(\frac{1}{T} - \frac{1}{T_{\mathrm{ref}}}\right)\right]. \tag{5.49}$$

Number density $N$, absorption length $l$, and partition functions $Q(T)$ cancel in this representation. The parameters $E_{\eta_i}$, line strength $S(T_{\mathrm{ref}})$, and in case of a height analysis, also the linewidth need to be known if the temperature is to be derived from equation (5.49). Literature data on these parameters is collected in the HITRAN database [49]. But typical uncertainties for the line strengths are of the order of 1%. The line intensity ratio $R_{\mathrm{ref}}$ should be known 'in the order of $10^{-4}$ to obtain the temperature with $\sim$ 12 mK uncertainty' [44]. Furthermore, the whole mathematical apparatus also assumes that all photons pass the medium and are accounted for. In practice, beam clipping and wavelength-dependencies in the optical set-up are often critical, and a careful calibration of the 'reference ratio' $R_{\mathrm{ref}} \equiv S_1(T_{\mathrm{ref}})/S_2(T_{\mathrm{ref}})$ is necessary to achieve sub-Kelvin accuracies. Alternatively, Tomberg *et al* [45] argue that in the case of atmospheric thermometry it is possible to work with a single transition only. For once, the oxygen contents (i.e. the number density $N$) in ambient air is well characterised and constant. Furthermore, polynomial expressions for partition functions $Q(T)$ are available. This 'single-line oxygen thermometry' reduces the experimental effort significantly, with a limited expense in accuracy.

Hieta *et al* [44, 48] have demonstrated the high potential of this method. They monitored a mechanical length of 30 m with a commercial interferometer. The temperature along the beam path was monitored in parallel by a network of eight distributed Pt-100 sensors and by a spectroscopic sensor aligned 10 cm away from the interferometer beam. Over a time span of more than 60 hours, local temperature variations were induced by two radiators and a fan. While the sensor-compensated lengths show abrupt changes in lengths of approximately 7 μm, the spectroscopic compensation remains smooth (see figure 5.8). This can be explained by a difference of sensor from spectroscopic temperature in the order of 10 mK [44]. Spectroscopically, a compensation at $1 \times 10^{-7}$ level is achieved. This accuracy requires a careful individual calibration of the system.

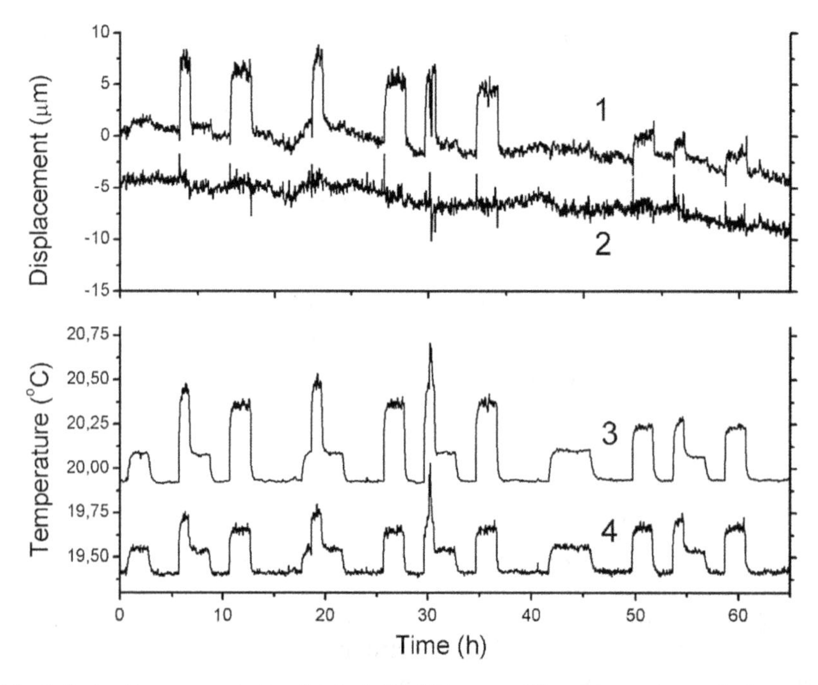


**Figure 5.8.** Optical path compensated by distributed Pt-100 sensors (1) and by spectroscopic thermometry (2). The average temperature of the Pt-100 sensors (3) is approximately 10 mK higher than the spectroscopically obtained (4). Dataset (2) has been shifted by −5 μm, dataset (4) by −500 mK for clarity. Reprinted from reference [44]. Copyright (2011) Optical Society of America.

One advantage of the method is the fact that the approach is scalable with length. The lines have to be chosen so that their absorption strength is suitable. Hieta *et al* [44] and Tomberg *et al* [45] have performed outdoor measurements, Tomberg *et al* [45] up to 864 m. After on-site calibration of the spectroscopic thermometers, they can reproduce temperature changes observed by the reference sensor network on sub-Kelvin level. They also see indications for systematic deviations of the sensor from the spectroscopic temperature of the order of 0.5 K which they attribute to radiative heating of the sensors as well as incorrect sampling of gradients by the sensors (see section 5.2.1).

Quantitative absorption spectroscopy can also be used to determine species concentration along the beam path. Multiple spectroscopic humidity sensors for well-defined measurement volumes have been developed since the availability of suitable tunable diode laser sources in the 1990s, from fundamental studies [50] to application solutions, e.g. for airborn hygrometry [51] or atmospheric measurements [42, 43]. Pollinger *et al* [52] report on a study comparing spectroscopic humidity sensors to the performance of capacitive sensor or dew point mirror networks. The spectroscopic sensors convince with their fast response time, but on-site calibration against the sensor system is needed to achieve accuracies of the order of 1% relative humidity or better.

Finally, it should be noted that the experimental effort for a spectroscopic parameter compensation can be substantial. When implemented by tunable diode laser spectroscopy, two almost disjoint experiments, a spectrometric and an interferometric set-up need to be combined. Frequency combs and femtosecond lasers are excellent sources both for interferometry (see chapters 6 and 7 of this book), as well as for spectroscopy (see e.g. reference [53, 54]). They can provide an elegant basis to combine both experiments efficiently. The first steps in this direction have recently been taken by Hänsel *et al* [55].

## 5.3 Dispersive intrinsic refractivity compensation

The compensation methods presented in section 5.2 have in common that the environmental parameters are measured in some way to apply the refractive index formula developed in section 5.1. In the following, a method is introduced that applies the refractivity models more indirectly, and is almost independent of auxiliary sensor information.

### 5.3.1 The $A$ coefficient

Assume a measurement of the same geometric path length $l$ with two different vacuum wavelengths $\lambda_1$ and $\lambda_2$. If performed in a dispersive medium, the determined optical path lengths $d_1$ and $d_2$ differ by their difference in refractivity $[(n(\lambda_2) - 1) - (n(\lambda_1) - 1)] \times l$. Using $d_i = n(\lambda_i) \times l$, it is obvious that the geometric path length could be derived from these two measurements if the indices of refraction were known according to

$$l = \frac{1}{n(\lambda_2) - n(\lambda_1)} \times (d_2 - d_1). \tag{5.50}$$

Formally, the geometric path length can also be retrieved by subtracting the refractivity-induced length change $(n(\lambda_1) - 1) \times l$ from the optical path length $l$

$$l = d_1 - (n(\lambda_1) - 1) \times l. \tag{5.51}$$

Substituting expression (5.50) in the right-hand side of equation (5.51), one obtains

$$l = d_1 - \frac{n(\lambda_1) - 1}{n(\lambda_2) - n(\lambda_1)} \times (d_2 - d_1). \tag{5.52}$$

The scaling factor which contains the influence of the refractivity before the optical difference path difference is usually referred to as the $A$ coefficient,

$$A(\lambda_1, \lambda_2) \equiv \frac{n(\lambda_1) - 1}{n(\lambda_2) - n(\lambda_1)}, \tag{5.53}$$

shortening equation (5.52) to

$$l = d_1 - A(\lambda_1, \lambda_2) \times (d_2 - d_1). \tag{5.54}$$

A straightforward application of equation (5.54) can be to calculate the $A$ coefficient for a given set of parameters and then keep it constant for the refractivity compensation in the following. In the measurement, it is sufficient to determine the optical paths only. Indeed, the sensitivity to environmental changes decreases for a length compensation in comparison to a standard single-colour Edlén based compensation. In table 5.2, relative changes of the length due to changes in the environmental parameters were calculated assuming standard conditions [56]. In general, the sensitivity to environmental parameters decreases by one order of magnitude, with the exception of the humidity. It is interesting to note that the influence of the temperature measurement depends strongly on whether relative humidity $h_{\mathrm{rel}}$ or water vapour pressure $p_{\mathrm{w}}$ are observed. In the former case, the temperature information is needed to derive the water vapour pressure. The required knowledge of the environmental conditions for the same level of relative accuracy is hence reduced if the optical path length is known for two different wavelengths.

The reliability of this approach was shown by Minoshima and co-workers [57, 58]. They designed a highly sophisticated two-colour pulse-to-pulse interferometer which is sensitive to changes in the optical path length $d_1$ and the optical path difference $(d_2 - d_1)$. A mechanically stable optical path length $l$ of 61 m was observed under thermal isolation for over 10 h. The data of reference [57] is reproduced in figure 5.9. The relative optical path $d_1/l$ varies due to air pressure variations of the order of $10^{-6}$. This change could be compensated for by scaling the variation of the optical path distance $\Delta(d_2 - d_1)$ with the $A$ coefficient according to equation (5.54) to a standard deviation of $\sigma = 1.5 \times 10^{-8}$. This is of the same order of magnitude as the uncertainty the Ciddor equation used. Similar results were obtained by Wu and co-workers using two-colour dispersive interferometry [59].

**Table 5.2.** Sensitivity of the relative length measurement against changes of the environmental parameters. Relative changes were calculated for a refractivity compensation using the Bönsch–Potulski compensation for a single measurement, and for a dispersive two-colour refractivity compensation based on a 532/1064 nm wavelength experiment (sensitive to the phase refractive index). Adapted from reference [56]. With permission from VDE Verlag GmbH Berlin.

| Environmental parameter variation by $\Delta X$ | Edlén relative length change | 532/1064 nm relative length change |
|---|---|---|
| $\Delta\vartheta = +1°C$ ($\Delta h_{\mathrm{rel}}$= const) | $-9.6 \times 10^{-7}$ | $-8.2 \times 10^{-8}$ |
| $\Delta\vartheta = +1°C$ ($p_{\mathrm{w}}$ = const) | $-9.3 \times 10^{-7}$ | $-4.2 \times 10^{-9}$ |
| $\Delta p = +1$ hPa | $+2.7 \times 10^{-7}$ | $+1.2 \times 10^{-9}$ |
| $\Delta x_{CO_2} = +1$ ppm | $+1.5 \times 10^{-10}$ | $+6.5 \times 10^{-13}$ |
| $\Delta h_{\mathrm{rel}} = +1\%$ | $-8.6 \times 10^{-9}$ | $-2.4 \times 10^{-8}$ |

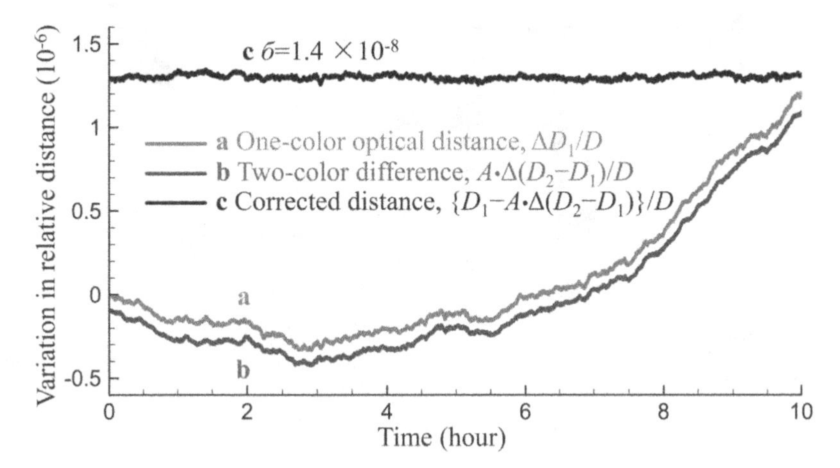


**Figure 5.9.** Demonstration of the potential of the two-colour refractive index compensation. The relative variation of the optical path of single-colour $\Delta D_1/D$ of the order of $10^{-6}$ can be compensated for by subtracting the variation of the optical path length differences $\Delta(D_2 - D_1)/D$ scaled by the $A$ coefficient to a standard deviation of $\sigma = 1.5 \times 10^{-8}$, the order of magnitude of uncertainty of the Ciddor equation. Reprinted with permission from reference [57]. With permission of Springer.

### 5.3.2 Uncertainty scaling of dispersive refractivity compensation

The implementation of this measurement principle poses some severe challenges. This can be seen when investigating the measurement uncertainty influences of the length derived according to equation (5.54). Following the 'Guide to the expression of uncertainty in measurement' (GUM) [30], the uncertainty of the length $u(l)$ can be calculated according to

$$u(l) = \sqrt{\left(\frac{\partial l}{\partial d_1} \times u(d_1)\right)^2 + \left(\frac{\partial l}{\partial d_2} \times u(d_2)\right)^2 + \left(\frac{\partial l}{\partial A} \times u(A)\right)^2}. \tag{5.55}$$

Using $(A + 1)^2 \approx A^2$, this can be evaluated to

$$u(l) = \sqrt{A^2(u(d_1)^2 + u(d_2)^2) + (d_2 - d_1)^2 u(A)^2}. \tag{5.56}$$

The first term in the square root indicates that any uncertainty $u(d_{1,2})$ in the optical path measurement will be scaled by the $A$ coefficient. As the latter is proportional to $1/(n(\lambda_2) - n(\lambda_1))$ (cf equation (5.53)), it is recommended to work with optical wavelengths limiting a spectral region of large dispersion. In this context it should be noted that equations (5.53) and (5.54) are applicable for both, phase and group refractive index. In general, $A$ factors based on group refractive indices are smaller than their phase equivalents for the same spectral regions. Measurement principles based on the air group velocity can be expected to behave more forgivingly with respect to uncertainty amplification. Figure 5.10 shows the wavelength-dependence of the $A$ coefficient, both for group refractivity and phase refractivity sensitive measurements. The coefficients are calculated for set-ups based on fundamental and second-harmonic measurement beam pairs.

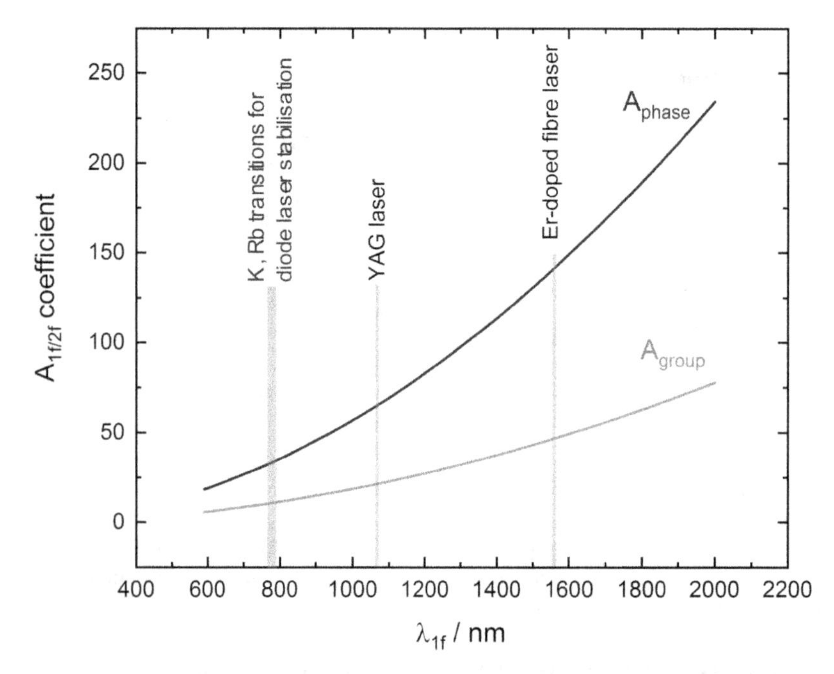


**Figure 5.10.** Wavelength-dependence of the $A$ coefficient. The $A$ coefficient for group and phase refractive distance measurements was calculated according to definition (5.53) using Bönsch's and Potulski's revision of the Edlén equation (see equation (5.36)). As probe beams, pairs of fundamental ($\lambda_{1f}$) and second-harmonic ($\lambda_{2f}$) wavelength were assumed. Typical fundamental wavelength sources for two-wavelength interferometers are indicated.

### 5.3.3 Independence from prior knowledge of the environment

The discussion so far relies only on the fundamental definition of the index of refraction given by equation (5.1). Therefore, equation (5.54) can be considered universally valid (though apparently of little added value except complexity in comparison to equation (5.50)). Remembering section 5.1, it is clear that the $A$ coefficient depends in general on the wavelengths $\lambda_1$ and $\lambda_2$, as well as temperature $\vartheta$, ambient pressure $p$, carbon dioxide content $x_{CO_2}$, and water vapour pressure $p_w$. Earnshaw and Owens [60], however, realised an intriguing property of the $A$ coefficient in dry air. Entering the general refractivity model expression (5.34) in the $A$ coefficient definition (5.53), the water dispersion factor $g_w$, and subsequently, also the multiplicative density factor $D$ cancel for $\rho_w \to 0$. The $A$ coefficient then only depends on the value of the dispersion factor $K$ at the respective vacuum wavelengths

$$A(\lambda_1, \lambda_2, \vartheta, p, x_{CO_2}, \rho_w) \xrightarrow{\rho_w \to 0} A(\lambda_1, \lambda_2) = \frac{K(\lambda_1)}{K(\lambda_2) - K(\lambda_1)}. \tag{5.57}$$

As a consequence, it is sufficient to measure two optical path lengths $d_{1,2}$ for known vacuum wavelengths $\lambda_{1,2}$ to derive the geometric path length $l$ in dry air from equation (5.54).

The prerequisite of dry air is relatively harsh and seems to limit the applicability of this measurement strategy. One possible extension is the measurement with an additional vacuum wavelength $\lambda_3$. In complete analogy to the two-colour derivation (5.50) to (5.54), one can show that the geometric length $l$ can then be expressed by [61]

$$l = d_1 - B(d_1 - d_2) - C(d_1 - d_3), \tag{5.58}$$

with $B$ and $C$ being constant for a given vacuum wavelength triplet $\lambda_{1,2,3}$. A few measurement schemes have been realised, e.g. by Thompson [62] or Golubev and Checkhovsky [61], but the complexity of these systems is intimidating.

A compromise between a full sensor environmental parameter measurement and a highly complex three-colour measurement was suggested by Meiners-Hagen and Abou-Zeid [63]. In fact, it is sufficient to measure the partial pressure of water $p_w$ in addition to deriving the geometric path length $l$ from two optical path length observations $d_{1,2}$ in humid air. This can be seen relatively easily. The starting point is the (slightly simplified) algebraic structure (5.35) for the index of refraction

$$\begin{aligned} n(\lambda, p, \vartheta, x_{CO_2}, p_w) - 1 = K(\lambda) \times D_{sda}(p, \vartheta) \\ \times g_{CO_2}(x_{CO_2}) - p_w/\mathrm{Pa} \times g_w(\lambda). \end{aligned}$$

This explicit expression for the index of refraction can be inserted into the equation system describing the two measurement observations

$$\begin{aligned} d_1 &= n(\lambda_1, p, \vartheta, x_{CO_2}, p_w) \times l \\ d_2 &= n(\lambda_2, p, \vartheta, x_{CO_2}, p_w) \times l. \end{aligned} \tag{5.59}$$

Combining these two equations, the product of density and $CO_2$ factor $D_{sda}(p, \vartheta) \times g_{CO_2}(x_{CO_2})$ can be eliminated. The resulting equation can be solved straightforwardly for the geometric path length $l$, leading to the Meiners-Hagen equation [63]:

$$l = \frac{K(\lambda_1)d_2 - K(\lambda_2)d_1}{K(\lambda_1) - K(\lambda_2) + p_w/\mathrm{Pa}(g_w(\lambda_1)\mathrm{K}(\lambda_2) - g_w(\lambda_2)\mathrm{K}(\lambda_1))}. \tag{5.60}$$

The uncertainty contribution $u_{p_w}(l)$ to the overall uncertainty of the resulting length measurement of the uncertainty of the water vapour measurement $u(p_w)$ can be calculated by

$$u_{p_w}(l) = \frac{\partial l}{\partial p_w} \times u(p_w). \tag{5.61}$$

For the popular wavelength combination 532 nm/1064 nm, this expression can be approximated to [63]

$$u_{p_w}(l) \approx 1.04 \times 10^{-9} \times l \times u(p_w)/\mathrm{Pa}. \tag{5.62}$$

The relative uncertainty of the length measurement is hence limited only to $10^{-7}$ if the water vapour pressure is known to be at least better than 100 Pa. Such an uncertainty in vapour pressure corresponds to an uncertainty of approximately 4% relative humidity for standard conditions ($\vartheta$ = 20 °C, $p$ = 1013 hPa). For many scenarios, such a measurement uncertainty seems feasible.

The challenge remains, however, that the optical path lengths $d_{1,2}$ need to be determined with high accuracy. For extended distances, synthetic wavelength interferometry can achieve this (see chapter 1). For this purpose, the Meiners-Hagen equation (5.60) can be generalised to pairs of synthetic wavelengths $\Lambda_{pq} = \lambda_p\lambda_q/(\lambda_p - \lambda_q)$. In this case, the optical path lengths obtained by synthetic wavelengths measurements can be intrinsically refractivity compensated using [64]:

$$l_s = \frac{K_s(\Lambda_1)d_2 - K_s(\Lambda_2)d_1}{K_s(\Lambda_1) - K_s(\Lambda_2) + p_w/\text{Pa} \times \Gamma_s(\Lambda_1, \Lambda_2)}, \tag{5.63}$$

where

$$\begin{aligned} \Gamma_s(\Lambda_1, \Lambda_2) &= g_s(\Lambda_1)K_s(\Lambda_2) - g_s(\Lambda_2)K_s(\Lambda_1) \\ K_s(\Lambda_i) &= K(\lambda_{i1}) - \frac{K(\lambda_{i2}) - K(\lambda_{i1})}{\lambda_{i2} - \lambda_{i1}}\lambda_{i1} \\ g_s(\Lambda_i) &= g(\lambda_{i1}) - \frac{g(\lambda_{i2}) - g(\lambda_{i1})}{\lambda_{i2} - \lambda_{i1}}\lambda_1 \end{aligned} \tag{5.64}$$

are defined in complete analogy to their single-wavelength equivalents.

### 5.3.4 Application to laser interferometry

Several applications of intrinsic dispersive refractivity compensation are presented in this section. The technique has been used from metre distances over kilometres up to thousands of kilometres. But the examples also show that there is still a need for technological development in this field.

*From proof-of-principle to coordinate measurement machine volume*

The experimental challenges for the implementation of the dispersive measurement principle in interferometry are substantial. Pioneer work was performed in the late 1980s and early 1990s by Ishida [66], and Matsumoto and Honda [67]. Ishida set up a two-colour interferometer based on a fundamental and frequency doubled $Ar^+$-ion laser. The resulting wavelength pair of 488 and 244 nm is very favourable in terms of $A$ factor magnitude. Ishida compensated displacements of a few hundred nanometres for the influence of the index of refraction [66]. Matsumoto and Honda tracked variations in the index of refraction over one metre by a YAG-based two-colour interferometer. They achieved relative agreement of $2 \times 10^{-7}$ with sensor-based data [67]. It is clear, however, that intrinsic dispersive refractivity compensation is not the method of choice for the refractivity compensation on these shorter scales. The unfavourable uncertainty scaling with the $A$ coefficient is a tremendous challenge when (sub-)nanometre uncertainties are the target. The instrumental effort is very high, and (experimentally and financially simpler) sensor-based compensation can provide competitive or even superior solutions.

In case of localised sources of environmental inhomogeneity, dispersive compensation is also advantageous in the case of measurement volumes of several metres. Applications can be imagined in large-scale measurements in industrial environments. Such a scenario was studied, e.g. by Kang and co-workers for the compensation of localised pressure variations due to air flow [68]. In a proof-of-principle experiment, intrinsic dispersive compensation was able to compensate the index of refraction approximately $5 \times 10^{-8}$ under air flow, while environmental sensors indicated a change three times larger [68].

One example to be discussed in more detail is the calibration of coordinate measurement machines (CMM) with measurement volumes of several cube metres. For this purpose, an interferometer basically follows the movement of an optical reflector mounted on the head of the CMM, and the lengths derived from the CMM internal scale are compared to the length derived from the interferometer. Typical measurement volumes are up to several cubic metres. Temperature gradients in the measurement volume are unavoidable, and since the CMM head is moved within the volume, temperature sensors cannot be placed along the beam. Dispersive intrinsic refractivity compensation seems an ideal tool for such a measurement.

Meiners-Hagen *et al* have developed a tracking two-colour interferometer for this purpose [65]. It is based on a two-colour heterodyne interferometer working with a YAG laser source at 1064 and 532 nm wavelengths. The design of the optical head is depicted in figure 5.11 together with an image of the realisation. The performance of the two-colour interferometer was investigated at a specially prepared interference comparator bench (lower image in figure 5.11). A fixed distance over 19 metres was monitored by the two interferometers and a network of distributed temperature, humidity and pressure sensors. In the beam path two localised heat sources were implemented. During the measurement, the apparent change upon heating was investigated. When compensated for using the sensor data, the baseline seemed to shrink by more than 25 μm. Analysing the data according to equation (5.60), however, the mean value of the distance remained fixed although the standard deviation increased substantially (upper dataset in figure 5.11) to micrometre level. The difference in geometric lengths derived from the two different compensation methods corresponds to a difference in temperature $\Delta\vartheta$ of approximately 1 K (lower dataset in figure 5.11). The sensor data is obviously not capable of monitoring the temperature change quickly and accurately enough.

So far, all measurement results were obtained by frequency doubled continuous wave laser sources. Recently, high-accuracy compensation results on dispersive refractivity compensation with femtosecond-comb-based light sources have also been demonstrated for distances of up to several metres [57–59]. These design principles might lead to further application scenarios in the field of precision engineering.

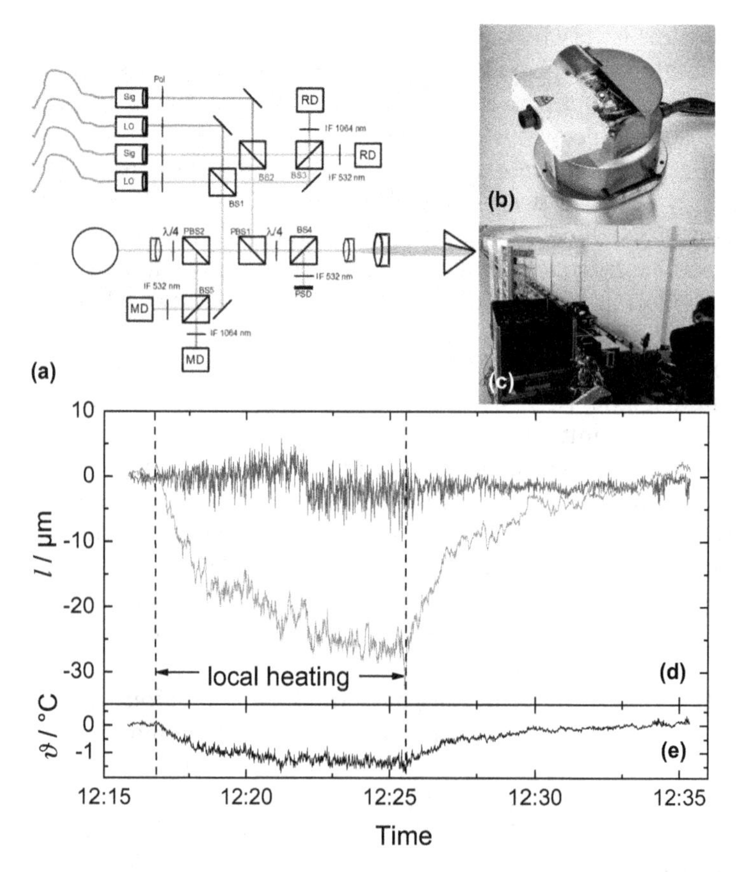


**Figure 5.11.** Heterodyne two-colour interferometer implemented in a tracking interferometer for coordinate measurement applications [65]. The sketch (a) shows the layout of the heterodyne two-colour interferometer as implemented in the tracking interferometer (b). For performance verification, a fixed distance of 19 m was monitored at an interference comparator equipped with two housings for localised heating (c). While conventionally compensated data indicates a shrinking of the length by more than 25 µm, the two-colour refractivity compensated results remain constant (d). The lower dataset translates the length difference in the corresponding difference $\Delta\vartheta$ between optical and sensor temperature (e). Adapted with permission from reference [65]. © [2016] Optical Society of America.

### *Compensation over hundreds of metres*

Surveying and geodesy are other fields in which high-accuracy measurements need to be performed over extended distances in uncontrolled environments (see e.g. [71, 72]). Many of the developments in this field go back to geodetic applications. In the early 1980s, an electronic distance metre (EDM) based on two-colour compensation was already commercially available, the so-called 'Terrameter' [73]. It depended on rough pre-values of the environmental parameters, determined at the start and end positions of the measurement. As optical sources, a HeNe and a HeCd laser were used, emitting at 633 and 442 nm. A modulation technique with a scale of 3 GHz was applied, and ranges up to 4 km were achieved [73]. Stabilities below $10^{-6}$ were successfully demonstrated. Handling issues and the advent of GNSS-based distance metrology seems to have ended this early impressive application work. Recent advances in fibre-based high-frequency modulation might have opened the possibility for implementing dispersive intrinsic refractivity compensation in modulation-based ranging technology [74].

Frequency doubling by second harmonic generation relaxes the requirements on the optical source a bit. Given enough primary power of the fundamental laser light, a second light beam with a fixed frequency distance can be generated with smaller effort than in the 1960s or 1970s. Minoshima and Matsumoto performed a two-colour distance measurement using a femtosecond laser and a second harmonic as the light source [75]. The measurement principle used the high stability of the repetition frequency of the femtosecond laser to generate modulations in the radio frequency bands by intermode beats. Although the accuracy of the single-colour measurement is limited, they could demonstrate a sensor-independent two-colour distance measurement up to 240 m with a relative accuracy of the order of $8 \times 10^{-6}$ [75]. Ideas to extend this measurement principle to three or more colours are currently being pursued [76].

As mentioned already in section 5.3.1, the $A$ coefficient in optical frequency domain scales up any uncertainty in two-colour refractive index compensation. Hence, the optical path needs to be determined with high relative accuracy for a satisfactory uncertainty below $10^{-6}$ for the geometric path measurement. Interferometric accuracy is difficult to obtain over extended distances of several hundreds of metres. Meiners-Hagen and co-workers from the Physikalisch-Technische Bundesanstalt (PTB), the German national metrology institute, developed the TeleYAG system as primary standard for the calibration of geodetic baselines [69]. It is based on heterodyne synthetic wavelength interferometry, leading to a relatively complex optical set-up [64]. The optical scheme is summarised in figure 5.12. By PLL stabilisation, two YAG lasers are stabilised to a frequency difference of $f_{\mathrm{PLL}} = 20.01$ GHz. This generates a synthetic wavelength of $\Lambda^1_{1064} \approx 15$ mm in the infra-red, and after frequency doubling, of $\Lambda_{532} = 7.5$ mm in the green. Longer synthetic wavelengths, and the frequency separation of local oscillator and measurement beams are generated by acousto-optic frequency shifters. The beam preparation unit and the interferometer head are separated. The heterodyne two-colour interferometer head as shown in figure 5.12 was compactly built with a focus on thermal stability. The system is characterised by an unforgiving uncertainty scaling. The $A$ coefficient amounts roughly to a factor of approximately 21.2. The synthetic wavelength interferometry, however, adds another uncertainty scaling factor of $\Lambda_{532}/\lambda \approx 14\,000$. Together, any uncertainty in the optical path measurement is scaled by a factor of almost 300 000 in the geometrical path result.

The system has been validated indoors, at the PTB 50 m interference comparator [64], as well as outdoors, at the Nummela reference baseline of the Finnish Geospatial Research Institute (FGI) [69]. At Nummmela, distances can be compared to reference distances obtained from the Vaisala white light interferometer [77]. Results of a one-week measuring campaign are depicted in figure 5.13(a). The mean value of the successful measurements is in good agreement with the reference value. The standard deviation for the longest distances approaches 0.6 mm. This

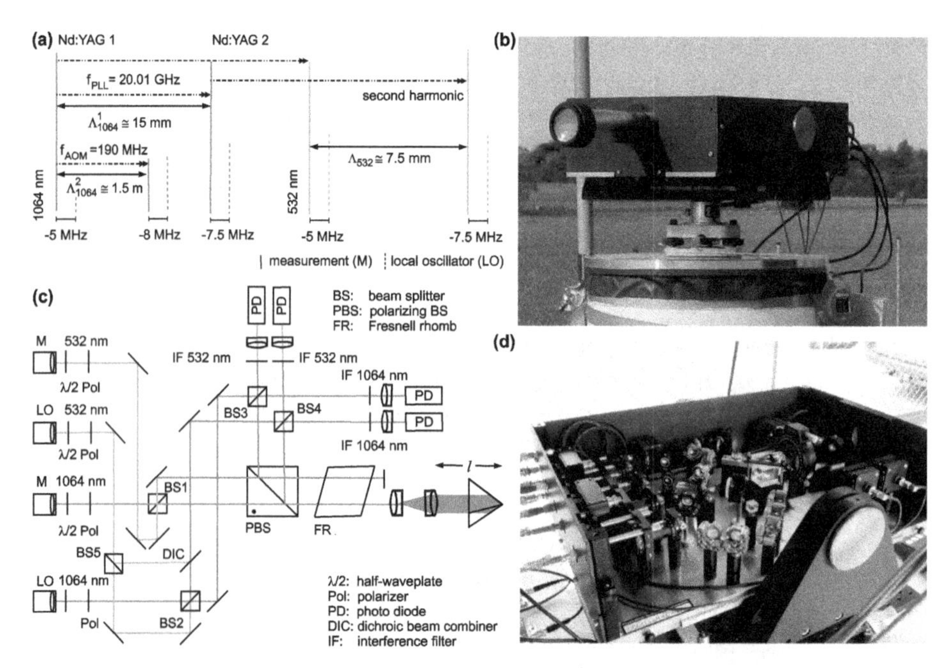


**Figure 5.12.** Refractivity compensated-absolute interferometer TeleYAG. (a) Frequency scheme for the heterodyne absolute interferometer. (b) Interferometer head mounted on a geodetic tribrach. (c) Scheme of the interferometer optical set-up. (d) View inside the interferometer head. (a) and (c) adapted under CC BY 4.0 license from reference [69], (d) reprinted with permission from reference [70]. Reprinted with permission from Proceedings of the 3rd Joint International Symposium on Deformation Monitoring, 2016.

corresponds to a relative uncertainty contribution of $7 \times 10^{-7}$, achieved without any auxiliary environmental sensor except a single humidity sensor. Nevertheless, the standard deviation seems relatively large. One explanation could be temperature-sensitive polarisation mixing in beam splitters, and thus time-dependent cyclic errors. In an air-conditioned laboratory, the uncertainty was indeed considerably lower [64]. Recent experiments support this assumption. Figure 5.13(b) shows the advantage of the two-colour approach in comparison with conventional refractive index compensation. The longest distance (realised by two geodetic pillars) at the Nummela baseline is monitored for 16 min. The two interferometer signals, infra-red (1064 nm) and green (532 nm), can be analysed independently as single-colour interferometers using environmental sensor data for compensation. Both signals indicate a systematic drift $\Delta l$ of the order of −0.6 mm. The intrinsically-compensated two-colour length change $\Delta l$ does not indicate any systematics, but shows scatter of the same order of magnitude due to the additional uncertainty scaling. Since the Nummela baseline is known for its stability, drifts of such an order of magnitude are not to be expected. Furthermore, the apparent length change is correlated with a simultaneous change of illuminance $E_v$. This indicates changing environmental conditions at this time that were not captured quickly enough by the sensor system. The time-resolved two-colour system, despite its increased scatter, hence provides a

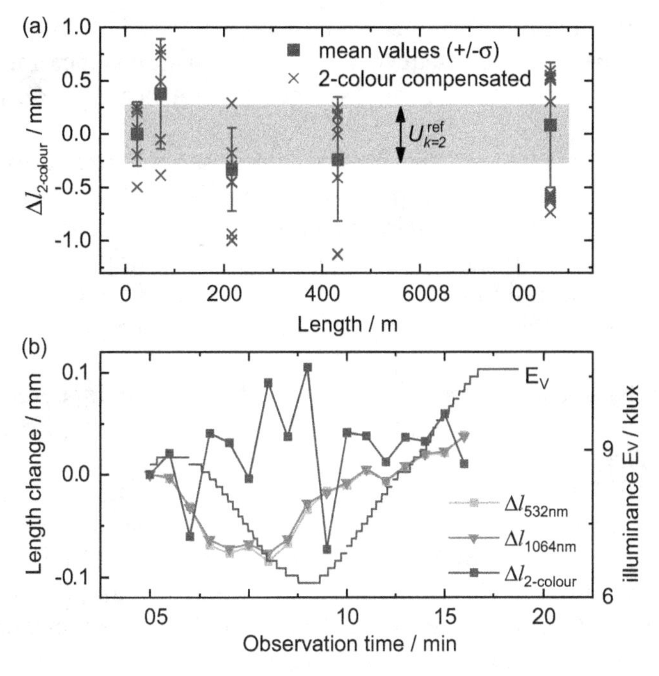


**Figure 5.13.** (a) Deviation of two-colour refractivity compensated heterodyne absolute interferometer from reference values obtained by white light interferometry. (b) Apparent position change of the 864 m pillar. Red and green curves were obtained with sensor data refractive index compensation, blue dots by two-colour compensation. The apparent position change follows the change in illuminance $E_V$. Adapted under CC BY 4.0 license from reference [69]. Reprinted with permission from AIP. Copyright 2016, American Institute of Physics.

better estimate of the actual measurement uncertainty than the 'low-pass-filtered' point sensor-compensated solution.

*Two-colour compensation in space geodesy*
Finally, though not really an interferential measurement, it should be noted that dispersive two-colour refraction compensation is also investigated for application in satellite and lunar laser ranging (SLR, LLR) (see, e.g. [78–80]). The distances are derived from the travel times of short light pulses in this measurement. Dispersion in the troposphere and ionosphere needs to be modelled for a single-colour measurement. The required standard deviation for this measurement to improve the overall accuracy for the difference in optical path lengths $\sigma(d_1 - d_2)$ is of the order of a few micrometres [80]. The instrumental realisation of these accuracies in green and infrared remains a challenge [81].

### 5.3.5 Limitations and prospects of intrinsic dispersive compensation

The TeleYAG and the two-colour system by Minoshima and Matumoto demonstrate the merciless uncertainty scaling of the intrinsic two-colour compensation. Satisfactory results in accuracy can only be obtained by a tremendous effort in the optical path determination. All uncertainty sources need to be controlled and mitigated with care. As a consequence, an easy-to-use distance meter based on two-colour interferometry for longer distances has not been demonstrated to this date.

It is important to note that two-colour compensation, despite its indisputable advantages, does not resolve the dispersion problem in length metrology completely. The more heterogeneous the environmental conditions become, the more the influence of turbulence grows. As can be seen, e.g. in figure 5.11, this leads to an enhanced standard deviation of the compensated result. The achievable resolution when applied, e.g. for real-time control in precision engineering, will be limited by this effect.

Finally, the basic assumption of dispersive refractivity compensation that both optical beams traverse the identical geometrical path cannot withstand a closer examination. The traversed air is not a simple homogeneous medium, but characterised by so-called 'bubbles' of equal local indices of refraction. Due to dispersion, beams of different wavelengths can be expected to follow different macroscopic paths (see e.g. the treatment by Wijaya and Brunner [80] for SLR). The magnitude of the effect depends of course on the turbulence condition and on the optical path length. The TeleYAG measurements over 864 m provides data indicating this effect. Figure 5.14 depicts time-resolved amplitude and length change data for the two infra-red and two green light beams. The amplitudes of the two red and green beams are almost identical for each colour pair. Red and green time-dependencies resemble each other, but appear shifted by approximately 50 ms. This is a strong indication

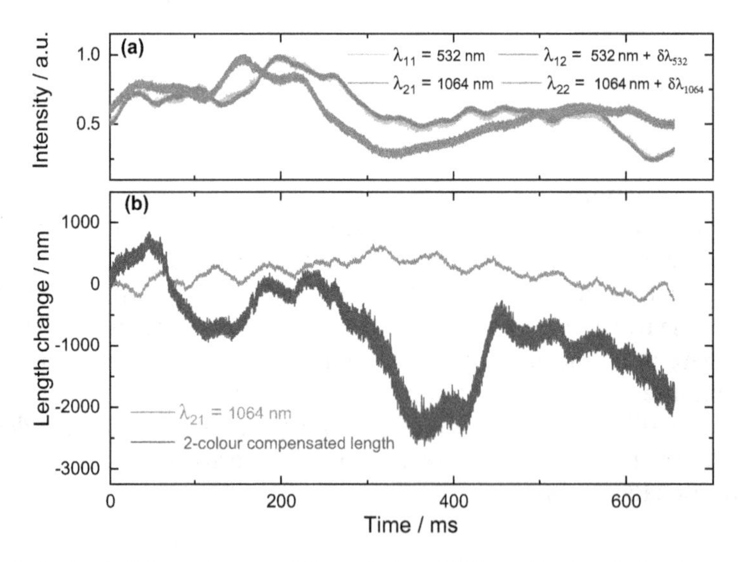


**Figure 5.14.** Optical path difference between 532 nm ($d_{532}$) and 1064 nm ($d_{1064}$) optical paths normalised to an absolute distance of 864 m for different signal averaging times. The fluctuations indicate a magnitude of refractive index changes on short time scale, but could also be caused in part by other noise sources. Adapted under CC BY 4.0 license from reference [69]. Reprinted with permission from AIP. Copyright 2017, American Institute of Physics.

for different beam paths. The magnitude of the associated uncertainty contribution can be derived from figure 5.14(b). The time-resolved interference data of a single-wavelength (here an infra-red one) varies only with a maximum of 500 nm. The two-colour compensated length generated from the optical paths of all four beams, however, shows a distance variation of up to 2000 nm. For a distance of 864 m, this corresponds to a relative uncertainty of the order of $2.3 \times 10^{-9}$, i.e. two orders of magnitude smaller than the overall uncertainty. In the case of moderate turbulence, this effect is still negligible, yet already identifiable.

The examples discussed in the previous sections show, however, that two-colour refractive index compensation has considerable advantages in uncontrolled environments for measurements on larger scales. It is not straightforward to implement, requiring a substantial effort in the high-accuracy determination of optical path lengths. Growing demands in flexibility in high accuracy engineering of larger parts will promote the further development of this technology.

## 5.4 Air wavelength stabilisation

The compensation approaches discussed in the previous chapters follow basically the same concept. The frequency and thus the vacuum wavelength of the optical source is stabilised. The measured phase is interpreted in terms of the wavelength in air. The air wavelength is therefore calculated from the (somehow measured) index of refraction and the known vacuum wavelength. This approach is very flexible in terms of measurement geometry and volume. But it is very difficult to achieve measurement uncertainties in the lower $10^{-7}$ or even in the $10^{-8}$ as the previous sections have shown.

There are applications in the semiconductor industry, e.g., in which lower uncertainties are targeted but operation in vacuum is not feasible. The measurement task is well-defined in these cases, and the environment is necessarily well-controlled. Temperature gradients are only to be expected in the order of 10 mK, for example [83]. Under these circumstances, the stabilisation of the wavelength in air against a mechanical reference can be an option to achieve relative uncertainties to $10^{-8}$ or below. For this, the standard needs to be manufactured by a material of very low thermal expansion like Zerodur or ULE.

If the measurement volume is well-defined, Lazar *et al* [82, 84] propose the use of a bidirectional interferometer as sketched in figure 5.15. The metrological frame is generated by optical elements fixed on an ultra-low expansion plate. The measurement path of both interferometers is generated by a moving mirror. Each interferometer probes one side. The sum of the phase signals of both interferometers is assumed to be constant due to the mechanical stability of the metrological frame. It is used to adjust the laser frequency to keep the air wavelength constant. In a reasonably stable environment, an uncertainty of the order of $5 \times 10^{-8}$ has been demonstrated [82]. In better controlled environments, a smaller 'tracking interferometer' can be applied the air wavelength stabilisation [83, 85]. In a very well-controlled environment, Quoc *et al* claim a compensation of refractive index

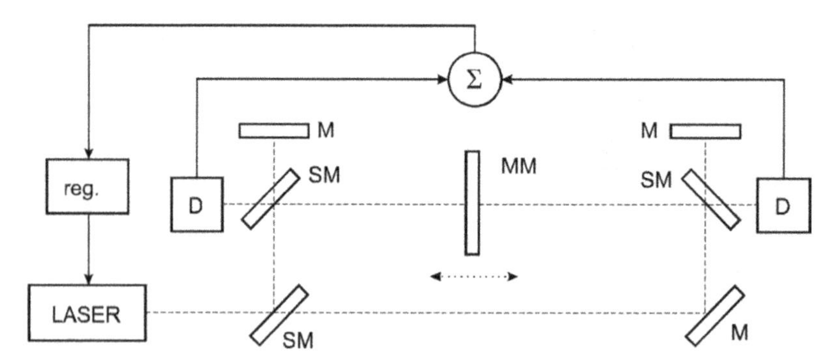


**Figure 5.15.** Example for air wavelength stabilised interferometry as proposed by Lazar and co-workers [82]. Two interferometers observe the length of the moving mirror (MM) in opposite directions. The sum of the optical path differences of both interferometers is constant since all optics are mounted on low-expansion material. The sum (Σ) of the detected phases is used to control the laser frequency. (M: mirror, SM: semi-transparent mirror, D: detector, reg.: frequency controler). Reprinted under CC BY 3.0 license from reference [82]. © 2011 by the authors; licensee MDPI, Basel, Switzerland.

fluctuation to $10^{-9}$ level, using a Fabry–Perot interferometer of low expansion material as length reference [85].

## 5.5 Conclusions

The correct interpretation of an interference measurement in air will always require a sensible approach to refractive index compensation.

First, it should be clear that relative uncertainties better than $10^{-8}$ can only be achieved for very special experimental designs in air, requiring a smart environmental control. Air wavelength stabilisation can then be a tool to compensate for residual refractive index fluctuations. However, for these tasks, a measurement in vacuum should be considered as a smart alternative if possible.

The empirical formulae that have been derived over the last 120 years have proven their validity to $10^{-8}$ level. The models benefit from extremely sophisticated refractivity experiments in the past. One can argue that given the growing importance of infra-red optical sources, e.g. Er-doped fibre lasers, in interferometry, more high-quality refractivity data in this region could be useful to test and improve the current empiric formulae. Nevertheless, practice indicates that they are also remarkably reliable in this spectral region.

In this chapter, various different ways to apply these refractivity models have been presented. The discussion shows that the application determines the optimum compensation method. In general, if gradients in environmental conditions can be avoided, the primary effort should be invested there, e.g. in good air conditioning and smart location of heat sources. In this case, smartly-distributed sensor networks can achieve relative uncertainties below $10^{-6}$ for the refractivity compensation. If gradients cannot be avoided, integrating methods are superior to sensor networks. For distances up to approximately 10 m, acoustic refractivity compensation shows a convincing performance to a moderate expense in experimental complexity. The method is, however, prone to echoes and unwanted reflections. This needs to be taken into account in the experimental design. More flexible, but also more complex is spectroscopic oxygen thermometry. When carefully optimised, refractivity compensation with uncertainty of $1 \times 10^{-7}$ has been demonstrated. It can be applied to larger volumes, but it needs to be adjusted to the measurement length. In addition, auxiliary information on the effective pressure is needed. Dispersive intrinsic refractivity compensation seems the most flexible and independent method. Except of a rough humidity measurement, no other auxiliary environmental data is needed for the refractivity compensation. However, uncertainty scaling by the $A$ coefficient is highly demanding on the quality of the length measurement. Furthermore, the integration of two interferometers of different colours in the same beam path is non-trivial. In the case of longer distances, as well as in uncontrolled industrial environments, the high flexibility and the potential justify the high effort if relative uncertainties below $10^{-6}$ are needed.

In conclusion, refractive index compensation will remain a major challenge in interferometry. No solution presented in this chapter combines all: lowest uncertainty, flexibility, and low cost for implementation. Given the growing demands for accuracy in precision engineering of big parts, e.g. in the aerospace industry, the topic will remain in the focus of dimensional metrology.

## Acknowledgments

The author would like to thank Petr Balling, Virpi Korpelainen, Kaoru Minoshima, and Marco Pisani for provision of data. The author is very grateful to Ahmed Abou-Zeid, Radu Doloca, Joffray Guillory, Thomas Fordell, Tuomas Hiéta, Jorma Jokela, Mikko Merimaa, Tobias Meyer, Jutta Mildner, Günther Prellinger, René Schödel, Jean-Pierre Wallerand, Martin Wedde, and Massimo Zucco for almost ten years of scientific collaboration in this field and for multiple fruitful and educating discussions. Finally, the author would like to acknowledge recently deceased Karl Meiners-Hagen. His insight and contributions will be missed.

## References

[1] Barrel H and Sears J E 1939 The refraction and dispersion of air and dispersion of air for the visible spectrum *Philos. Trans. R. Soc. Lond. A: Math. Phys. Eng. Sci.* **238** 1–64

[2] Born M and Wolf E 1999 *Principles of Optics* 7th edn (Cambridge: Cambridge University Press)

[3] Edlén B 1966 The refractive index of air *Metrologia* **2** 71

[4] Ciddor P E 1996 Refractive index of air: new equations for the visible and near infrared *Appl. Opt.* **35** 1566–73

[5] Balling P, Mašika P, Křen P and Doležal M 2012 Length and refractive index measurement by Fourier transform interferometry and frequency comb spectroscopy *Meas. Sci. Technol.* **23** 094001

[6] Edlén B 1953 The dispersion of standard air *J. Opt. Soc. Am.* **43** 339–44

[7] Bönsch G and Potulski E 1998 Measurement of the refractive index of air and comparison with modified Edlénas formulae *Metrologia* **35** 133–9

[8] Giacomo P 1982 Equation for the determination of the density of moist air (1981) *Metrologia* **18** 171

[9] Hyland R W and Wexler A 1973 The second interaction (cross) virial coefficient for moist air *J. Res. Natl. Bur. Stand. A Phys. Chem.* **77A** 133–47

[10] Davis R S 1992 Equation for the determination of the density of moist air (1981/91) *Metrologia* **29** 67

[11] Birch K P and Downs M J 1993 An updated Edlén equation for the refractive index of air *Metrologia* **30** 155

[12] Ciddor P E 2002 Refractive index of air: 3. The roles of $CO_2$, $H_2O$, and refractivity virials *Appl. Opt.* **41** 2292–8

[13] Schödel R, Walkov A and Abou-Zeid A 2006 High-accuracy determination of water vapor refractivity by length interferometry *Opt. Lett.* **31** 1979–81

[14] Owens J C 1967 Optical refractive index of air: Dependence on pressure, temperature and composition *Appl. Opt.* **6** 51–9

[15] Dvořáček F 2018 Survey of selected procedures for the indirect determination of the group refractive index of air *Acta Polytech.* **58** 9–16

[16] Hardy B 1998 ITS-90 formulations for vapor pressure, frostpoint temperature, dewpoint temperature, and enhancement factors in the range -100 to 100 C, *The Proc. of the Third Int. Symp. on Humidity and Moisture, Teddington, London April 1998*

[17] Egan P F, Stone J A, Hendricks J H, Ricker J E, Scace G E and Strouse G F 2015 Performance of a dual Fabry-Perot cavity refractometer *Opt. Lett.* **40** 3945–8

[18] Andersson M, Eliasson L and Pendrill L R 1987 Compressible Fabry-Perot refractometer *Appl. Opt.* **26** 4835–40

[19] Eickhoff M L and Hall J L 1997 Real-time precision refractometry: new approaches *Appl. Opt.* **36** 1223–34

[20] Khélifa N, Fang H, Xu J, Juncar P and Himbert M 1998 Refractometer for tracking changes in the refractive index of air near 780 nm *Appl. Opt.* **37** 156–61

[21] Fox R W, Washburn B R, Newbury N R and Hollberg L 2005 Wavelength references for interferometry in air *Appl. Opt.* **44** 7793–801

[22] Badr T, Azouigui S, Wallerand J P and Juncar P 2010 Absolute refractometry using helium *CPEM 2010* pp 506–7

[23] Egan P and Stone J A 2011 Absolute refractometry of dry gas to ± 3 parts in $10^9$ *Appl. Opt.* **50** 3076–86

[24] Pikálek T and Buchta Z 2015 Air refractive index measurement using low-coherence interferometry *Appl. Opt.* **54** 5024–30

[25] Kruger O and Chetty N 2016 Robust air refractometer for accurate compensation of the refractive index of air in everyday use *Appl. Opt.* **55** 9118–22

[26] Yang L, Wu X, Wei H and Li Y 2017 Frequency comb calibrated frequency-sweeping interferometry for absolute group refractive index measurement of air *Appl. Opt.* **56** 3109–15

[27] Ciddor P E and Hill R J 1999 Refractive index of air. 2. Group index *Appl. Opt.* **38** 1663–7

[28] Yang L J, Zhang H Y, Li Y and Wei H Y 2015 Absolute group refractive index measurement of air by dispersive interferometry using frequency comb *Opt. Express* **23** 33597–607

[29] Prellinger G, Meiners-Hagen K and Pollinger F 2015 Spectroscopically in situ traceable heterodyne frequency-scanning interferometry for distances up to 50 m *Meas. Sci. Technol.* **26** 084003

[30] JCGM 2008 *Evaluation of measurement data—Guide to the expression of uncertainty in measurement* http://www.bipm.org/en/publications/guides/gum.html.

[31] Pollinger F, Meyer T, Beyer J, Doloca N R, Schellin W, Niemeier W, Jokela J, Häkli P, Abou-Zeid A and Meiners-Hagen K 2012 The upgraded PTB 600 m baseline: a high-accuracy reference for the calibration and the development of long distance measurement devices *Meas. Sci. Technol.* **23** 094018

[32] Heinonen M 2006 Uncertainty in humidity measurements: Publication of the EUROMET Workshop P758 *MIKES J Series*, J4 (Espoo: Mittatakniikan kesus) https://www.vtt.fi/inf/pdf/MIKES/2006-J4.pdf

[33] Bonfig K W 1997 *Temperatursensoren—Prinzipien und Applikationen* (Renningen-Malmsheim: Expert-Verlag)

[34] Childs P R N 2001 *Practical Temperature Measurement* (Oxford: Butterworth-Heinemann)

[35] Eschelbach C 2009 Refraktionskorrekturbestimmung durch Modellierung des Impuls- und Wärmeflusses in der Rauhigkeitsschicht *PhD Thesis* Karlsruher Institut für Technologie

[36] Lewis A J, Campbell M and Stavroulakis P 2016 Performance evaluation of a cheap, open source, digital environmental monitor based on the Raspberry Pi *Measurement* **87** 228–35

[37] Pisani M, Astrua M and Zucco M 2018 An acoustic thermometer for air refractive index estimation in long distance interferometric measurements *Metrologia* **55** 67

[38] Korpelainen V and Lassila A 2004 Acoustic method for determination of the effective temperature and refractive index of air in accurate length interferometry *Opt. Eng.* **43** 2400–9

[39] Cramer O 1993 The variation of the specific heat ratio and the speed of sound in air with temperature, pressure, humidity, and $CO_2$ concentration *J. Acoust. Soc. Am.* **93** 2510–6

[40] Morfey C L and Howell G P 1980 Speed of sound in air as a function of frequency and humidity *J. Acoust. Soc. Am.* **68** 1525–7

[41] Howell G P and Morfey C L 1987 Frequency dependence of the speed of sound in air *J. Acoust. Soc. Am.* **82** 375–6

[42] Underwood R, Gardiner T, Finlayson A, Few J, Wilkinson J, Bell S, Merrison J, Iverson J J and Podesta M 2015 A combined non-contact acoustic thermometer and infrared hygrometer for atmospheric measurements *Meteorol. Appl.* **22** 830–5

[43] Underwood R, Gardiner T, Finlayson A, Bell S and de Podesta M 2017 An improved non-contact thermometer and hygrometer with rapid response *Metrologia* **54** S9

[44] Hieta T, Merimaa M, Vainio M, Seppä J and Lassila A 2011 High-precision diode-laser-based temperature measurement for air refractive index compensation *Appl. Opt.* **50** 5990–8

[45] Tomberg T, Fordell T, Jokela J, Merimaa M and Hieta T 2017 Spectroscopic thermometry for long-distance surveying *Appl. Opt.* **56** 239–46

[46] Silver J A and Kane D J 1999 Diode laser measurements of concentration and temperature in microgravity combustion *Meas. Sci. Technol.* **10** 845

[47] Chang A Y, DiRosa M D, Davidson D F and Hanson R K 1991 Rapid tuning cw laser technique for measurements of gas velocity, temperature, pressure, density, and mass flux using NO *Appl. Opt.* **30** 3011–22

[48] Hieta T and Merimaa M 2010 Spectroscopic measurement of air temperature *Int. J. Thermophys.* **31** 1710–8

[49] Gordon I E *et al* 2017 The HITRAN2016 molecular spectroscopic database *J. Quant. Spectrosc. Radiat. Transf.* **203** 3–69 HITRAN2016 Special Issue

[50] Arroyo M P and Hanson R K 1993 Absorption measurements of water-vapor concentration, temperature, and line-shape parameters using a tunable InGaAsP diode laser *Appl. Opt.* **32** 6104–16

[51] Buchholz B, Kühnreich B, Smit H G J and Ebert V 2013 Validation of an extractive, airborne, compact TDL spectrometer for atmospheric humidity sensing by blind intercomparison *Appl. Phys.* B **110** 249–62

[52] Pollinger F, Hieta T, Vainio M, Doloca N R, Abou-Zeid A, Meiners-Hagen K and Merimaa M 2012 Effective humidity in length measurements: comparison of three approaches *Meas. Sci. Technol.* **23** 025503

[53] Diddams S A, Hollberg L and Mbele V 2007 Molecular fingerprinting with the resolved modes of a femtosecond laser frequency comb *Nature* **445** 627

[54] Coddington I, Newbury N and Swann W 2016 Dual-comb spectroscopy *Optica* **3** 414–26

[55] Hänsel A, Reyes-Reyes A, Persijn S T, Urbach H P and Bhattacharya N 2017 Temperature measurement using frequency comb absorption spectroscopy of $CO_2$ *Rev. Sci. Instrum.* **88** 053113

[56] Meiners-Hagen K, Terra O and Abou-Zeid A 2006 Two colour interferometry, *Sensoren und Messsysteme 2006: Vorträge der 13. ITG/GMA-Fachtagung vom 13. bis 14.3.2006 in Freiburg/Breisgau* p 249

[57] Wu G H, Takahashi M, Arai K, Inaba H and Minoshima K 2013 Extremely high-accuracy correction of air refractive index using two-colour optical frequency combs *Sci. Rep.* **3** 1894

[58] Minoshima K, Arai K and Inaba H 2011 High-accuracy self-correction of refractive index of air using two-color interferometry of optical frequency combs *Opt. Express* **19** 26095–105

[59] Wu H, Zhang F, Liu T, Li J and Qu X 2016 Absolute distance measurement with correction of air refractive index by using two-color dispersive interferometry *Opt. Express* **24** 24361–76

[60] Earnshaw K and Owens J 1967 9.6-A dual wavelength optical distance measuring instrument which corrects for air density *IEEE J. Quant. Electron.* **3** 544–50

[61] Golubev A N and Chekhovsky A M 1994 Three-color optical range finding *Appl. Opt.* **33** 7511–7

[62] Thompson M C 1968 Space averages of air and water vapor densities by dispersion for refractive correction of electromagnetic range measurements *J. Geophys. Res.* **73** 3097–102

[63] Meiners-Hagen K and Abou-Zeid A 2008 Refractive index determination in length measurement by two-colour interferometry *Meas. Sci. Technol.* **19** 084004

[64] Meiners-Hagen K, Bosnjakovic A, Köchert P and Pollinger F 2015 Air index compensated interferometer as a prospective novel primary standard for baseline calibrations *Meas. Sci. Technol.* **26** 084002

[65] Meiners-Hagen K, Meyer T, Prellinger G, Pöschel W, Dontsov D and Pollinger F 2016 Overcoming the refractivity limit in manufacturing environment *Opt. Express* **24** 24092–104

[66] Ishida A 1989 Two-wavelength displacement-measuring interferometer using second-harmonic light to eliminate air-turbulence-induced errors *Jpn. J. Appl. Phys.* **28** L473–75

[67] Matsumoto H and Honda T 1992 High-accuracy length-measuring interferometer using the two-colour method of compensating for the refractive index of air *Meas. Sci. Technol.* **3** 1084

[68] Kang H J, Chun B J, Jang Y-S, Kim Y-J and Kim S-W 2015 Real-time compensation of the refractive index of air in distance measurement *Opt. Express* **23** 26377–85

[69] Meiners-Hagen K, Meyer T, Mildner J and Pollinger F 2017 Si-traceable absolute distance measurement over more than 800 meters with sub-nanometer interferometry by two-color inline refractivity compensation *Appl. Phys. Lett.* **111** 191104

[70] Pollinger F, Mildner J, Köchert P, Yang R, Bosnjakovic A, Meyer T, Wedde M and Meiners-Hagen K 2016 Si-traceable high-accuracy EDM based on Multi-Wavelength Interferometry, *Proc. of the 3rd Joint Int. Symp. on Deformation Monitoring*

[71] Pollinger F *et al* 2015 Metrology for long distance surveying: A joint attempt to improve traceability of long distance measurements *IAG 150 years—Proc. of the 2013 IAG Scientific Assembly, Potsdam, Germany, 1–6 September 2013, Int. Association of Geodesy Symp.* vol 143 ed C Rizos and P Willis (Berlin: Springer) pp 651–6

[72] Rüeger J M 1996 *Electronic Distance Measurement* (Berlin: Springer)

[73] Huggett G R 1981 Two-color terrameter *Tectonophysics* **71** 29–39

[74] Guillory J, Šmid R, Garcia-Marquez J, Truong D, Alexandre C and Wallerand J-P 2016 High resolution kilometric range optical telemetry in air by radio frequency phase measurement *Rev. Sci. Instrum.* **87** 075105

[75] Minoshima K and Matsumoto H 2000 High-accuracy measurement of 240-m distance in an optical tunnel by use of a compact femtosecond laser *Appl. Opt.* **39** 5512–7

[76] Salido-Monzú D and Wieser A 2018 Simultaneous distance measurement at multiple wavelengths using the intermode beats from a femtosecond laser coherent supercontinuum *Opt. Eng.* **57** 044107

[77] Jokela J 2014 *Length in Geodesy—On Metrological Traceability of a Geospatial Measurand, Publications of the Finnish Geodetic Institute* vol 154 (Kirkkonummi: Finnish Geodetic Institute)

[78] Zagwodzki T W, McGarry J F, Degnan J J and Varghese T 1997 Two-color SLR experiments at the GSFC 1.2-m telescope *Proc. SPIE Laser Ranging and Atmospheric Lidar Techniques* vol 3218

[79] Böer A and Hessels U 1999 Two color laser ranging with the TIGO SLR System: status and first results *Mitteilungen des Bundesamtes für Kartographie und Geodäsie* vol 10 (Frankfurt am Main: Bundesamt für Kartographie und Geodäsie) pp 353–8

[80] Wijaya D and Brunner F 2011 Atmospheric range correction for two-frequency SLR measurements *J. Geod.* **85** 623–35

[81] Courde C *et al* 2017 Lunar laser ranging in infrared at the Grasse laser station *Astron. Astrophys.* **602** A90

[82] Lazar J, Čip O, Čižek M, Hrabina J and Buchta Z 2011 Suppression of air refractive index variations in high-resolution interferometry *Sensors* **11** 7644–55

[83] Ishige M, Aketagawa M, Quoc T B and Hoshino Y 2009 Measurement of air-refractive-index fluctuation from frequency change using a phase modulation homodyne interferometer and an external cavity laser diode *Meas. Sci. Technol.* **20** 084019

[84] Lazar J, Holá M, Čip O, Čížek M, Hrabina J and Buchta Z 2012 Displacement interferometry with stabilization of wavelength in air *Opt. Express* **20** 27830–7

[85] Quoc T B, Ishige M, Ohkubo Y and Aketagawa M 2009 Measurement of air-refractive-index fluctuation from laser frequency shift with uncertainty of order $10^9$ *Meas. Sci. Technol.* **20** 125302

**IOP** Publishing

Modern Interferometry for Length Metrology
Exploring limits and novel techniques
**René Schödel (Editor)**

# Chapter 6

# Frequency comb based spectral interferometry and homodyne many-wavelength interferometry for distance measurements

**Nandini Bhattacharya and Steven van den Berg**

## 6.1 Introduction to frequency comb lasers and their applications to dimensional metrology

The invention of the femtosecond frequency comb at the beginning of this century has been a step change for optical frequency measurements. It has boosted academic research in many fields, like high-accuracy optical spectroscopy for fundamental research [19, 42], gas spectroscopy [1], the development of optical clocks [29], the search for exo-planets via more accurate calibration of astrographs [40] and many others [15, 25, 31]. The frequency comb also quickly found its way into national metrology institutes, because it provides an accurate and convenient way to calibrate optical frequency standards traceable to an atomic clock and thus to the SI second, or, in other words, to provide a direct coupling between the meter and the second. The huge impact of this invention was recognized in 2005 when the Nobel prize was awarded to two founding fathers of the field of laser frequency stabilization and frequency combs, John Hall and Theodor Hänsch [17, 18].

### 6.1.1 Principles of operation of an optical frequency comb

The invention of the frequency comb laser is based on two key developments in the fields of laser physics and optics. First, the development of mode-locked, ultrashort lasers, especially the Ti:sapphire laser in the 1990s [39]. The laser oscillator forms the heart of the frequency comb. A second key development is the invention of microstructured optical fibers [26, 33]. These highly nonlinear fibers enable broadening of the optical spectrum [9]. Spectral broadening covering one octave is

doi:10.1088/2053-2563/aadddcch6

regularly achieved, i.e. the high optical frequencies within the generated spectrum exceed the low optical frequencies by at least a factor of two. This range is especially important for locking the frequency comb to a reference clock using self-referencing and it allows the measurement of optical frequencies over a wide spectral range.

The word 'frequency comb' refers to the spectrum of a femtosecond laser. This spectrum consists of a large number of optical modes, showing a comb-like structure because of the fixed frequency difference between neighboring modes. The mode spacing is determined by the free spectral range (FSR) of the laser cavity. Since the modes are phase locked, the mutual phase between the modes is fixed, leading to the emission of an ultrashort pulse. The optical frequency of an individual mode can be written as $f_m = f_0 + m \times f_{\text{rep}}$, with $m$ an integer number, $f_{\text{rep}}$ the repetition rate of the laser, which is equal to the FSR and $f_0$ an offset frequency. Both $f_{\text{rep}}$ and $f_0$ can be stabilized against a frequency standard, like a cesium atomic clock. Having these two frequencies stabilized, stabilizes the full frequency comb, which means that all optical frequencies are directly traceable to the SI second. The stability of an individual optical frequency depends on the quality of the reference clock. A typical relative stability for a commercial cesium clock is $10^{-11}$ in 1 s. As a result of the stabilized repetition frequency of the laser, the pulse-to-pulse distance is automatically stabilized as well with the same level of accuracy, in vacuum. The periodic train of pulses emitted from the laser thus has its counterpart in the frequency domain as a comb of equidistant modes with a mutual separation equal to the repetition frequency $f_{\text{rep}}$ as can be seen in figure 6.1. The difference between the group velocity and phase velocity in the laser cavity gives rise to a pulse-to-pulse phase shift $\Delta\phi$ between the carrier wave and the envelope. This phase shift results in the offset frequency $f_0$, generally referred to as the carrier–envelope offset (CEO) frequency. The CEO and the repetition frequency are related by $f_0 = (\Delta\phi/2\pi)f_{\text{rep}}$.

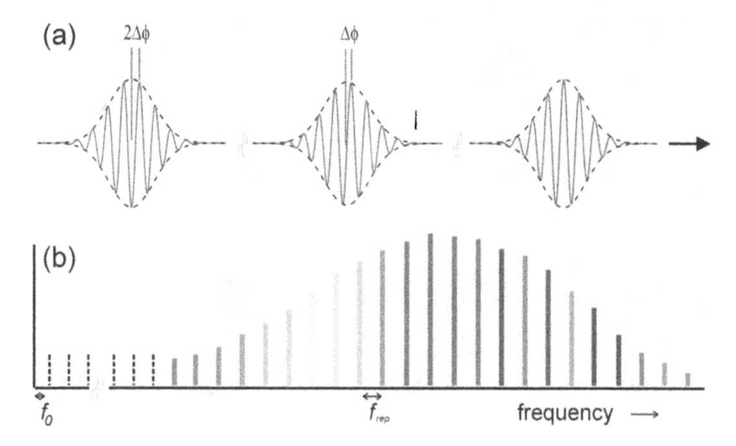


**Figure 6.1.** (a) Illustration of the carrier–envelope phase shift in the time domain. A pulse-to-pulse phase shift $\Delta\phi$ is observed. (b) Corresponding optical frequency spectrum with $f_0$ the offset frequency due to the carrier–envelope phase shift, and $f_{\text{rep}}$ the pulse repetition rate.

Typically, the spectral width emitted by a femtosecond laser of 10–100 fs pulse duration, is tens of nm. The spectrum is further broadened by sending the pulse through a nonlinear fiber. Due to large index contrast that arises from the fiber structure the light is confined within a very small core with a diameter of about 2 $\mu$m. For ultrashort pulses this leads to a large power density, even for modest pulse energies. The resulting strong nonlinear interaction of the electric field with the fiber gives rise to supercontinuum generation. Supercontinuum generation is based on several nonlinear processes like self-phase modulation, stimulated Raman scattering and soliton dynamics. Which process is dominant depends on the specific fiber design. In figure 6.2 an example of spectral broadening in a fiber is shown.

Having a spectrum covering a full octave means that optical radiation is generated from the fiber at both $f_m = f_0 + m \times f_{\text{rep}}$ and $f_{2m} = f_0 + 2m \times f_{\text{rep}}$. The offset frequency can then be measured by frequency doubling $f_n$ by second harmonic generation and measuring the frequency difference with $f_{2m}$. This frequency difference represents $f_0$. The repetition frequency can be measured easily with a photodiode that is sufficiently fast. To reference the frequency comb to a time standard both $f_0$ and $f_{\text{rep}}$ need to be locked to a reference value. These reference values are provided by signal generators referenced to a time base. Locking of the repetition frequency takes place by providing feedback to the cavity length, usually by driving a cavity mirror on a piezo-electric transducer (PZT). Feedback to the PZT is provided to stabilize the repetition rate via the optical cavity length to a reference value. Feedback to stabilize $f_0$ is usually provided via the power level of the pump laser driving the ultrafast laser. The carrier–envelope phase is sensitive to the pump

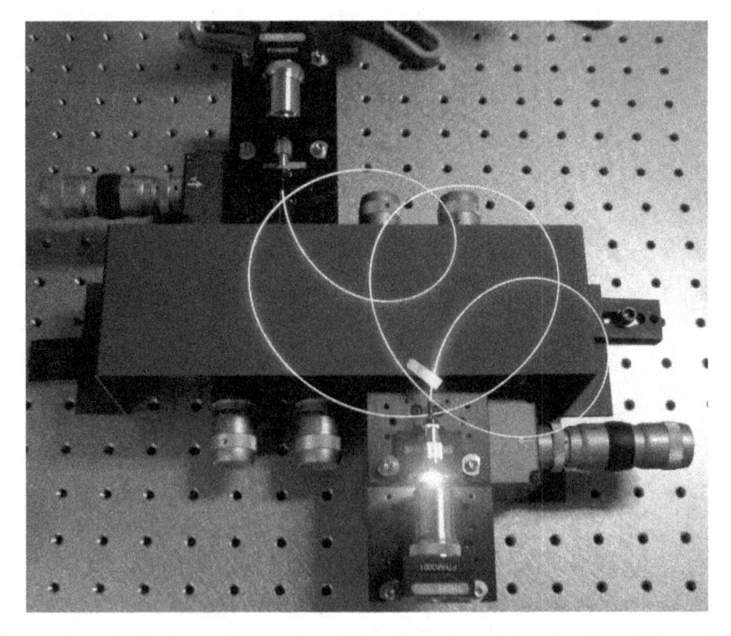

**Figure 6.2.** Spectral broadening in a microstructured fiber for pulses with a duration of about 40 fs and a pulse energy of 0.3 nJ as generated by a Ti:sapphire laser.

power because of the Kerr effect in the gain medium, e.g. the Ti:sapphire crystal. Once both $f_0$ and $f_{\mathrm{rep}}$ are phase locked to a time standard, the complete optical frequency comb spectrum is stabilized with a relative stability close to the clock stability.

### 6.1.2 Dimensional metrology and distance measurement with frequency combs

Since 1983 the meter has been defined as the distance traveled by light in 1/299 792 458 s in vacuum [16]. This definition is based on the fact that the speed of light in vacuum is a constant, which has been defined as $c$ = 299 792 458 m s$^{-1}$. Stabilized lasers like iodine-stabilized HeNe lasers are widely used as reference for length measurements [32], having relative wavelength stability of typically $10^{-10}$–$10^{-11}$, depending on the measurement time. A frequency comb offers the possibility to easily calibrate the laser frequency $f_{\mathrm{laser}}$ and thus the vacuum wavelength $\lambda_{\mathrm{laser}}$ via $\lambda_{\mathrm{laser}} = c/f_{\mathrm{laser}}$ against the SI second [46]. This is a huge advantage compared to earlier schemes based on a phase locked frequency chain [35]. The optical frequency measurement takes place by measuring the beat frequency $f_{\mathrm{beat}}$ between the optical frequency standard and the frequency comb. Here, some prior knowledge of the optical frequency is required in order to determine the value of $n$ and the signs of the various frequencies, since

$$f_{\mathrm{laser}} = n \times f_{\mathrm{rep}} \pm f_0 \pm f_{\mathrm{beat}}. \tag{6.1}$$

An accurate wavemeter is able to measure the optical frequency with an uncertainty of tens of MHz, which is sufficient to determine $f_{\mathrm{laser}}$ unambiguously. In figure 6.3 an example of a laser frequency measurement is shown for an iodine stabilized HeNe laser and for a Zeeman stabilized green HeNe laser. The latter is one of the wavelength standards that is typically used for multi-wavelength gauge block calibration.

Soon after the first applications of the frequency comb for optical wavelength measurement, length and distance measurement with or assisted by frequency combs were demonstrated. In the first application of a frequency comb for distance measurement, the comb served as a stable frequency modulated laser source [30]. Distance measurement was based on the synthetic wavelength corresponding to

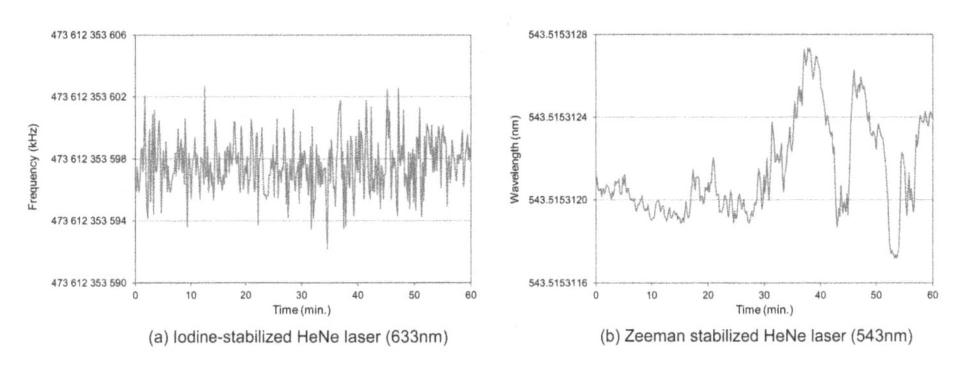


**Figure 6.3.** Calibration of an iodine stabilized HeNe laser (left) and Zeeman stabilized green laser (right).

higher harmonics of the repetition frequency. Another early scheme for distance measurement exploiting the frequency comb, was the application of cw lasers for distance measurement, while they were simultaneously referenced to a frequency comb [24]. In a similar experiment two lasers were referenced to different modes of the frequency comb for heterodyne measurements [37]. This scheme was further expanded by introducing more wavelengths by frequency doubling [34]. The authors reported a relative measurement accuracy of $10^{-8}$ when measuring optical path differences of up to 800 mm. Another multi-wavelength interferometry experiment was reported [20] which involved the continuous tuning and locking of an external cavity diode laser to the modes of the frequency comb.

The first proposal for distance measurement based on interferometry with the frequency comb itself was published by Ye in 2004 [45]. In the proposed measurement scheme a distance is retrieved from correlation functions that are obtained after propagation through a Michelson interferometer. Distance information can also be retrieved by spectral analysis of the interferometer output, as demonstrated in the first experimental paper on comb-based interferometry [22]. Another experiment reported by the same research group in 2006 implemented the absolute length calibration of gauge blocks using the frequency comb [21].

In this chapter both distance determination based on cross-correlation measurement and several forms of distance measurement based on spectral interferometry will be discussed. The measurements based on cross correlations are discussed in section 6.2 and numerical study of the evolution of the cross-correlation patterns for long distances will be discussed in section 6.3. In section 6.4, the dispersive method using a grating spectrometer for analysis will be introduced. In section 6.5 an extension of this measurement approach is discussed, showing the power of mode-resolved interferometry with a high-resolution spectrometer. An advanced scheme on mode-resolved interferometry will be introduced in section 6.6 utilizing very high repetition rate frequency combs.

## 6.2 Distance measurement based on cross-correlation

Distance measurement schemes can generically be described in terms of a Michelson interferometer illuminated with a light source, where the short arm is the reference length and the long arm is the unknown length to be determined. For incoherent non-interferometric detection schemes pulses of light sent down both arms are detected and the differences in their arrival times are used to calculate the unknown length, a technique traditionally known as time of flight. For coherent detection schemes the interferometric fringes are detected and extremely high resolution measurements can be done. The serious disadvantage for a single wavelength measurements using this principle is the distance ambiguity, leading to the use of this technique only for displacement measurements. Using multiple wavelengths for the measurement the ambiguity problem can be mitigated and beat signals between several frequency components can be used for absolute distance measurement. A frequency comb with its broad spectrum with extremely stable frequencies thus is an ideal source which could be fully exploited to measure long distances with high resolution. The extremely stable frequencies that the femtosecond laser is locked to determine $f_0$ and $f_{\mathrm{rep}}$, the characteristic frequencies which define the frequency comb as described earlier. This stability allows for hi-fidelity interference fringes to be established between pulse pairs which have traveled different delay distances. The first work on absolute distance measurement used a mode-locked femtosecond laser, was reported by Minoshima and Matsumoto [30], as mentioned in the earlier section. The first published experimental proposal for using the carrier–envelope stabilized frequency comb for distance measurement was published by Jun Ye [45] in 2004. In this paper Ye proposes using a frequency comb source in conjunction with a Michelson interferometer for absolute distance measurement of the long arm which was the unknown distance. Using this pulsed laser source the scheme involves changing the inter-pulse distance that is the repetition frequency of the frequency comb to measure the unknown distance. Measuring the timing difference between the pulses returning from each arm at different repetition frequencies one can determine a coarse estimate of the length. The finer measurement is done using an optical cross-correlator which can resolve the fringes when the $f_{\mathrm{rep}}$ is such that the pulses overlap. In figure 6.4 a schematic representation of the measurement scheme can be seen.

An unpublished report by Joseph Braat [6] described an alternative scheme using the frequency comb and Michelson interferometer. He proposed scanning the short arm of the Michelson interferometer till the cross-correlation fringes could be seen between the pulses traveling in both arms The short arm could be measured independently. The unknown distance was to be determined with high accuracy when the global path length difference is known in terms of the number of synthetic

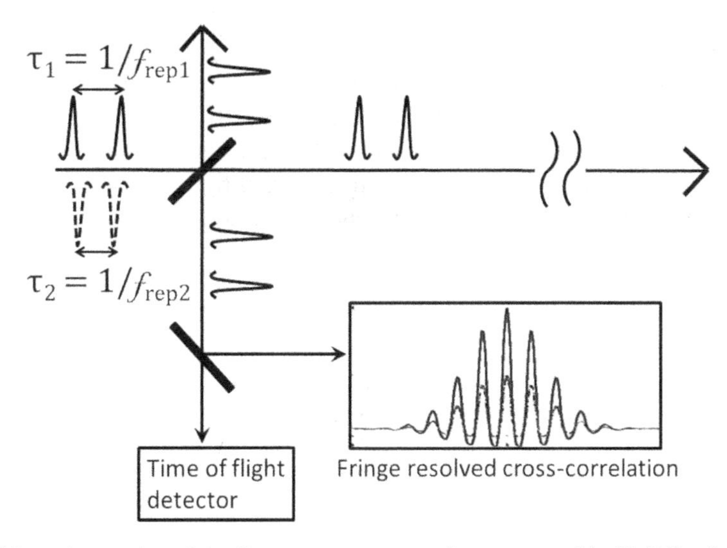


**Figure 6.4.** Schematic overview of the distance measurement scheme proposed by Ye [45], using a frequency comb as light source. Here the repetition frequency $f_{\mathrm{rep}}$ is varied to overlap the femtosecond pulses coming from two arms of the Michelson interferometer.

wavelengths. The only constraint on the experimental setup was that the minimum optical path length of the reference arm needed to be the same optical path length as the femtosecond laser cavity. Measurements based on recording inter-pulse cross-correlation patterns was first reported in 2008 by Cui *et al* in an experiment [10] which was an implementation of the experiment described by Joseph Braat. The experimental setup can be visualized in figure 6.5 as a schematic. In this experiment the authors used intensity correlation fringes to measure displacement of the long arm of the Michelson interferometer.

The optical path length which could be measured by the long arm was longer than the cavity length of the pulsed laser so that two correlation patterns could be obtained when scanning. These correlation patterns are created by pulses which are separated by the distance $L_{pp}$, which is the length of the laser cavity. By fitting the profiles of the two recorded cross-correlations and comparing the envelopes of the correlation patterns as shown in figure 6.5 the optical path length between them could be extracted. The frequency comb-based measurements were referenced with a stabilized calibrated fringe counting HeNe displacement meter. It could be shown that the measurements between the two interferometric systems agreed to within half a wavelength.

Similar experiments where interferometric distance measurement was performed with a frequency comb and compared with a fringe counting displacement interferometer was reported [2, 12] in 2009 by two groups. In the first experiment [2] the authors also reported developing a numerical model of pulse propagation in air. In the model the authors propagated the individual frequencies which constituted the pulse emanating from the frequency comb and propagated them with the Ciddor formula or the modified Edlén formula [4, 7, 8] which describes the refractive index of air. This model will be further described in the following section. This model was then compared with the experimental results and an agreement up to $5 \times 10^{-8}$ could be shown where the calculation was done for the actual spectra of the pulses used in the experiment. This is discussed further in the next section. In the measurements [12] reported by Cui and colleagues the experimental setup was in principle similar to the previous experiment reported by them except that they now used first order field cross-correlations instead of intensity cross-correlations and had

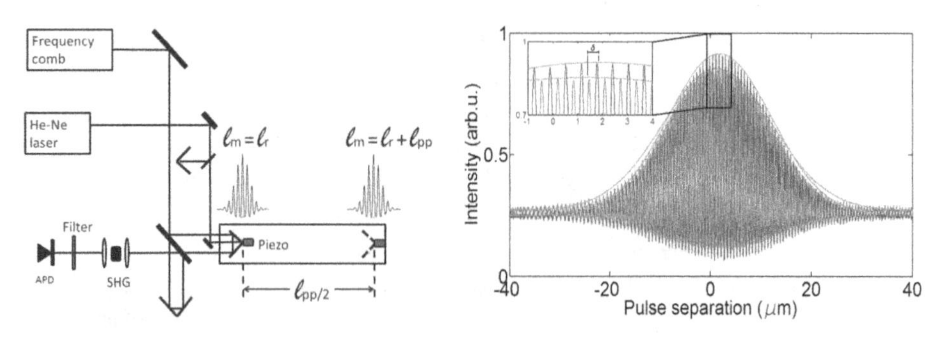


**Figure 6.5.** A schematic overview of the experimental setup used by Cui [10] for distance measurement. Two measured cross-correlation patterns in red and blue are shown with the corresponding fitted envelopes in green.

extended their long arm to 50 m. Figure 6.6 is a photograph of the long arm of the interferometer which uses the 50 m bench in VSL at Delft in the Netherlands. They demonstrated that even with dispersion and pulse broadening the measurements of long distances in air using this technique was possible. A numerical model similar to the one mentioned above was used by the authors to compensate for the dispersion and pulse broadening in air. The comparison of the frequency comb measurements with the fringe counting laser interferometer as seen in figure 6.7 showed an agreement with the measured distance to $4 \times 10^{-8}$ at 50 m. The intrinsic uncertainty in the updated Edlén equation is $3 \times 10^{-8}$ and in practice an uncertainty of 0.2 °C in temperature or 0.5 hPa in pressure already leads to an uncertainty in the refractive index determination of about $5 \times 10^{-7}$. This corresponds to about 25 $\mu$m at 50 m. But the environmental influences on the refractive index of air for the wavelengths of the Ti:sapphire laser based frequency comb and the HeNe laser cancel in the first order. This is the key to why the experimental results between the two interferometric measurements agree much better than the estimated 25 $\mu$m.

**Figure 6.6.** Photograph of the long arm of the interferometer which uses the 50 m bench in VSL at Delft in the Netherlands.

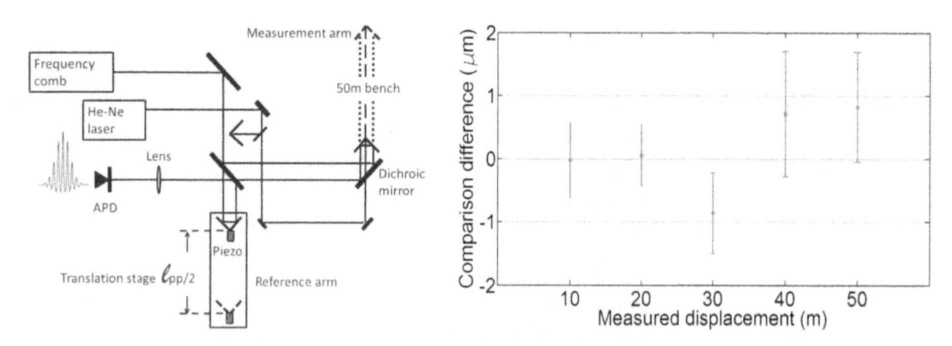


**Figure 6.7.** A schematic overview of the experimental setup used by Cui [12] for distance measurement on the 50 m bench at VSL. The comparison of the frequency comb measurements with the fringe counting laser interferometer. An agreement of $4 \times 10^{-8}$ is found between the two measurements for measured distances up to 50 m.

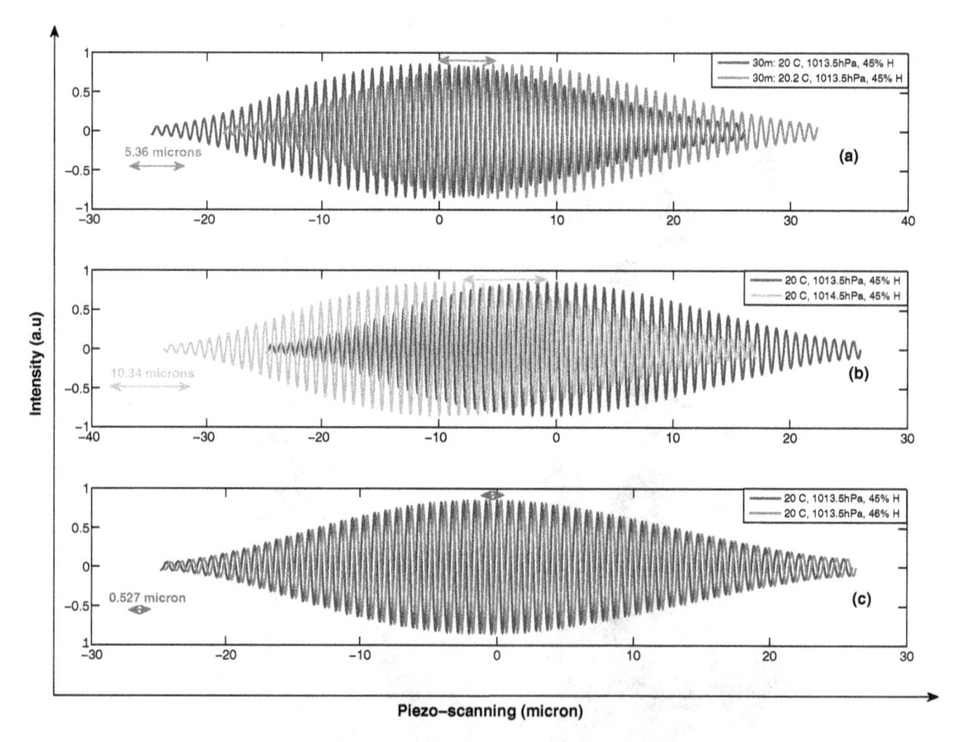


**Figure 6.8.** The effect of variation in the environmental parameters (a) temperature, (b) pressure and (c) humidity on the cross-correlation pattern. Figure reproduced with permission from [48]. Copyright (2010) by the American Physical Society.

The effects of the environmental parameters i.e. temperature, pressure, and humidity of the air was also investigated using the numerical model which is described in the next section, since environmental fluctuations have the largest influence on the accuracy of the measurements. The results of the numerical simulations are summarized in figure 6.8, for a pulse propagation distance of

60 m. The reference is considered to be the position of the fringe at the maximum of the correlation pattern obtained for 20 °C, 1013.25 hPa and 45% humidity and this position is centered at '0'. From the figure 6.8(a) it can be seen that for a 0.2 °C temperature variation the maximum of the correlation pattern will shift by 5.36 $\mu$m from the reference. Similarly a pressure increase of 1 hPa results in 10.34 $\mu$m shift from the reference as seen in figure 6.8(b). A humidity variation of 1% shows a shift of 0.527 $\mu$m as seen in figure 6.8(c). The fact that the correlation patterns are effected by the variation of the environmental parameters is known. From the simulations one observes that the patterns only shift without any extra broadening or chirp.

## 6.3 Evolution of cross-correlations at longer pulse propagation distances

The propagation of femtosecond pulses from the frequency comb laser for long distances in air leads to considerable dispersion and thus distortion in the pulse profile. Therefore, for length measurements based on cross-correlation between pulses emerging from both arms of the interferometer a better understanding of the pulse propagation is necessary to be able to extract data from distorted correlation profiles. For this purpose numerical pulse propagation models were developed by several groups as mentioned in the earlier section. In the reported works [2, 12, 48] the total electric field at a position was considered as a superposition of all the modes which constituted the pulse and were separated in frequency by $f_{\text{rep}}$. Expressing this in angular frequency the field of the pulse emitted by the laser, propagating in the direction of positive $x$, at $x = 0$ can be written as

$$\mathcal{E}(0, t) = \sum_{m=0}^{\infty} A_m \cos[(m\omega_r + \omega_0)t + \phi_m] \tag{6.2}$$

where $A_m$ is a real amplitude and $\phi_m$ is the phase. Here the frequency reference for the individual comb components was obtained from the property of the frequency comb which has the relation

$$\omega_m = m\omega_r + \omega_0, \tag{6.3}$$

where $\omega_0$ is the common offset frequency, $m$ is an positive integer and $\omega_r$ is the repetition frequency $f_{\text{rep}}$ expressed in angular notation

$$\omega_r = 2\pi f_r = \frac{2\pi}{T_r}, \; T_r = \frac{1}{f_{\text{rep}}}. \tag{6.4}$$

Here $T_r$ is the time distance between the pulses. The offset frequency $\omega_0$ is caused by the difference between the group velocity and the phase velocity inside the laser cavity. Both $\omega_0$ and $\omega_r$ are stabilized to an atomic clock. For exact simulation of experiments the electric field of the comb component was generated using the spectral width and shape. In figure 6.9 one can see a spectrum measured by Cui [12] from a Ti:sapphire laser based frequency comb. The central frequency of the comb is $\omega_c = 2.3254 \times 10^{15}$ Hz, corresponding to a wavelength of 810 nm in vacuum, the

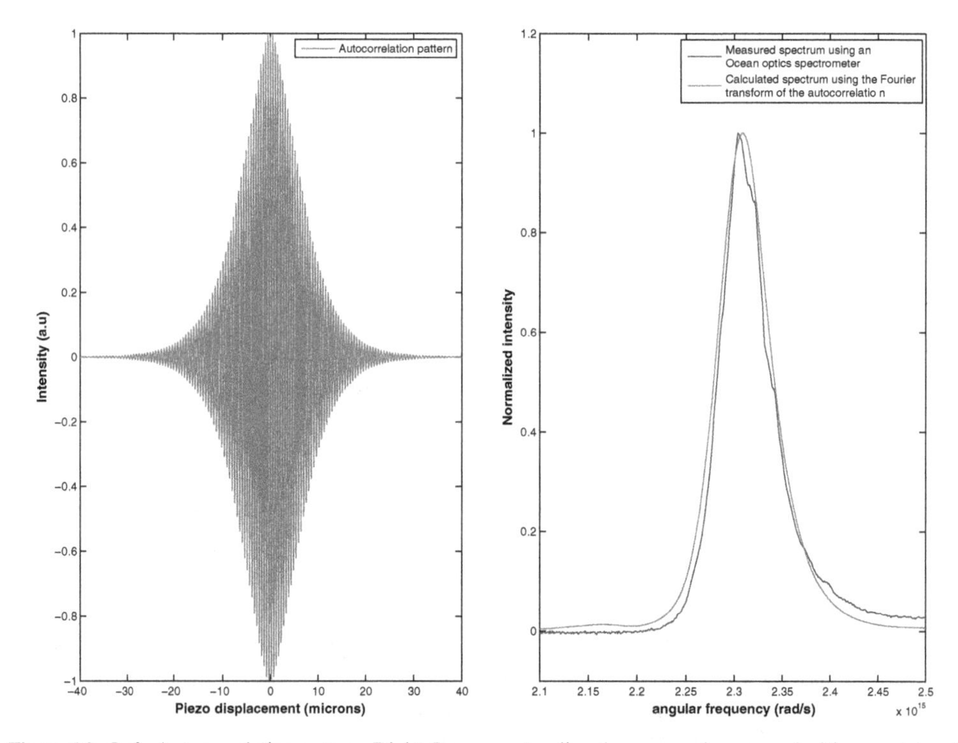


**Figure 6.9.** Left: Autocorrelation pattern. Right: Laser spectra directly measured compared with one retrieved from the autocorrelation. Figure reproduced with permission from [48]. Copyright (2010) by the American Physical Society.

bandwidth is typically $\Delta\omega \approx 5 \times 10^{13}$ Hz, which correspond to a pulse width of $\Delta x \approx 12$ $\mu$m and a pulse duration of 40 fs. The frequency offset is $\omega_0 \approx 2\pi \times 180 \times 10^6$ rad s$^{-1}$ and the repetition frequency is $\omega_r = 6.28 \times 10^9$ rad s$^{-1}$, corresponding to a cavity length $L_{pp} = 30$ cm and period $T_r \approx 1$ ns. This spectral profile consisted of $8 \times 10^3$ spectral lines. The equations for the refractive index of air, the modified Edlén formula and the Ciddor formula [4, 7, 8] at the environmental parameters specified were used to extract the phase refractive index for each of these frequencies. Plane wave propagation of the individual comb components to the position $x$, is done using the corresponding phase refractive index of the components. The electric fields of the individual components are then summed to generate the total electric field at the intended position. Thus the entire pulse can be generated for a position index $x$. The range of the position index for the pulses is appropriately chosen such that the electric field outside the range is only a negligible tail. The increment in the position index $x + \delta x$ is chosen to be much smaller than the wavelength for the shape and chirp of the pulse to be reproduced correctly.

The cross-correlations or the interferograms are formed by the overlap of two pulses from a frequency comb which have propagated in two arms of the Michelson interferometer when the difference between the pulse propagation distances is an integer multiple of the distance between two consecutive pulses emitted by the laser. In the numerical simulation this is then generated from the convolution of the electric fields of both pulses. Neglecting the DC-background the cross-correlation is, given by

$$\Gamma(X) = \sum_{m=0}^{\infty} |a_m|^2 \cos\left[(m\omega_r + \omega_0)n(m\omega_r + \omega_0)\frac{X}{c}\right] \quad (6.5)$$

where $|a_m|^2 \equiv$ power spectral density (PSD) and $X = x_1 - x_2$ the difference between the distances traversed by the two pulses forming the correlation. The details of the calculations can be found in the work of Zeitouny and colleagues [48]. Here too the range and increment have to be taken carefully into account since for long distances one pulse is substantially more chirped than the other leading to a chirped interferogram. When a realistic or actual experimental spectral profile is input into the numerical simulation the entire length measurement experiment can be numerically simulated thus helping in the interpretation of experimentally measured correlation patterns.

Cross-correlation patterns are calculated for different pulse propagation distances of the long arm of the interferometer and compared with the measured interferograms. To analyze the experimental data to extract the distances from the interferograms the brightest fringe position is used since a maximum fringe intensity indicates the occurrence of a maximum temporal coherence between superposed pulses. When the measurements are done in vacuum the position of the maximum coherence implies that the path length difference is $q \times L_{pp}$ where $L_{pp} = 2\pi c/\omega_r$ and $q$ is an integer. The measurements in air are more complex since the dispersion relation has to be known in detail. Using the Edlén's equation to obtain the refractive index $n(\omega)$ the phase delay is given by $v_p(\omega) = \omega/k(\omega) = c/n(\omega)$ where $k(\omega)$ is the wavenumber. The pulse envelope carries the separate frequency oscillations which form the group oscillation. The group delay representing the delay of the pulse envelope after propagation is the rate of change of the total phase shift with respect to the angular frequency $v_g = d\omega/dk(\omega) = c/n_g$ where $n_g$ is the group refractive index defined at the carrier frequency of the pulse, $\omega_c$. The carrier frequency of the pulse, $\omega_c$, is conventionally accepted as the frequency with the maximum intensity in the spectrum. The group refractive index is represented by

$$n_g = n(\omega_c) + \omega_c\left[\frac{d(n(\omega))}{d\omega}\right]_{\omega_c}. \quad (6.6)$$

The maximum fringe visibility is reached at a delay of $L$ for interferometric measurements which in vacuum is $L = q \times L_{pp}$ whereas in a dispersive medium $L_n = q \times L_{pp}/n_g$. The maximum of the cross-correlation pattern does not coincide with the position of $L_n$ unless the medium is linearly dispersive over the entire spectrum of the pulse.

The formation of a cross-correlation after the pulse propagates in the dispersive media does not only depend on the individual pulse path in one arm, but rather on the interference between pulses which have encountered different delays. Simulated and corresponding measured cross-correlation patterns are shown in figure 6.10. The

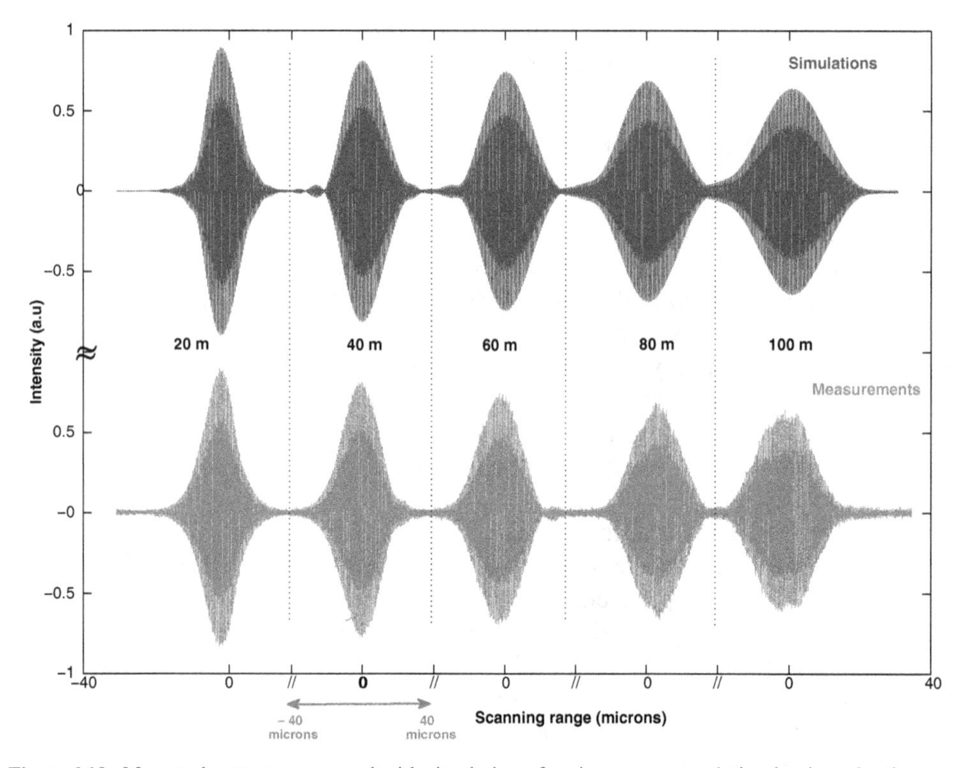


**Figure 6.10.** Measured patterns compared with simulation of various cross-correlation in air under the same environmental conditions. Figure reproduced with permission from [48]. Copyright (2010) by the American Physical Society.

simulated and measured cross-correlation patterns correspond to path length differences of 20, 40, 60, 80, and 100 m in air. As can be seen from figure 6.10 the simulated and measured patterns both show similar chirp and broadening. The numerical model accounts quite well for the nonlinear dispersion effects on the pulses which have an asymmetric frequency spectrum. To characterize the broadening of the dispersed pulses the full width at $1/e$ of the maximum ($FW_{1/e}$) of measured and simulated correlation patterns are compared in figure 6.11, for 0 up to 200 m propagation in air. The $FW_{1/e}$ of the simulated and measured patterns agree very well for short distances, less than 80 m. After that distance the $FW_{1/e}$ of the patterns do not agree very well and this was mainly attributed to unpredictable effects of vibrations in the interferometer and air turbulence in the measurement room. From studying the simulated and measured pattern it was inferred that the chirp of the pulses in the long arm of the interferometer reaches its maximum at a distance of ~30 m; beyond this distance the broadening effects are linear.

Further study of long distance pulse propagation and distortion in the cross-correlations formed by these pulses can be found in the work by Zeitouny and colleagues [48, 49], where the importance of the spectral content of the pulse to these effects is also discussed. An observation was that a large part of the spectral content of the pulse contributes to the formation of the correlation patterns and in particular the

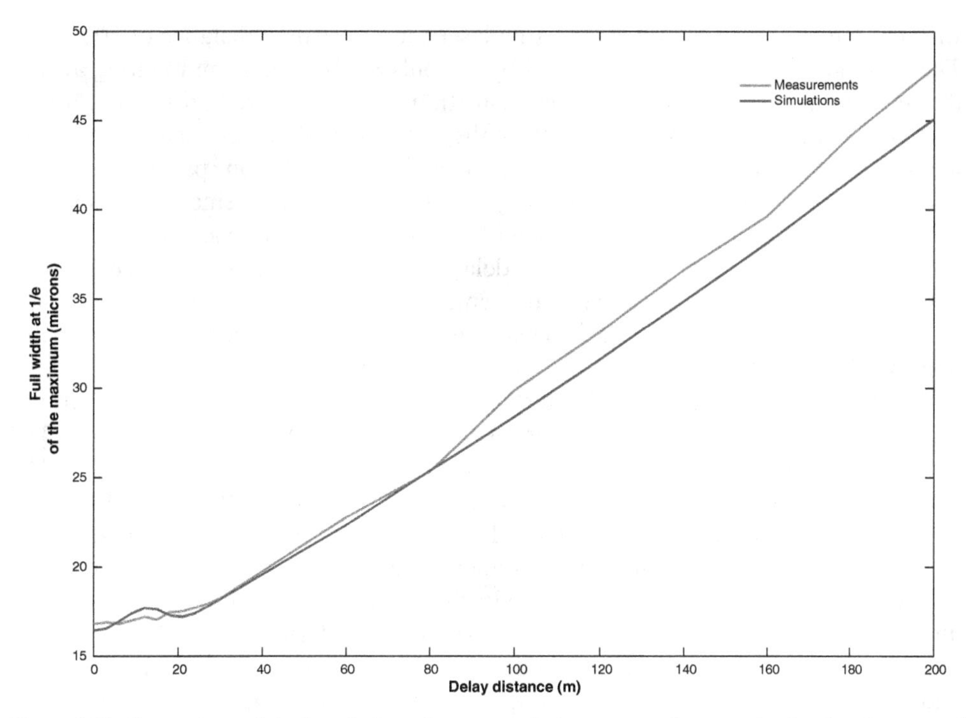


**Figure 6.11.** Comparison of the broadening of cross-correlation patterns between simulated and measured data. Figure reproduced with permission from [48]. Copyright (2010) by the American Physical Society.

brightest fringe for short path length differences. There they also extend the discrete numerical model to a continuous semi-analytical model to further investigate the many aspects of the formation of correlation patterns after propagation in dispersive media. The main observations are that the cross-correlation patterns are subject to nonlinear broadening at short path length differences and linear broadening at larger path length differences as mentioned before. This has also been observed in experiments. The change in refractive index varies nonlinearly for different frequencies as can be seen from the Edlén's equation. The phase accumulation of the different frequencies in the spectral content contributing to the correlation pattern is also nonlinear. This leads to the position of the maximum coherence varying nonlinearly and also nonlinear broadening of the entire pattern for the initial path length differences. For the large path length differences the nonlinear phase accumulation plays a smaller role and very few frequencies of the spectra contribute to the fringes of the correlation pattern, so both the variation in the position of the maximum coherence and the broadening are seen to be linear. These effects were used for analysis of the correlation patterns for the distance measurements done on the 50 m bench.

## 6.4 Spectral interferometry

In the previous sections we discussed long distance measurement by detecting the cross correlations formed in a Michelson interferometer when illuminated by a femtosecond frequency comb. The measurements could achieve the agreement to within 2 $\mu$m in comparison to a counting laser for a measured distance of 50 m [12]. The nonlinear broadening experienced by the pulses returning from the long arm of the interferometer led to considerable distortion in the measured cross-correlations and careful analysis was necessary before the distance could be extracted from these measurements. To extract the distance from the correlation patterns detailed knowledge of the PSD of the frequency comb was necessary since each point of the cross-correlation pattern is a coherent sum of all the frequency components in each pulse with the appropriate phase delay. Besides, measurement of correlation patterns requires constant periodic movement of the reference arm to build the interferogram. The technique of dispersive interferometry or spectral interferometry does not have these limitations.

A typical setup using spectral interferometry uses a spectrometer at the output arm of the Michelson interferometer to record the spectral interferograms. The phase is unwrapped from the spectral interferogram to determine the distance. The technique is strongly reminiscent of white light interferometry with dispersive interferometers [36] before the arrival of ultrashort pulsed lasers. The first experimental work to report distance measurement with the frequency comb source [22], in 2006, used dispersive interferometry. In this paper Joo *et al* measured a spectral interferogram at the output arm of the Michelson interferometer. For this they placed a grating at the output arm to disperse the inter-pulse interference spectrally and captured it using a line camera. Further analysis of this spectrogram reveals the distance which was to be measured. The experiment reports measuring 0.89m with an accuracy of $10^{-5}$. The authors reported in 2008 another experiment [23] where they had further expanded their setup with time of flight detection in addition to dispersive interferometry in their detection arm. Here they concluded that the maximum measurable range is $1.5 \times 10^{7}$ m as determined by the temporal coherence length derived from the linewidth of the modes of the comb to be 10 Hz when the frequency comb is locked and stabilized [19] to such precision. These experiments are further discussed in chapter 7. In this chapter we focus on the experiments done by Cui *et al* [11] using the technique of spectral interferometry on the 50 m bench at VSL in Delft.

In figure 6.12 an artists impression of the experimental setup with a visualization of spectral interferometry can be seen. The pulses from the frequency comb being phase-coherent make interferometry possible for path length differences close to multiples of the inter-pulse distance. To evaluate this technique we consider the combined spectrum of the two arms as described by equation:

$$S(\omega) = 2\,|\hat{E}_r(\omega) \cdot \hat{E}_m(\omega)|\left[1 + \cos\left(\varphi_r(\omega) - \varphi_m(\omega)\right)\right], \tag{6.7}$$

where $|\hat{E}_r(\omega)|$ and $\varphi_r(\omega)$ are the electric amplitude and phase of the pulse from the reference arm. $|\hat{E}_m(\omega)|$ and $\varphi_m(\omega)$ are the electric amplitude and phase from the measurement arm. The phase difference $\varphi_r - \varphi_m$ is embedded in the interferogram thus permitting the reconstruction of the path length difference. Above we have neglected absorption and assumed that the intensities of the reflected beams from the two arms are equal. In case of propagation in air we can write

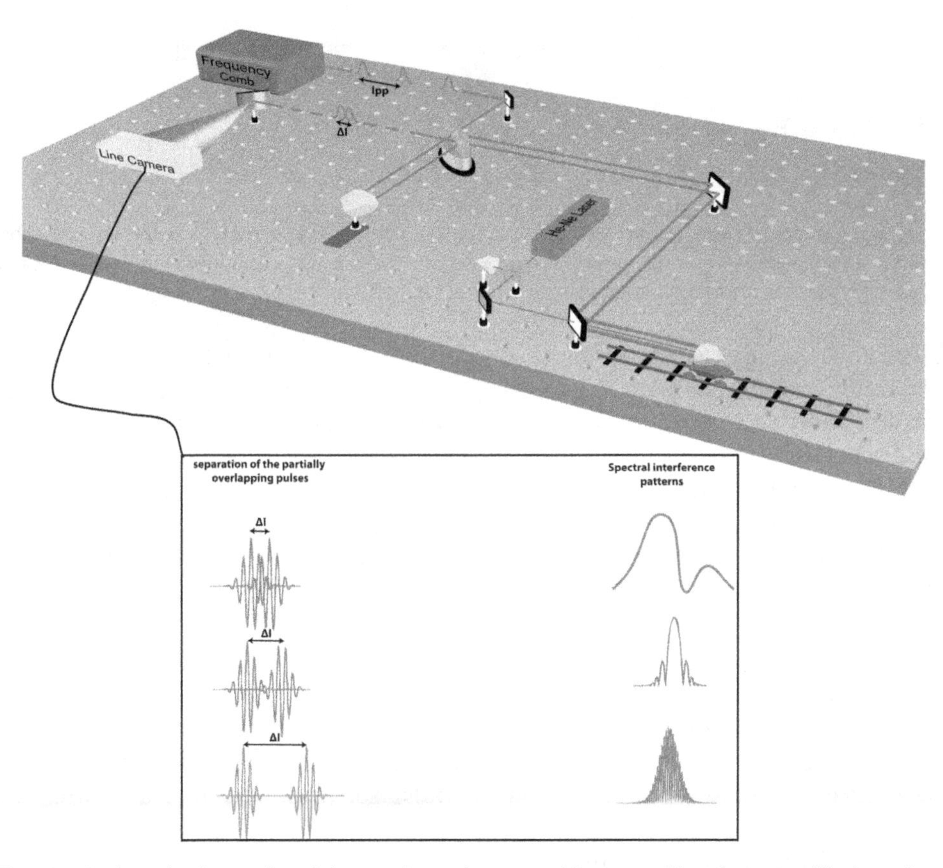


**Figure 6.12.** An artists impression of the experimental setup used for spectral interferometry. The inset shows the evolution of the spectral interferogram with different pulse overlap.

$$\varphi(\omega) = \varphi_r(\omega) - \varphi_m(\omega) = n(\omega)\omega L/c \tag{6.8}$$

where $n$ and $c$ are the refractive index of air and the speed of light in vacuum, respectively. The pulse separation between the interfering pulses after the beam splitter is represented by $L$. The measurement is initially set up such that the path length difference between the two arms is relatively small but not zero and a modulated spectrum with interference fringes can be observed as seen in figure 6.13(a). As the retroreflector in the measurement arm is displaced further from the beam splitter the number of fringes in the spectrum increases since the separation between the two pulses increases. The modulation depth becomes much shallower as the fringe density is beyond the resolution of the spectrometer. This can be seen in figure 6.13(b). When the path length difference between the two arms becomes so large that more than one fringe is imaged to a single pixel of the camera the fringes disappear. The fringes re-appear as in figure 6.13(c), when the measurement arm is displaced about half of the inter-pulse distance since then there is enough pulse overlap to reduce the fringe density enough so that the spectrometer can resolve the fringes. The 'twin image' problem arises where the spectrum is indistinguishable

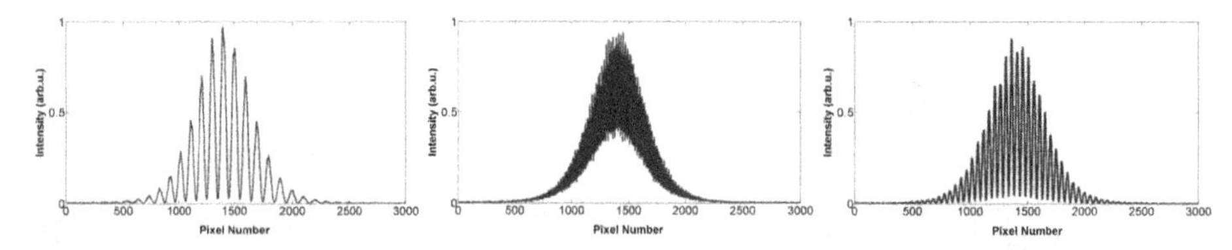


**Figure 6.13.** Spectral interferograms recorded for different pulse separations. (a) Two pulses at a short separation. (b) Two pulses at a large separation. (c) After one cavity length, the fringes re-appear, where one pulse is interfering to the next. Reproduced from [11]. CC BY 4.0.

from its twin image with the same pulse separation but an opposite sign. This can be seen in the cosine term in equation (6.7). This issue can be resolved by moving one arm a little bit and observing how the spectrum changes to determine the sign of the pulse separation.

Half of the inter-pulse distance $L_{pp}/2$ can be viewed as the synthetic wavelength and around multiples of this length there is enough overlap between the pulses to record a spectral interferogram. The total displacement of the measurement arm $\Delta l$ can be expressed by

$$\Delta l = m \cdot L_{pp}/2 + L_1/2 - L_2/2 \tag{6.9}$$

with $m$ an integer and $L_{pp}$ the inter-pulse distance, calculated by $L_{pp} = c/n_g f_{\mathrm{rep}}$. Here $c$ is the speed of light in vacuum, $f_{\mathrm{rep}}$ is the repetition frequency and $n_g$ is the group refractive index calculated by using the wavelength at the maximum of the PSD of the pulse. $L_1$ and $L_2$ are the pulse separations before and after the movement of the measurement arm respectively. The factor of two arises due to back and forth propagation and $L_1$ and $L_2$ are positive when the pulse from the measurement arm is ahead the pulse from the reference arm. Ignoring the jitter in $f_{\mathrm{rep}}$ the absolute uncertainty of the method does not increase due to the increase in the integer $m$ thus implying that the maximum distance measured by this technique is only limited by the coherence length of the laser source. If $f_{\mathrm{rep}}$ is locked and stabilized within 1 Hz, an uncertainty of 1 $\mu$m in 1 km measured distance is possible in vacuum.

The spectral interferogram captured by the CCD is then Fourier transformed, the AC peak is extracted and inverse Fourier transformed as seen in figure 6.14. The phase obtained from the inverse Fourier transform is then unwrapped and the distance can be calculated from the derivative of the unwrapped phase,

$$L(\omega) = \frac{c}{n_g}\left(\frac{d\varphi}{d\omega}\right). \tag{6.10}$$

Here $\omega$ is the angular frequency and $n_g$ is the group refractive index of air. The spectrometer needs careful calibration and this can be done using calibration sources in the appropriate spectral region or using an independent displacement meter like the HeNe fringe counting laser interferometer as described by the authors. In this experiment the authors chose a different $f_{\mathrm{rep}}$ for each of the five groups of long distance measurements that were performed. Each group consisted

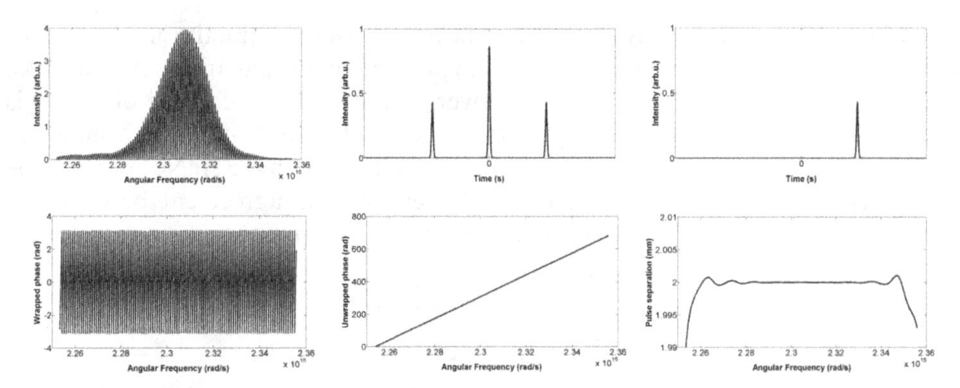


**Figure 6.14.** Data processing procedure for measurement of $L$. (a) The spectral interferogrami.e. dispersed interference intensity captured by the CCD line. (b) Fourier transform of the measured spectral interferogram. (c) The DC peak and one AC peak are band pass filtered. (d) The wrapped phase. (e) The unwrapped phase. (f) The pulse separation obtained from the derivative of the unwrapped phase. Reproduced from [11]. CC BY 4.0.

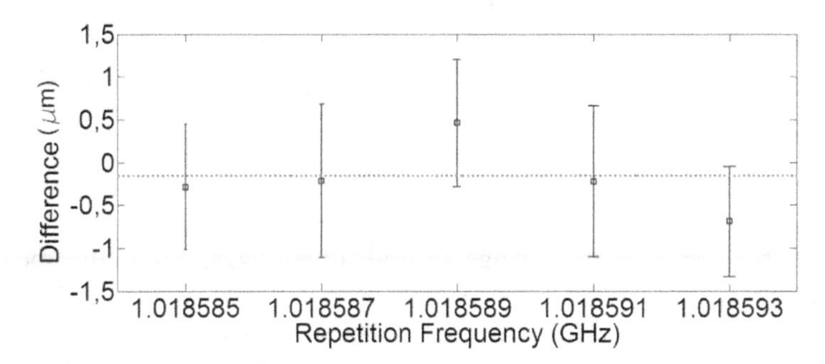


**Figure 6.15.** Comparison measurement of displacements of around 50 m. The error bars indicate the standard uncertainty, derived from measurement reproducibility. The average of all measurements is shown by the dotted line. Reproduced from [11]. CC BY 4.0.

of six independent measurements. In each measurement $L_1$ and $L_2$ in equation (6.9) was chosen arbitrarily, but at positions where the spectral fringes could be resolved. The spectral interferogram is recorded five times at each position to statistically minimize the uncertainty. One 50 m measurement took about 10 min, which mostly consisted of transporting the retroreflector down the 50 m bench. The measurements done on the 50 m bench were compared to a calibrated counting laser interferometer for validation and the comparison is shown in the figure 6.15. All five measurements agree within one wavelength, with the standard deviation of around 1 $\mu$m. An agreement within 200 nm is seen in the average value of all measurements for 100 m pulse propagation in air, as shown by the dotted line in figure 6.15. The authors attribute the residual differences and uncertainty to vibrations and air turbulence. Evaluating the environmental parameters for example an uncertainty of 1 hPa in pressure leads to an uncertainty of $2.7 \times 10^{-7}$, corresponding to 26 $\mu$m at 100 m propagation. Again a typical uncertainty of

0.1 °C in temperature already leads to an uncertainty of the refractive index of air of about $1 \times 10^{-7}$, corresponding to 10 $\mu$m at 100 m propagation. An intrinsic uncertainty of $1 \times 10^{-8}$, implying 1 $\mu$m over a path length difference of 100 m is inherent to the Edlén equation itself. But here again the effect of the environmental parameters on the refractive index at the wavelengths of both the HeNe laser and the frequency comb cancel in first order and therefore the agreement between the experimental results is much better than 10 $\mu$m. Thus the measurement result is not limited by the chosen method.

## 6.5 Mode-resolved homodyne interferometry

As discussed in the previous section spectral measurements of the Michelson interferometer output provide distance information without the need for moving elements, which is an advantage compared to the cross-correlation measurements. Another advantage of the spectral measurements compared to cross-correlation measurements is that the measurement range is not restricted to exact multiples of $L_{pp}$, but allows for measurements of distance with some extension around multiples of $L_{pp}$. For the spectral interferometry measurements the distance is derived from the phase change as a function of optical frequency. Here a well-calibrated wavelength scale of the spectrometer is required. The measurement range is still relatively small and depends on the resolution of the spectrometer. Washing out of the interference fringes arises from phase variation of the comb components within the resolution of the system. Ideally, the spectrometer is able to resolve the individual modes of the frequency comb, because this brings several advantages. First, the measurement range is no longer restricted by the resolution of the spectrometer. Since all modes are resolved individually, fringes do not wash out anymore, even for the largest distance between the pulses ($L_{pp}/2$). Second, the wavelength scale needed for spectral interferometry is immediately available, because the spacing between the modes is equal to the known $f_{\text{rep}}$. Third, having all these optical frequencies available from a single source provides a unique opportunity for optical interferometry, since all these modes can be individually exploited for interferometry. In other words: each mode can be considered as an individual single-mode laser, with a wavelength accuracy close to the reference clock.

In this section, we discuss the application to distance measurement of a very high resolution spectrometer that is able to resolve individual frequency comb modes [43, 44]. This is possible with a spectrometer based on a virtually imaged phased array (VIPA) and a grating. A VIPA is an etalon which generates an angular dispersion of the optical wavelengths. Originally, the VIPA was developed and applied for wavelength division and multiplexing in telecommunication [38]. The first application to resolve frequency comb modes was reported in 2007, showing high-resolution molecular fingerprinting [14]. A detailed background on the VIPA-based spectrometer can be found in references [3, 41]. The thickness of the etalon determines the free spectral range of the VIPA, which is typically about 50 GHz. In order to separate the overlapping orders from the VIPA, the light is also dispersed with a grating in the horizontal plane. The dispersed light is imaged on a CCD

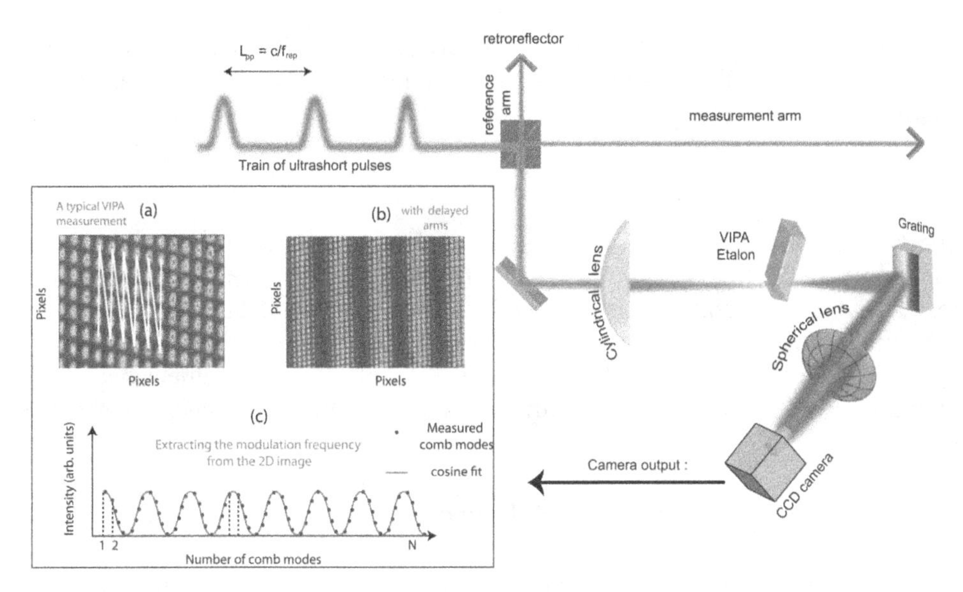


**Figure 6.16.** Schematic overview of the spectral analysis of the output of a Michelson interferometer, using a frequency comb as a light source. Reprinted with permission from [43], Copyright (2012) by the American Physical Society.

camera, resulting in an image as shown in figure 6.16. Each dot represents an individual wavelength from the frequency comb. Along the vertical axis the optical frequency difference between neighboring dots is equal to $f_{rep}$. Along the horizontal axis it is about 50 GHz, as determined by the FSR of the VIPA. From the 2D dotted pattern the full spectrum of the comb can be reconstructed. Here it is necessary to have a reference frequency to know which dot is which and to assign absolute values of the optical frequency to each individual dot. For this purpose a tunable single-mode reference laser is used, having a wavelength close to the central wavelength of the comb spectrum. The absolute wavelength of the laser is measured with an uncertainty <100 MHz with a wavemeter, which is accurate enough to assign absolute values to an individual dot i.e. to determine the integer value $m$ for an individual dot in the formula $f_m = mf_{rep} \pm f_0$. The light of the frequency comb is sent to the VIPA spectrometer via a single-mode fiber. Using the same fiber for delivering the light from the single-mode laser to the VIPA spectrometer ensures perfect overlap of both laser beams. The light from the single-mode laser appears as a single dot on the CCD image, serving as a known reference marker. If needed the laser can be tuned using current or temperature, to generate extra reference markers, which provides additional information about the direction along which the dots need to be stitched.

In order to exploit the presence of many different wavelengths of the comb two methods have been developed. First, a distance can be derived from the phase change as a function of wavelength, which is called spectral interferometry or sometimes dispersive interferometry. Secondly, the knowledge of the wavelengths of the individual modes can be used. Each wavelength interferes with itself. With the many comb modes present homodyne interferometry takes place with all these modes simultaneously, providing a wealth of information on the distance to be determined. Both methods are worked out in the next two subsections.

### 6.5.1 Spectral interferometry

Spectral interferometry for distance measurement has been demonstrated in various configurations [11, 22]. With the repetition frequency being stabilized against the cesium atomic clock, an almost perfect frequency scale is available for determining the slope of the phase change with wavelength. The phase change as a function of frequency can be obtained from a cosine fit through the measurement data, or by unwrapping the phase by fast Fourier transform. The interference term can be written as:

$$I(\lambda) = I_0 \cos\left(\frac{2\pi \cdot 2L \cdot n}{\lambda}\right), \tag{6.11}$$

with $I_0$ the intensity of the light sent into the interferometer, $L$, the path length difference of the interferometer arms (single path), $\lambda$ the vacuum wavelength and $n$ the refractive index of the medium, usually air. The total accumulated phase can be written as:

$$\phi = \frac{4\pi Ln}{\lambda} = \frac{4\pi Lnf}{c}, \tag{6.12}$$

with $f$ the optical frequency and $c$ the speed of light in vacuum.

### 6.5.2 Spectral interferometry in dispersionless media

Two cases are considered. First the case in which dispersion can be neglected, i.e. $n \neq n(\lambda)$. In this case the distance can be written as:

$$L = \frac{\mathrm{d}\phi}{\mathrm{d}f}\frac{c}{4\pi n}. \tag{6.13}$$

The phase is determined a cosine fit as, described by $A + B\cos(C(p - D))$. The fitted phase is given by

$$\phi_{\mathrm{fit}} = C(p - D), \tag{6.14}$$

with $C$ and $D$ the fitting parameters and $p$ a label associated to a specific comb frequency via:

$$f_p = f_{\mathrm{rep}}(Q - p) + f_0, \tag{6.15}$$

with $Q$ a (large) integer number (several 100 000 for a GHz repetition rate). $p$ ranges from 1 to about 6000, corresponding to the number of comb modes taking into account. Considering that $f_{\mathrm{rep}}$ is nominally 1 GHz, a wavelength range of 813.5–827.5 nm is covered in this way. Note that the optical frequency decreases with increasing $p$. By expressing

$$p = Q - \left[\frac{f(p) - f_0}{f_{\mathrm{rep}}}\right], \tag{6.16}$$

and taking into account equation (6.14) the distance $L$ is related to the fitting parameter $C$:

$$L = \frac{\mathrm{d}\phi}{\mathrm{d}f}\frac{c}{4\pi n} = -\frac{C}{f_{\mathrm{rep}}} \cdot \frac{c}{4\pi n}, \tag{6.17}$$

with $\mathrm{d}\phi/\mathrm{d}f = -C/f_{\mathrm{rep}}$. Note that the sign for $C$ still needs to be determined. As can be directly seen from the cosine term in equation (6.11), it is not possible to distinguish positive and negative values of $L$. The interference patterns are identical in this case. This will be discussed in section 6.5.4.

### 6.5.3 Spectral interferometry in media with linear dispersion

In case dispersion cannot be neglected, e.g. for longer distances in air, the description needs to be extended a bit. Dispersion is taken into account via:

$$\frac{\mathrm{d}\phi}{\mathrm{d}f} = \frac{4\pi L}{c}\left[n + f\frac{\mathrm{d}n}{\mathrm{d}f}\right] = \frac{4\pi L}{c}\left[n + f\frac{\mathrm{d}n}{\mathrm{d}\lambda}\frac{\mathrm{d}\lambda}{\mathrm{d}f}\right]. \tag{6.18}$$

Using $\lambda = c/f$ and $\mathrm{d}\lambda/\mathrm{d}f = -c/f^2$, this can be rewritten as

$$\frac{\mathrm{d}\phi}{\mathrm{d}f} = \frac{4\pi L}{c}\left[n - \lambda\frac{\mathrm{d}n}{\mathrm{d}\lambda}\right] = \frac{4\pi L}{c}n_g. \tag{6.19}$$

The equation above is expressed in terms of the group index, which is given by:

$$n_g = n - \lambda\frac{\mathrm{d}n}{\mathrm{d}\lambda}. \tag{6.20}$$

This leads to the following expression for the distance $L$:

$$L = \frac{\mathrm{d}\phi}{\mathrm{d}f}\frac{c}{4\pi n_g} = -\frac{C}{f_{\mathrm{rep}}} \cdot \frac{c}{4\pi n_g}. \tag{6.21}$$

Note that if $n_g$ depends on $\lambda$, i.e. if there is group velocity dispersion, $\mathrm{d}\phi/\mathrm{d}f$ is not constant. If group velocity dispersion cannot be neglected the approach based on cosine fitting, having a fixed periodicity, cannot be applied anymore, since some chirp will occur as a result from the nonlinearity. In the time domain this leads to a deformation of the pulse shape. For the distance measured and the targeted uncertainty, group velocity dispersion can be neglected. This is equivalent to the assumption that the refractive index linearly depends on wavelength. It can easily be shown that in that case $n_g$ is wavelength-independent. As demonstrated by Cui [11] the effect of group velocity dispersion is far below 1 $\mu$m for a 50 m distance, so this assumption is valid for this range.

### 6.5.4 Determination of the sign of $L$

In the description of spectral interference discussed so far, the distance $L$ is purely determined from spectral interferometry, which leads to solutions:

$$-\frac{1}{2}L_{pp} < L < \frac{1}{2}L_{pp} \tag{6.22}$$

here $L_{pp}$ is the pulse-to-pulse distance in the medium, being defined as:

$$L_{pp} = \frac{c}{f_{\text{rep}} n_g}. \tag{6.23}$$

The fact that the range of $L$ is restricted to the range as given in equation (6.22), is a direct result of the periodicity of the pulse train. Here the method is different from conventional white light interferometry, as will be discussed later. As already indicated above, additional information is needed for the determination of the sign of $L$. There are several ways to obtain this information. For example, the sign can be retrieved from the observation of the change of the $\mathrm{d}\phi/\mathrm{d}f$ when moving the measurement arm towards the measurement position. Suppose that the measurement position is approached in the direction of increasing the interferometer arm length. When the fringe density decreases when approaching the measurement position $L_{pp}$ is negative; for increasing fringe density $L_{pp}$ is positive. Alternatively, one may change $f_{\text{rep}}$ (and thus $L_{pp}$) and observe the change of $\mathrm{d}\phi/\mathrm{d}f$. A third possibility is that $|L/2|$ exceeds the uncertainty of a rough measurement that is performed with another method. In that case the information from the rough measurement is sufficient to determine the sign of $L$. Considering equation (6.21), one finds that for $L > 0$, the coefficient $C < 0$.

### 6.5.5 Measuring distances exceeding the pulse-to-pulse distance

Due to the periodicity of the pulse train, the interference patterns are repeating themselves. Obviously, pulse overlap occurs when the total path length difference of the interferometer is at a multiple of the pulse-to-pulse distance:

$$L_t = \frac{1}{2}q\frac{c}{f_{\text{rep}} n_g} = \frac{1}{2}qL_{pp}, \tag{6.24}$$

with $q$ an integer and $L_t$ the one way path length difference. Filling this into equation (6.11), the phase in vacuum can then be written as:

$$\phi = 1/2\frac{4\pi q}{c} \cdot \frac{c}{f_{\text{rep}}} f = 2\pi q \frac{p f_{\text{rep}} + f_0}{f_{\text{rep}}} = 2\pi q p + 2\pi q \frac{f_0}{f_{\text{rep}}}. \tag{6.25}$$

From this equation one sees that for overlapping pulses the phase difference between neighboring modes i.e. the modes $k$ and $k + 1$ equals $2\pi q$. Hence at $L_t = qL_{pp}$, the phase difference between all modes is a multiple of $2\pi$ and modes are thus in phase. Note that the phase is only equal to 0 if $f_0$ equals 0. For $f_0 \neq 0$, successive pulses are not fully identical due to the carrier–envelope phase slip. This leads to a phase of $2\pi q f_0/f_{\rm rep}$ for distances at a multiple of $L_{pp}$, which is thus equal for all modes. For distance measurement based on spectral interferometry, the distance is determined from the $\mathrm{d}\phi/\mathrm{d}f$, which equals 0 at distances that are multiples of $L_{pp}$. An arbitrary distance is written as:

$$L_t = \frac{1}{2}qL_{pp} + L. \tag{6.26}$$

To determine $L_t$, the integer $q$ needs to be known. Therefore it is necessary to know the distance to be measured with an accuracy better than $L_{pp}/2$. Furthermore it is necessary to determine the sign of $L$, as described above.

Similarly, the phase difference between neighboring modes can also be determined for propagation in a medium with linear dispersion. For simplicity and without loss of generality $f_0 = 0$ is considered here. In this case the phase difference between neighboring modes for overlapping pulses at an integer value of $q$, can be written as:

$$\phi_k - \phi_{k+1} = 2\pi 4\pi L\left(\frac{n_k}{\lambda_k} - \frac{n_{k+1}}{\lambda_{k+1}}\right) = 2\pi q. \tag{6.27}$$

In air the distance $L_t$ can then be written as:

$$L_t = \frac{q/2}{\frac{n_k f_k}{c} - \frac{n_{k+1} f_{k+1}}{c}} = \frac{q/2c}{n_k f_k - (n_k + \Delta n)(f_k + f_{\rm rep})} \tag{6.28}$$

$$= \frac{qc}{-2[n_k f_{\rm rep} + \Delta n f_k + \Delta n f_{\rm rep}]} = \frac{qc}{2[f_{\rm rep}(n_k + f_k\frac{\Delta n}{f_{\rm rep}} + \Delta n)]} \tag{6.29}$$

$$= -\frac{qc}{2f_{\rm rep}(n_g + \Delta n)}, \tag{6.30}$$

with

$$n_g = n - \lambda\frac{\mathrm{d}n}{\mathrm{d}\lambda} = n + f\frac{\mathrm{d}n}{\mathrm{d}f}. \tag{6.31}$$

Since

$$\Delta n = f_{\rm rep} \cdot \frac{\mathrm{d}n}{\mathrm{d}f} \ll n_g \tag{6.32}$$

this can be written as:

$$L_t = \frac{qc}{2f_{\text{rep}}n_g} = \frac{q}{2}L_{pp}. \quad (6.33)$$

The value of $\Delta n$ is smaller than $10^{-9}$ for air, so can be considered negligible. Note that for spectral interferometry the value of $f_0$ does not need to be known for distance determination.

### 6.5.6 Frequency comb interferometry versus white light interferometry

Frequency comb interferometry can be considered as a special case of white light interferometry. However, the available spectrum is not continuous, but periodic (the comb). With conventional white light interferometry fringes get closer and closer for increasing path length difference. The resolution of the spectrometer determines the maximum distance that can be measured. However, since the spectrum of a comb is discrete, at some point a case of under sampling, or aliasing occurs. This happens for path length differences exceeding $L_{pp}/2$. When $L = L_{pp}/2$ the pulses have maximum separation in the time domain corresponding to a phase difference of $\pi$ between neighboring modes. The Nyquist frequency is exceeded when the pulse separation is further increased and the comb doesn't contain enough frequency components that would be needed to map the continuous white light interference spectrum. In fact, one profits here from the fact that the spectrum is discrete and aliasing occurs, since the measurement range is now no longer limited by the resolution of the spectrometer as would be the case with a continuous spectrum. In the time domain this can be understood from the fact that a pulse train is used, with the result that the maximum distance between pulses arriving at the spectrometer output will never exceed $L_{pp}/2$.

### 6.5.7 Homodyne interferometry

So far, only the repetition rate has been used to derive a distance from the interference pattern. However, since the frequency comb is fully stabilized i.e. both $f_{\text{rep}}$ and $f_0$ are locked to the atomic clock, all wavelengths are also accurately known. This is additional information that can be used for more accurate determination of the distance. By combining equations (6.12) and (6.16), the total accumulated phase $\phi_p$ for a particular comb wavelength $\lambda_p$ can be written as:

$$\phi_p = -\frac{2 \cdot 2\pi n_p L f_{\text{rep}}}{c}\left[p - (Q + \frac{f_0}{f_{\text{rep}}})\right], \quad (6.34)$$

with $n_p$ the (phase) refractive index corresponding to the wavelength $\lambda_p$. Here a cosine has been fitted through the spectral interferometry data using the parametrized phase as given in equation (6.14). Note the $\phi_p$ and $\phi_{\text{fit}} = C(p - D)$ are not identical. Only $\cos\phi_p = \cos\phi_{\text{fit}}$, which means that $[\phi_p \bmod 2\pi] = [C(p - D) \bmod 2\pi]$. Or, in other words:

$$\phi_p = 2\pi q_p + [C(p - D) \bmod 2\pi], \tag{6.35}$$

with $q_p$ an integer. Again note that for $L > 0$, i.e. measurement arm > reference arm, $C < 0$ should be chosen. This is to be expected, since the total accumulated phase $\phi_p$ decreases with increasing wavelength $\lambda_p$. When plotting the total phase against label $p$, one sees that the phase is decreasing with increasing $p$, which is correct, since wavelength increases with $p$.

To find the distance as measured by a particular wavelength $\lambda_p$, one uses $\phi_p = Cp - CD \bmod 2\pi$ as determined from the curve fit, combined with the integer number $q_p$, which is found from the distance based on spectral interferometry. The accuracy of the spectral interferometry measurement is good enough to determine $q_p$ unambiguously. This is confirmed by the fact that usually the integer value is clearly approximated (deviation < 0.2) and varying $q_p$ with ±1 leads to inconsistent results between different wavelengths. The total measured distance can then be written as:

$$L_p = \frac{\lambda_p}{2} q_p + \frac{1}{2\pi}[C(p - D) \bmod 2\pi]. \tag{6.36}$$

### 6.5.8 Summary of mode-resolved distance measurement

To summarize, the total distance determination consists of three measurement steps. First, a rough pre-measurement is needed to determine the distance within a fraction of the pulse-to-pulse distance $L_{pp}$. Secondly, the distance is determined more accurately using spectral interferometry, as described in the previous section.

The distance determination based on spectral interferometry is described in equation (6.21). This equation can be rewritten as: $L = -C/2\pi \cdot \Lambda/2n_g$. With $\Lambda$ the synthetic wavelength of neighboring comb modes, which is equal to $L_{pp} = \Lambda/n_g$.

The measurement of a total distance $L_t$ can be rewritten as:

$$L_t = \frac{\Lambda}{2n_g}(q - C/2\pi), \tag{6.37}$$

with $C$ negative for $L > 0$, as discussed in subsection 6.5.4 and $q$ the integer number of $L_{pp}$ as determined from the rough measurement. This rough measurement needs to provide information with an accuracy of $\Lambda/2$, which is 15 cm in our case.

The third step is to determine the distance $L$ by using the additional information of the accurately known optical wavelengths in the combs, giving

$$L_p = \frac{\lambda_p}{2n_p} q_p + \frac{1}{2\pi}[C(p - D) \bmod 2\pi], \tag{6.38}$$

which is equal to the equation above. Note that $q_p >> q$, since $\lambda_p << \Lambda$. As a result the measurement uncertainty resulting from equation (6.38) is expected to be smaller than the measurement uncertainty resulting from equation (6.37), since the uncertainty is mainly determined by the uncertainty on the fitting parameters $C$ and $D$. For a practical measurement other contributions to the measurement uncertainty arise, like wavelength stability and the stability of the interferometer itself. Furthermore, the knowledge of the refractive index of air can be a limit to the best achievable absolute accuracy.

### 6.5.9 Results of mode-resolved frequency comb interferometry

Figure 6.17 shows the experimental setup for mode-resolved interferometry in more detail and includes the reference HeNe laser, which is a counting interferometer and is used to compare the measurement results. In the experiment shown here, distances up to 50 m have been measured. In figure 6.18 typical results of mode-resolved interferometry with a frequency comb laser are shown (taken from [44]). For this experiment a Ti:sapphire laser has been used with a spectral range from 813.5 nm to 827.5 nm, containing about 6000 comb modes. The figure shows several patterns of spectral interference at various distances between the pulses. These patterns repeat for displacements exceeding $L_{pp}/2$.

Following the data analysis as described in the sections above, the distance has been retrieved based on the phase change as a function of frequency and the refractive index calculated from the environmental parameters using Edlén's formula [5]. Figure 6.19 shows the results of the distance measurements up to 50 m, comparing the comb and HeNe results.

### 6.5.10 Considerations on measurement uncertainty

As mentioned above, the dominant uncertainty contribution for absolute distance measurement is the limited knowledge of the refractive index of air. The uncertainty on the environmental parameters i.e. temperature, pressure and humidity and the

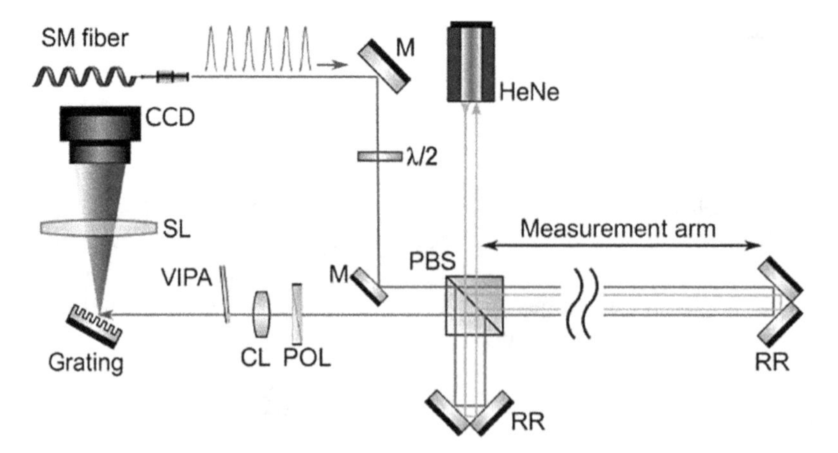


**Figure 6.17.** Schematic overview of the measurement setup for comparing distance measurement up to 50 m with a counting helium-neon laser and a frequency comb. The comb light is delivered to the setup with a single-mode (SM) fiber, providing a clean mode profile. Both the HeNe laser (orange line) and comb laser (red line) measure the displacement quasi-simultaneously. PBS: polarizing beam splitter, POL: polarizer, $\lambda/2$: half-waveplate, CCD: charge-coupled device camera, M: planar mirror, RR: hollow retroreflector, CL: cylindrical lens (100 mm focal length), SL: spherical lens (400 mm focal length). Figure taken from [44]. CC BY 4.0.

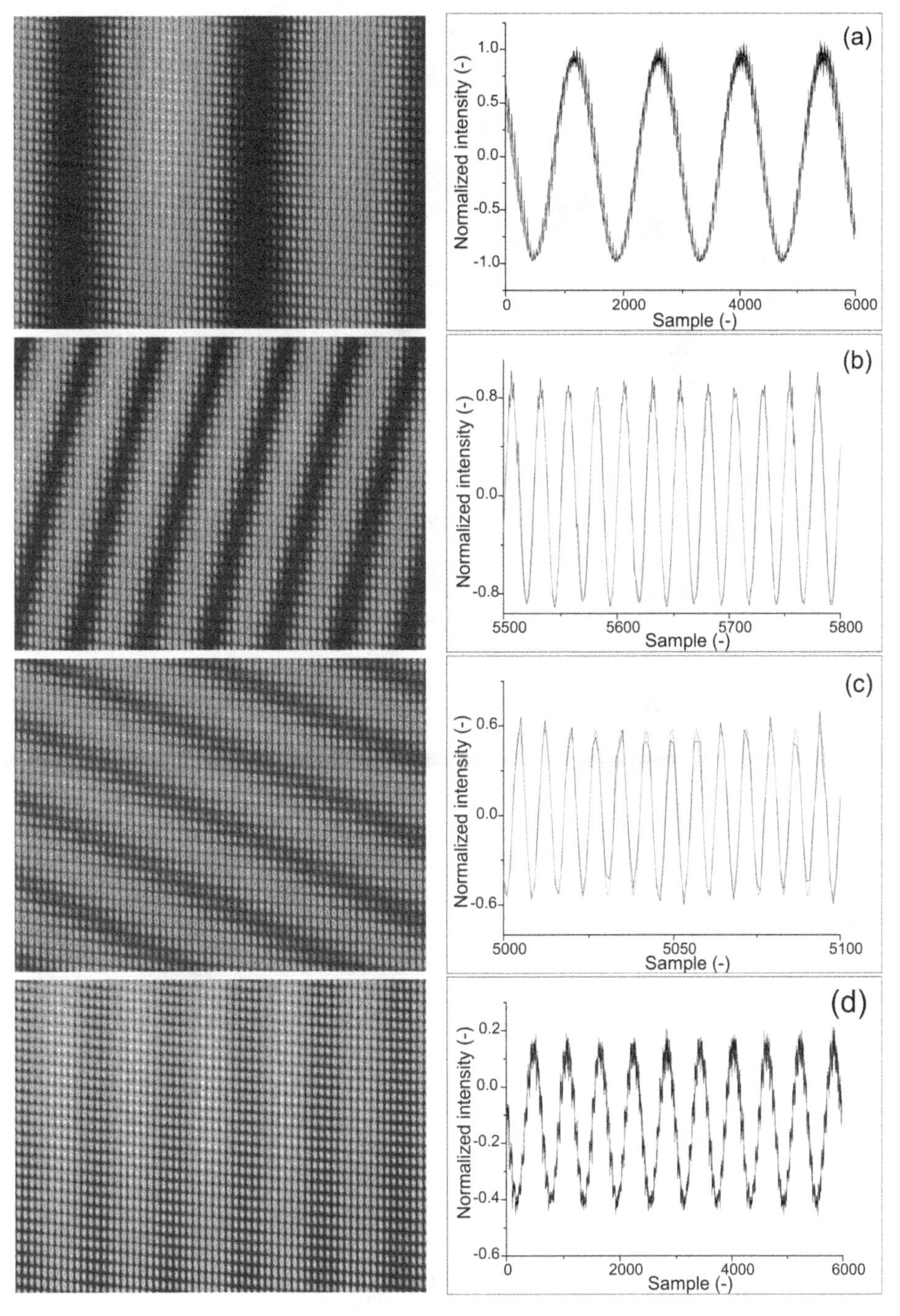


**Figure 6.18.** Spectral interference as measured with the VIPA spectrometer for distances of $\approx$ 0 m (a), 5 mm (b), 20 m (c) and 50 m (d), respectively. Spectral interferometry images are shown on the left. For clarity only 1/4 of the CCD chip area has been selected. On the right side the reconstructed spectra are shown. The $x$-axis with the sample number, has been scaled differently for these graphs to clearly visualize the fringes. Figure taken from [44]. CC BY 4.0.

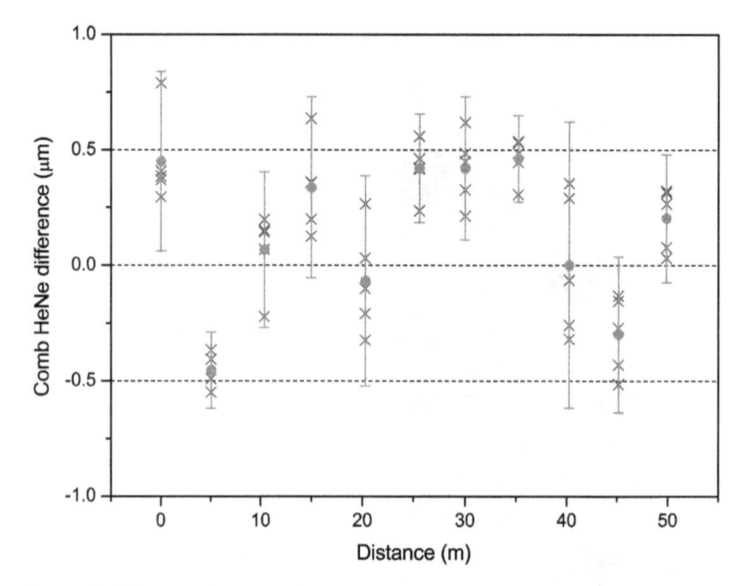


**Figure 6.19.** Observed differences between the frequency comb distance measurement, based on spectral interferometry and the counting laser interferometer. The error bars indicate twice the standard deviation over the five measurements. Figure taken from [44]. CC BY 4.0.

uncertainty of the Edlén formula itself. Even for accurate measurements of environmental parameters e.g. with uncertainties of 0.3 °C for air temperature, 7 Pa for pressure and 1.3% for relative humidity, respectively, the uncertainty on refractive index is $3.1 \times 10^{-7}$. In the comparison between comb and HeNe laser only the uncertainty on the refractive index *difference* for the comb and HeNe wavelengths is relevant. Based on the uncertainties mentioned above this uncertainty contribution is about $1.7 \times 10^{-9}$. In the experiments presented here, the dominant uncertainty contribution comes from vibration of the long interferometer and air turbulence. These come into play because the data-acquisition times of both methods are not equal, creating sensitivity to path length changes that occur within 0.1 s. The associated uncertainty is about 0.3 $\mu$m.

It is interesting to consider the potential of homodyne many-wavelength for quiet environments, like space. The uncertainty on the fitting parameter $C$ is typically $u_C < 10^{-6}$ for distances close to zero delay ($C \ll 1$). For spectral interferometry this leads to an uncertainty contribution to the measured distance of $10^{-7} \times \Lambda$ as seen in equation (6.37), which is about 30 nm. The other uncertainty contribution comes from the stability of the clock and is proportional to the distance measured as seen again in equation (6.37), which can be as small as 1 nm on a km distance for a high quality clock and sufficiently long measurement time. When considering equation (6.38), one sees that potentially much lower uncertainties can be achieved when the distance is determined based on the accurately known wavelengths of each comb mode. In this case the uncertainty contribution to the measurement due to the uncertainty of $C$ is much smaller: $10^{-7} \times \lambda \ll 1$ nm and the main limitation to the distance measurement comes from the wavelength accuracy, i.e. it is only based on the quality of the reference clock.

## 6.6 Recent developments and outlook

Although the measurement method as outlined in this chapter is very powerful, widespread practical implementation requires both cost reduction and easy field operation. The ongoing development of frequency comb technology will enable simplification of the mode-resolved homodyne measurement. It may be advantageous to work with a widely spaced frequency comb to relax the requirements on the resolution of the spectrometer. Note that a trade-off between range of non-ambiguity and spectral resolution needs to be made here. Having a large mode spacing of e.g. 10 s of GHz allows for the use of simple and affordable array spectrometers for detection. However, the range of non-ambiguity decreases to a few mm, so the requirements on initial knowledge of the distance to be measured gets tighter. As an example of distance measurement with a high repetition rate frequency comb, distance measurements have been performed with an 'astro-comb', i.e. a mode filtered comb to increase the mode spacing [28]. The name 'astro-comb' refers to the original application of such a widely spaced comb: the calibration of the frequency scale of an astronomical spectrograph. In the current example the frequency comb is generated with a 1 GHz Ti:sapphire laser, which is filtered with a cavity that transmits every 56th mode. A schematic overview of this setup is shown in figure 6.20. Again distance measurements up to 50 m have been performed, but with a linear array spectrometer, instead of a VIPA spectrometer. The measurement results are very similar, with an agreement between HeNe laser and frequency comb within 1 $\mu$m for distances up to 50 m.

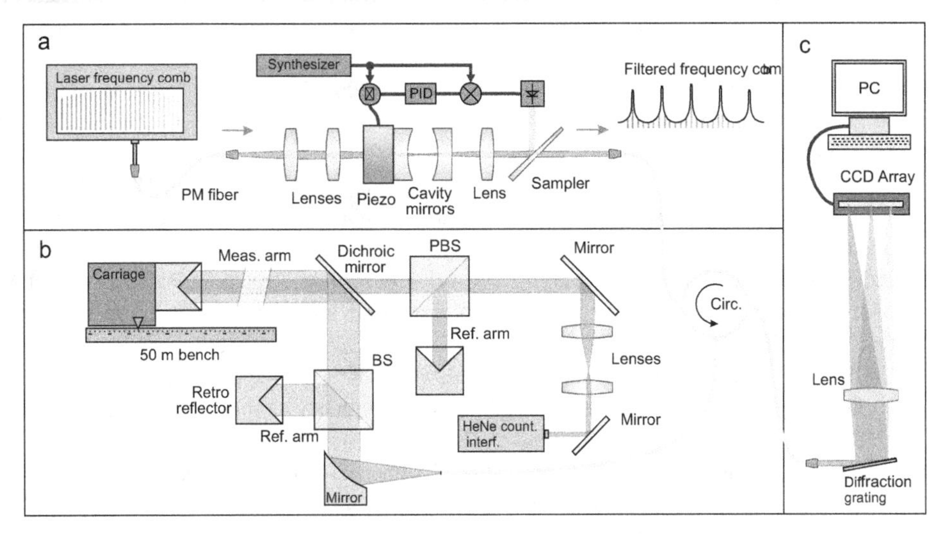


**Figure 6.20.** Experimental setup split into three sections. (a) The filtering part showing the frequency comb laser, cavity mode-matching optics, cavity with piezo actuator, collimation lens and beam sampler. The green parts are the detector and electronics for locking the cavity length. The filtered frequency comb is led through an optical circulator towards the 50 m bench part. (b) This part shows two Michelson interferometers with a common measurement arm for the filtered frequency comb and the HeNe laser. (c) Spectral analysis of the interferometer output with a diffraction grating and linear CCD camera. Taken from [20]. CC BY 4.0.

There has been a strong effort to develop miniature frequency comb sources, both based on semiconductors [27, 47] and based on optically pumped microcavities [13]. Such developments will enable field applications of multi-wavelength homodyne or heterodyne comb interferometry.

## References


[1] Adler F, Masłowski P, Foltynowicz A, Cossel K C, Briles T C, Hartl I and Ye J 2010 Mid-infrared fourier transform spectroscopy with a broadband frequency comb *Opt. Express* **18** 21861–72

[2] Balling P, Křen P, Mašika P and van den Berg S A 2009 Femtosecond frequency comb based distance measurement in air *Opt. Express* **17** 9300–13

[3] Balslev-Clausen D 2007 Broad band cavity enhanced direct frequency comb spectroscopy *Master's thesis* University of Copenhagen & University of Colorado

[4] Birch K P and Downs M J 1993 Correction to the updated edlén equation for the refractive index of air *Metrologia* **30** 155

[5] Bönsch G and Potulski E 1998 Measurement of the refractive index of air and comparison with modified Edlén's formulae *Metrologia* **35** 133

[6] Braat J J M 2003 Absolute distance measurement using a locked femto-second laser unpublished

[7] Ciddor P E 1996 Refractive index of air: new equations for the visible and near infrared *Appl. Opt.* **35** 1566–73

[8] Ciddor P E and Hill R J 1999 Refractive index of air. 2. group index *Appl. Opt.* **38** 1663–67

[9] Coen S, Chau A H L, Leonhardt R, Harvey J D, Knight J C, Wadsworth W J and Russel P St 2001 White-light supercontinuum generatrion with 60-ps pump pulses in a photonic crystal fiber *Opt. Lett.* **26** 1356

[10] Cui M, Schouten R, Bhattacharya N and Berg S 2008 Experimental demonstration of distance measurement with a femtosecond frequency comb laser *J. Eur. Opt. Soc. Rapid Publ.* **3** 08003

[11] Cui M, Zeitouny M G, Bhattacharya N, van den Berg S A and Urbach H P 2011 Long distance measurement with femtosecond pulses using a dispersive interferometer *Opt. Express* **19** 6549–62

[12] Cui M, Zeitouny M G, Bhattacharya N, van den Berg S A, Urbach H P and Braat J J M 2009 High-accuracy long-distance measurements in air with a frequency comb laser *Opt. Lett.* **34** 1982–84

[13] Del'Haye P, Arcizet O, Schliesser A, Holzwarth R and Kippenberg T J 2008 Full stabilization of a microresonator-based optical frequency comb *Phys. Rev. Lett.* **101** 053903

[14] Diddams S A, Hollberg L and Mbele V 2007 Molecular fingerprinting with the resolved modes of a femtosecond laser frequency comb *Nature* **445** 627–30

[15] Diddams S A 2010 The evolving optical frequency comb [invited] *J. Opt. Soc. Am.* B **27** B51–62

[16] Giacomo P 1984 News from the bipm *Metrologia* **20** 25

[17] Hall J L 2006 Nobel lecture: Defining and measuring optical frequencies *Rev. Mod. Phys.* **78** 1279–95

[18] Hänsch T W 2006 Nobel lecture: Passion for precision *Rev. Mod. Phys.* **78** 1297–309

[19] Holzwarth R, Udem T h, Haensch, Knight T, Wadsworth J C and Russell W 2000 Optical frequency synthesizer for precision spectroscopy *Phys. Rev. Lett.* **85** 2264–67

[20] Hyun S, Kim Y-J, Kim Y, Jin J and Kim S-W 2009 Absolute length measurement with the frequency comb of a femtosecond laser *Meas. Sci. Technol.* **20** 095302

[21] Jin J, Kim Y-J, Kim Y, Kim S-W and Kang C-S 2007 Absolute length calibration of gauge blocks using optical comb of a femtosecond pulse laser *Opt. Express* **14** 5968–74

[22] Joo K-N and Kim S-W 2006 Absolute distance measurement by dispersive interferometry using a femtosecond pulse laser *Opt. Express* **14** 5954–60

[23] Joo K-N, Kim Y and Kim S-W 2008 Distance measurements by combined method based on a femtosecond pulse laser *Opt. Express* **16** 19799–806

[24] Jost J D, Hall J L and Ye J 2002 Continously tunable, precise, single frequency optical signal *Opt. Express* **10** 515

[25] Kim S-W 2009 Combs rule *Nat. Photon* **3** 313–14

[26] Knight J C, Birks T A, Russell P J St and Atkin D M 1996 All-silica single-mode optical fiber with photonic crystal cladding *Opt. Lett.* **21** 1547–49

[27] Latkowski S, Moskalenko V, Tahvili S, Augustin L, Smit M, Williams K and Bente E 2015 Monolithically integrated 2.5 ghz extended cavity mode-locked ring laser with intracavity phase modulators *Opt. Lett.* **40** 77–80

[28] Lešundák A, Voigt D, Cip O and van den Berg S 2017 High-accuracy long distance measurements with a mode-filtered frequency comb *Opt. Express* **25** 32570–80

[29] Margolis H S, Barwood G P, Huang G, Klein H A, Lea S N, Szymaniec K and Gill P 2004 Hertz-level measurement of the optical clock frequency in a single 88sr. ion *Science* **306** 1355–58

[30] Minoshima K and Matsumoto H 2000 High-accuracy measurement of 240-m distance in an optical tunnel by use of a compact femtosecond laser *Appl. Opt.* **39** 5512–17

[31] Newbury N R 2011 Searching for applications with a fine-tooth comb *Nat. Photon.* **5** 186–88

[32] Quinn T J 2003 Practical realization of the definition of the metre, including recommended radiations of other optical frequency standards (2001) *Metrologia* **40** 103–33

[33] Russell P 2003 Photonic crystal fibers *Science* **299** 358–62

[34] Salvadé Y, Schuhler N, Lévêque S and Le Floch S 2008 High-accuracy absolute distance measurement using frequency comb referenced multiwavelength source *Appl. Opt.* **47** 2715–20

[35] Schnatz H, Lipphardt B, Helmcke J, Riehle F and Zinner G 1996 First phase-coherent frequency measurement of visible radiation *Phys. Rev. Lett.* **76** 18

[36] Schnell U, Zimmermann E and Dandliker R 1995 Absolute distance measurement with synchronously sampled white-light channelled spectrum interferometry *Pure Appl. Opt.: J. Eur. Opt. Soc. Part A* **4** 643

[37] Schuhler N, Salvadé Y, Lévêque S, Dändliker R and Holzwarth R 2006 Frequency-comb-referenced two-wavelength source for absolute distance measurement *Opt. Lett.* **31** 3101–03

[38] Shirasaki M 1996 Large angular dispersion by a virtually imaged phased array and its application to a wavelength demultiplexer *Opt. Lett.* **21** 366–68

[39] Spence D E, Kean P N and Sibbett W 1991 60-fsec pulse generation from a self-mode-locked ti:sapphire laser *Opt. Lett.* **16** 42–4

[40] Steinmetz T *et al* 2008 Laser frequency combs for astronomical observations *Science* **321** 1335–37

[41] Thorpe M 2009 Cavity-enhanced direct frequency comb spectroscopy *PhD Thesis* University of Colorado

[42] Udem T h, Reichert J, Holzwarth R and Hänsch T W 1999 Absoluted optical frequency measurement of the cesium $d_1$ line with a mode-locked laser *Phys. Rev. Lett.* **82** 3586
[43] van den Berg S A, Persijn S T, Kok G J P, Zeitouny M G and Bhattacharya N 2012 Many-wavelength interferometry with thousands of lasers for absolute distance measurement *Phys. Rev. Lett.* **108** 183901
[44] van den Berg S A, van Eldik S and Bhattacharya N 2015 Mode-resolved frequency comb interferometry for high-accuracy long distance measurement *Sci. Rep.* **5** 14661
[45] Ye J 2004 Absolute measurement of a long, arbitrary distance to less than an optical fringe *Opt. Lett.* **29** 1153–55
[46] Yoon T, Ye J, Hall J and Chartier J-M 2001 Absolute frequency measurement of the iodine-stabilized he-ne laser at 633 nm *Appl. Phys.* B **72** 221–26
[47] Zaugg C A *et al* 2014 Gigahertz self-referenceable frequency comb from a semiconductor disk laser *Opt. Express* **22** 16445–55
[48] Zeitouny M G, Cui M, Bhattacharya N, Urbach H P, van den Berg S A and Janssen A J E 2010 From a discrete to a continuous model for interpulse interference with a frequency-comb laser *Phys. Rev.* A **82** 023808
[49] Zeitouny M G, Cui M, Janssen A J E M, Bhattacharya N, van den Berg S A and Urbach H P 2011 Time-frequency distribution of interferograms from a frequency comb in dispersive media *Opt. Express* **19** 3406–17

**IOP** Publishing

Modern Interferometry for Length Metrology
Exploring limits and novel techniques
**René Schödel (Editor)**

# Chapter 7

# Distance measurements using mode-locked pulse lasers

**Seung-Woo Kim and Yoon-Soo Jang**

## 7.1 Introduction

Optical interferometry has long been leading the progress of precision length metrology using continuous-wave lasers [1]. Meanwhile, mode-locked lasers are being made available as a versatile light source capable of responding to the ever-growing demands on the measurement precision and functionality beyond the capability of conventional lasers [2]. Especially, the optical spectrum of a mode-locked laser, referred to as the frequency comb, acts as the ruler permitting ultra-stable wavelengths to be produced for precision interferometry with traceability to the atomic clock. In addition, mode-locked lasers are employed directly as the light source offering ultrashort pulses, of which the time-of-flight in air or space can be detected with unprecedented precision for long distance metrology. As a result, mode-locked lasers are now ready to take on the advanced role of length metrology by providing unique temporal and spectral benefits over conventional lasers or other broad spectrum light sources.

Most laser-based distance interferometers commercially available today are capable of providing a sub-nanometer resolution in measuring distances up to tens of meters [3, 4]. This superb capability is achieved by resolving the interference phase using the homodyne or heterodyne phase-measuring technique with the aid of high speed electronics [5, 6]. However, when a single-wavelength laser is used as the light source, the non-ambiguity range (NAR) of an interferometric measurement is confined to just half of the used light wavelength. This limitation is overcome by accumulating the interference phase continuously while moving the target mirror all the way to a given position from a zero-reference datum point. This incremental way of distance measurement is often not permissible in reality, particularly when a long distance has to be measured without a usable motion guideway for the target mirror.

doi:10.1088/2053-2563/aadddcch7  

The concept of absolute distance measurement aims to determine a distance with a single measurement by synthesizing an extended non-ambiguity range. As part of laser-based distance interferometry, much work has been done to develop effective methods of absolute distance measurement. As a result, quite a few methods have been made available, including the time-of-flight measurement using pulsed light [7], continuous-wave laser intensity modulation [8, 9], wavelength scanning using a frequency-tunable laser [10], and incorporation of multiple lasers of different wavelengths [11]. These methods permit measuring long distances with an extensive non-ambiguity range, but their performance is not yet comparable to that of single-wavelength interferometry, particularly in terms of the measurement precision and uncertainty.

From the viewpoint of optical interferometry, a mode-locked laser is distinguished from conventional lasers in its distinct temporal and spectral characteristics as illustrated in figure 7.1. In the time domain, a mode-locked laser is seen as a train of ultrashort pulses emitted at a radio-frequency repetition rate ($f_r$). The pulse duration falls into the picosecond or even femtosecond range. In the frequency domain, its optical spectrum is dubbed the frequency comb as it is comprised of numerous discrete frequency modes equally spaced with a spacing of $f_r$. The spectral

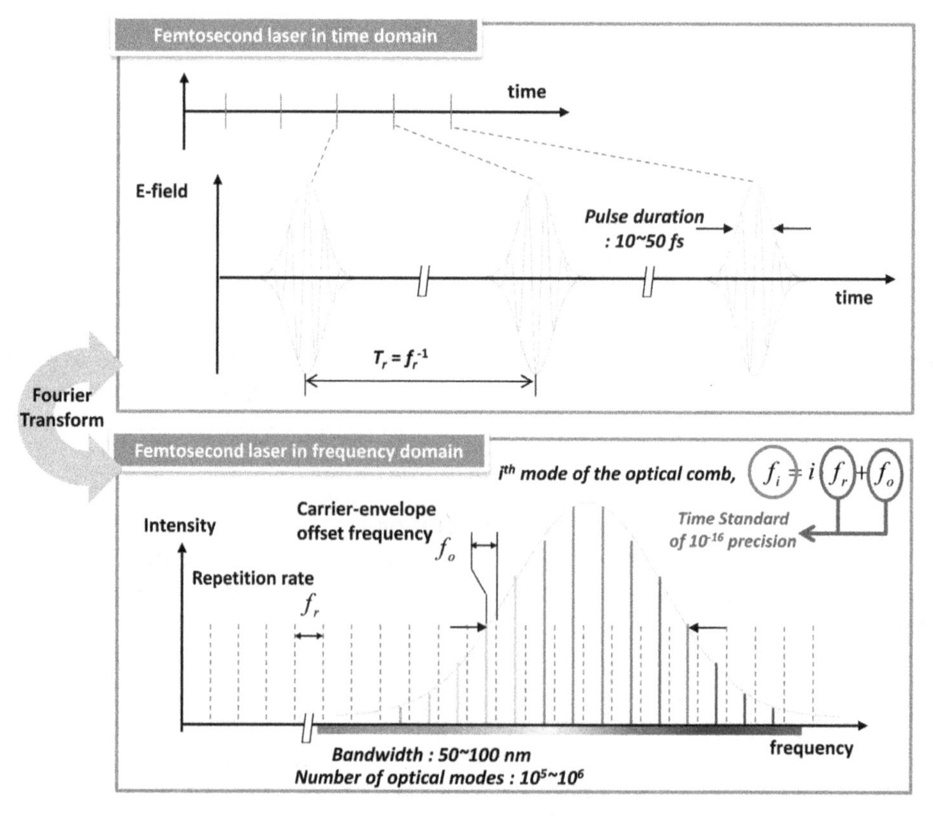


**Figure 7.1.** Temporal and spectral characteristics of a mode-locked laser emitting ultrashort pulses at a radio-frequency repetition rate.

width of the frequency comb relates to the pulse duration as a Fourier-transform pair. Each mode within the frequency comb has a center frequency expressed as $f_i = i \times f_r + f_o$ with the integer $i$ denoting the individual mode number and $f_o$ being the carrier-envelope offset frequency. The repetition rate $f_r$ is readily detected using a photo-detector, while $f_o$ is identified using a self-referencing $f$–$2f$ interferometer [12]. Both $f_r$ and $f_o$ fall in the radio-frequency regime, so they can be phase locked to the micro-wave atomic clock operating on the erbium or cesium absorption limes. This leads to a collective stabilization of the frequency comb, encompassing all the modes in it.

This chapter describes how the above-mentioned temporal and spectral characteristics of mode-locked lasers can be exploited for the advance of length metrology with particular focus on absolute distance measurements. In this regard, section 7.2 explains multi-wavelength interferometry performed by generating multiple wavelengths using the frequency comb as the ruler. Section 7.3 deals with dispersive interferometry to determine the absolute value of a distance by Fourier-transform-based analysis of the broad band interferogram obtained using a mode-locked laser as the light source. Section 7.4 presents time-of-flight measurement of ultrashort pulses by means of optical cross-correlation detection with unprecedented timing accuracy. Finally, section 7.5 give details how dual-comb interferometry can be exploited to measure the time-of-flight of ultrashort pulses in real time by optical frequency down-conversion to the radio-frequency region.

## 7.2 Comb-referenced multi-wavelength interferometer

Figure 7.2 illustrates the principle of multi-wavelength interferometry (MWI) to determine a given target distance by employing multiple wavelengths $\lambda_i$ for $i$ = 1, 2, …, $N$. The target distance $L$ for each wavelength is expressed as $L_i = (\lambda_i/2)(m_i + e_i)$

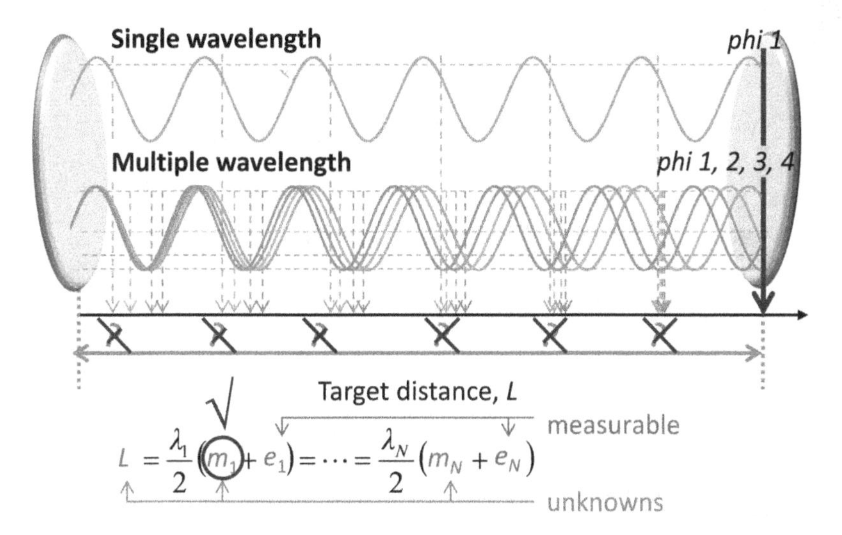


**Figure 7.2.** Basic concept of multi-wavelength interferometer (MWI) for absolute distance measurement. Phi; interference phase.

with $m_i$ and $e_i$ being an integer value and an excess fraction, respectively. The excess fraction $e_i$ is measured for each wavelength $\lambda_i$ individually in the same way as in single-wavelength interferometry. Then the unknown integer value $m_i$ is determined by numerical iteration so as to minimize the total error sum of $\Sigma|L - L_i|$ for all $\lambda_i$. This method of multi-wavelength interferometry permits the non-ambiguity range to be extended but requires each constituent wavelength to maintain high temporal stability with a well certified nominal value [13, 14]. Otherwise, the measured distance $L$ would result in a largely augmented amount of error fluctuating severely in the time domain.

Figure 7.3 a multi-wavelength interferometer configured by generating four wavelengths with reference to the frequency comb [15]. Each wavelength is produced by selecting a distributed feedback (DFB) diode laser of 10 mW optical power. By controlling the input current, the output frequency of the DFB diode laser is phase-locked to a mode within the frequency comb to produce its preassigned wavelength. The frequency comb is constructed from an Er-doped fiber oscillator of a 1550 nm center wavelength. As the ruler, the frequency comb is stabilized to the Rb atomic clock by locking the repetition rate $f_r$ to 100 MHz and the carrier-envelope-offset frequency $f_o$ to 30 MHz [16, 17]. The Rb atomic clock provides a fractional stability of $2.57 \times 10^{-11}$ at 1 s averaging, so the generated wavelength has a linewidth of $\sim 1.0 \times 10^{-6}$ nm. This is applied to all the wavelengths from four DFB diode lasers at 1530.279 693 nm, 1531.040 888 nm, 1554.179 409 nm and 1554.937 151 nm, respectively.

The four wavelengths contained in four DFB diode lasers are delivered through a single-mode fiber to the Michelson type interferometer. The interference phase of each wavelength is detected by providing a 40 kHz heterodyne beat using a pair of acousto-optic modulators (AOMs). The four wavelengths are separated using a fiber Bragg-grating (FBG) array, of which the interference phases are detected simultaneously using a multi-channel phase meter with 0.1° resolution. For comparison

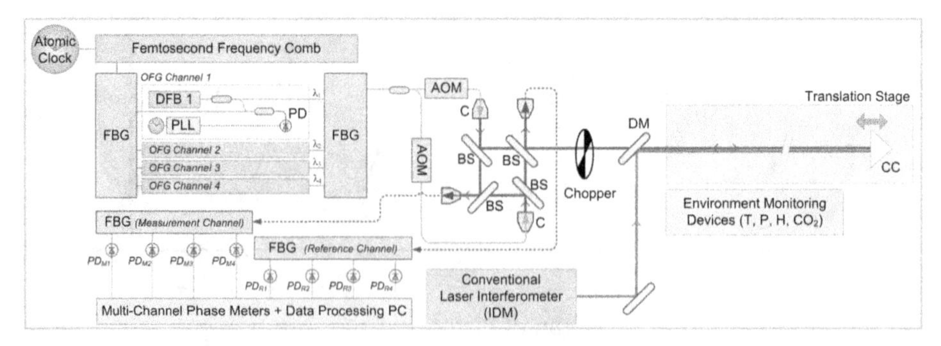


**Figure 7.3.** Hardware configuration to implement multi-wavelength interferometry with four wavelengths generated with reference to the frequency comb stabilized to the Rb atomic clock. AOM: acousto-optic modulator, DFB: distributed feedback diode laser, DM: dichroic mirror, C: fiber coupler, CC: corner cube retroreflector, PLL: phase-locked loop control, PD: photo-detector, FBG: fiber Bragg grating, BS: beam splitter, IDM: incremental distance measurement, PC: personal computer, T: temperature, P: pressure, H: relative humidity.

purpose, an incremental type commercial HeNe laser interferometer (IDM) is installed along the same optical path. The refractive index of air is estimated using the Ciddor's formula by monitoring the ambient air temperature, relative humidity, $CO_2$ concentration, and air pressure [18–20]. The target mirror is installed on a granite air-bearing stage to move over a 3 m translation range. The whole interferometer system is contained inside a 10 mm thick acrylic shield chamber to minimize the environmental change during measurement.

Figure 7.4 presents a test result of the multi-wavelength interferometer configured in figure 7.3, which demonstrates the real-time capability of absolute distance measurement (ADM). During the test, the target mirror was positioned at a 3.8 m distance while the measurement beamline was interrupted by activating a mechanical chopper on and off periodically at a 2 Hz switching rate. For comparison, the

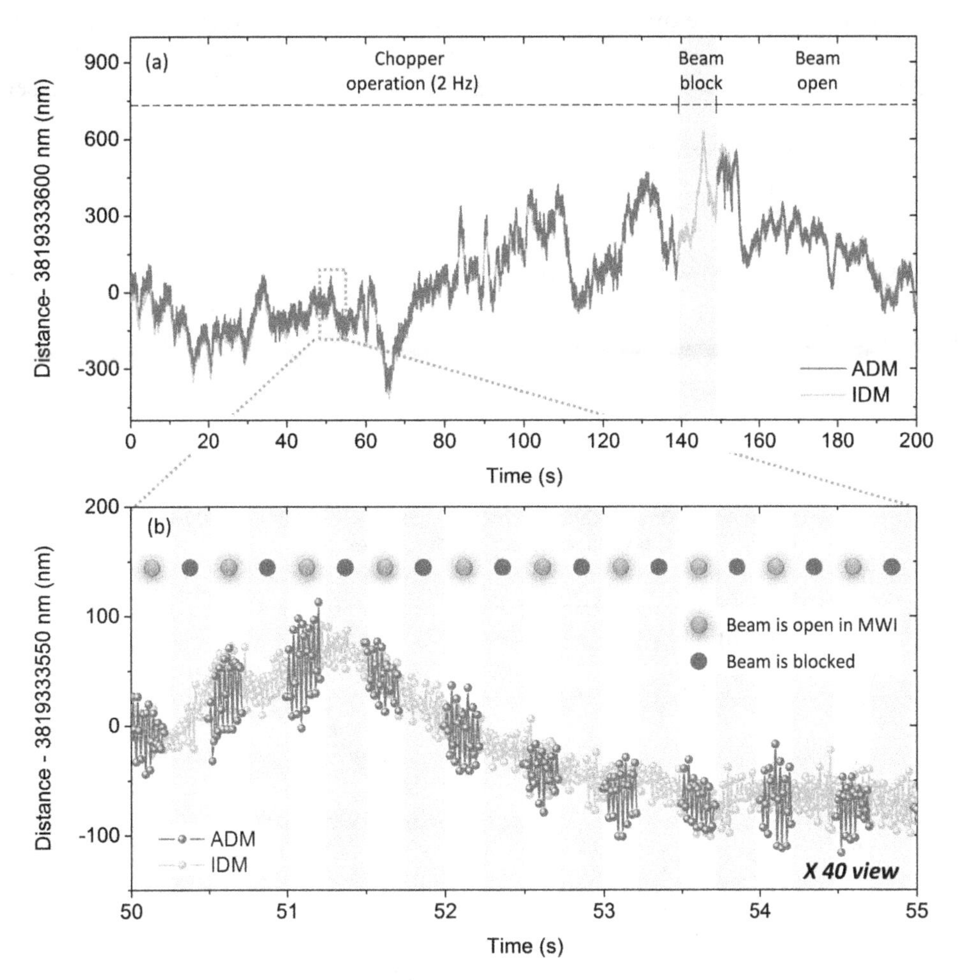


**Figure 7.4.** Real-time capability of multi-wavelength interferometry for absolute distance measurement (ADM) in comparison with a conventional laser interferometer performing incremental distance measurement (IDM). (Adapted from reference [21] with permission.)

incremental HeNe laser (IDM) was also continuously operated but without beamline interruption. When the resulting ADM (blue color) and IDM (pink color) measurements are overlapping as in the figure, no notable difference is observed. The test result clearly shows that the multi-wavelength interferometer is able to catch up with that of the IDM reading immediately every time when the ADM measurement beamline is recovered from blockage. This also verifies that the proposed multi-wavelength interferometer is able to perform ADM with the same resolution and speed as IDM.

Figure 7.5 shows another test conducted to evaluate the measurement linearity and stability while the target distance was incremented from 1.0 to 3.8 m consecutively with an equal interval of 150 mm. Note that $L_i$ indicates the absolute distance determined from $\lambda_i$. Then the linearity of $L_i$ was quantified in terms of the discrepancy from the IDM distance obtained from the HeNe laser interferometer installed along the same optical path. The discrepancy of $L_1$ is identified as 35.3 nm (±17.7 nm) in peak to valley (standard deviation), which corresponds to $\pm 4.6 \times 10^{-9}$ linearity error over the entire test range of 2.8 m. In a strict sense, the HeNe laser

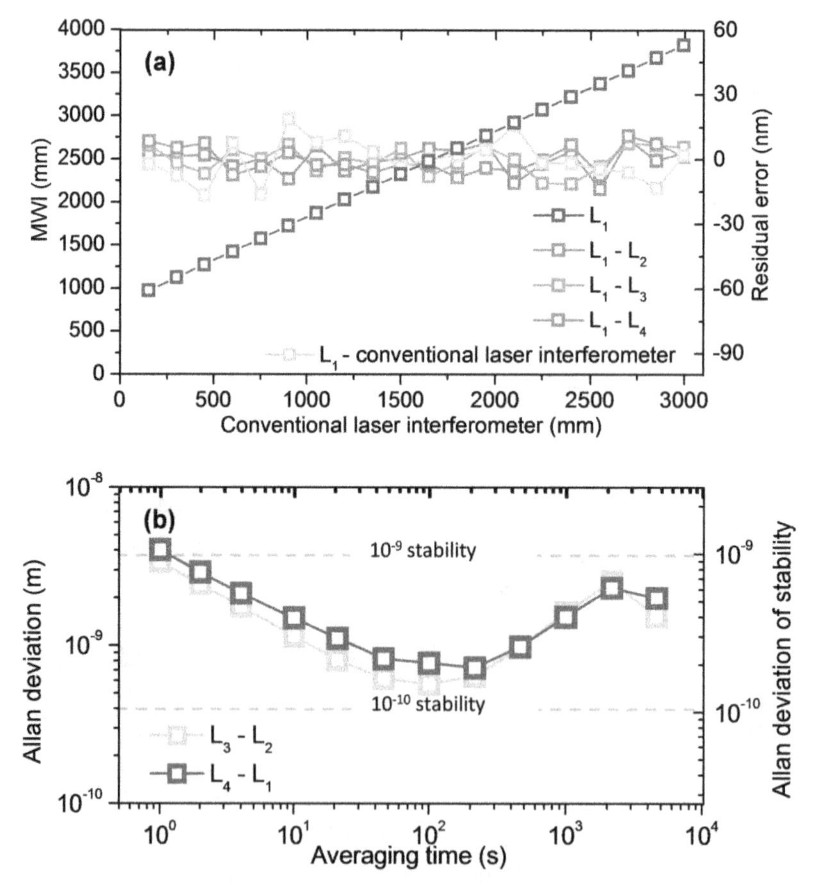


**Figure 7.5.** Performance evaluation in terms of (a) linearity and (b) stability of multi-wavelength interferometry. (Adapted from reference [22] with permission.)

interferometer cannot be a perfect reference of linearity evaluation. Thus, self-referencing evaluation of $L_1$ with respect to $L_2$, $L_3$ and $L_4$ of other constituent wavelengths of $\lambda_2$, $\lambda_3$, and $\lambda_4$ was made to reveal that their discrepancy is within $\pm 10$ nm in peak-to-valley, corresponding to $\pm 2.6 \times 10^{-9}$ linearity error as presented in figure 7.3(a).

In figure 7.3(b), the measurement stability is plotted in terms of the Allan deviation for a target distance of 3.8 m. Then the inter-wavelength distance differences of $L_3 - L_2$ and $L_4 - L_1$ are taken to calculate the measurement stability. The result is 0.57 nm for $L_3 - L_2$ and 0.78 nm for $L_4 - L_1$ at 100 s averaging, corresponding to $1.5 \times 10^{-10}$ and $2.1 \times 10^{-10}$ in terms of the fractional stability. This implies that albeit the combined uncertainty of distance measurement in air is limited to a $10^{-8}$ level due to the uncertainty of the refractive index of air, the measurement stability can be achieved to a $10^{-10}$ level, two orders of magnitude less.

## 7.3 Comb-based dispersive interferometry

In principle, dispersive interferometry exploits a broad spectrum light source to produce an abundance of interference over a large spectral range. Being also called spectrally-resolved interferometry, this principle has long been used in Fourier-transform spectroscopy to identify absorption lines of gaseous atoms or molecules. This principle began to be adopted for length metrology by making the most of the dispersive phenomenon that the interference phase maintains a linear distribution with the wavelength for a given target distance. The target distance can therefore be determined by identifying the linear dispersive slope of the interference phase distribution.

Dispersive interferometry has long been realized by inserting a scanning mechanism either in the interferometer's reference or measurement arm. This conventional way is not suitable for fast real-time distance measurement, so the scanning mechanism is replaced with a spectrometer to speed up the measurement time at the sacrifice of measurement precision and range due to the limited spectrometer's resolving power and spectral width.

Figure 7.6 illustrates a comb-based scheme of dispersive interferometry operating on a Michelson interferometer [23–26]. The frequency comb is produced from a Ti: Sapphire femtosecond laser emitting 10 fs pulses at a repetition rate of 75 MHz. The spectral width is 80 THz at a center frequency of 375 THz. Each mode within the frequency comb has a narrow linewidth, equivalent to a temporal coherence length of ~150 m. The target distance is defined as the optical length difference between the reference mirror ($M_R$) and the measurement mirror ($M_M$). The interference signal is captured using a spectrometer consisting of a line grating and a line array of 3648 line photo-detectors. A Fabry–Perot etalon (FPE) is inserted before the spectrometer to reduce the comb mode density so that only one comb mode is received by one photo-detector. The interference intensity captured by the spectrometer is a function of the optical frequency $\nu$ as $g(\nu) = s(\nu)\ \{1 + \cos \phi(\nu)\}$; $s(\nu)$ is the power spectral density of the light source and $\phi(\nu)$ is the interference phase that has a linear relationship with the target distance $L$, i.e. $\phi(\nu) = 4\pi nL\nu/c_0$; $n$ is the refractive index and $c_0$ is the speed of light in vacuum.

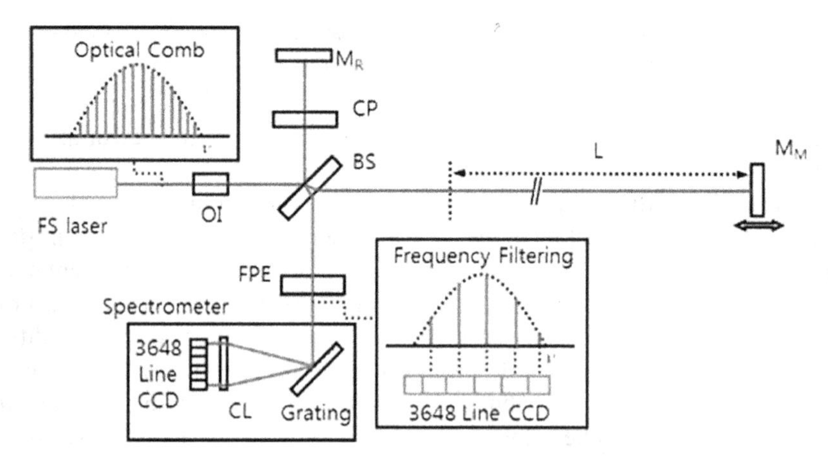


**Figure 7.6.** Optical system layout of comb-based dispersive interferometry together with a representative spectrally-resolved interferogram. L: target distance, MM: mreasurement mirror, MR: reference mirror, FPE: Fabry–Perot etalon, BS: beam splitter, OI: optical isolator, CL: collimating lens, CP: compensation plate, FS: femtosecond.

The interference signal $g(\nu)$ detected by the spectrometer is processed to determine the target distance $L$ through a series of calculation steps as illustrated in figure 7.7. The first step is to take the Fourier-transform of $g(\nu)$, i.e. $G(\tau) = \mathrm{FT}\{g(\nu)\} = S(\tau) \otimes \{(1/2)\delta(\tau + \alpha) + \delta(\tau) + (1/2)\delta(\tau - \alpha)\}$; $\tau$ represents the optical delay time, $S(\tau)$ is the Fourier transform of $s(\nu)$, and $\delta(\tau)$ is the Dirac delta function. The symbol $\otimes$ indicates the convolution operation. The calculated $S(\tau)$ is symmetrical about $\tau = 0$, having three peaks at—$\alpha$, 0, $\alpha$. The target distance $L$ may be calculated from the relation of $L = c_0\alpha/2n$. However, the value of α is not precisely identified since the peak at $\alpha$ is convoluted by $S(\tau)$ with a broadened width. Besides, the spectrometer's resolution is usually not fine enough. Thus, as an alternative way, the peak at $\alpha$ is extracted by means of band-pass filtering and subsequently Fourier-transformed inversely to the $\nu$ domain as $g'(\nu) = \mathrm{FT}^{-1}\{S(\tau) \otimes (1/2)\delta(\tau - \alpha)\} = (1/2)s(\nu)\exp[\mathrm{i}(2\pi\alpha\nu)] = (1/2)\exp[\mathrm{i}\phi(\nu)]$ with $i = \sqrt{-1}$. This permits calculating $\phi(\nu)$ directly as $\phi(\nu) = \tan^{-1}[\mathrm{Im}\{g'(\nu)\}/\mathrm{Re}\{g'(\nu)\}]$. Then the first-order slope of $\phi(\nu)$ is related to the target distance $L$ as $\mathrm{d}\phi(\nu)/\mathrm{d}\nu = 4\pi NL/c_0$ with $N$ being the group refractive index, i.e. $N = n + (\mathrm{d}n/\mathrm{d}\nu)\nu$ [27]. Thus, the target distance $L$ is finally obtained as $L = (c_0/4\pi NL)(\phi(\nu)/\mathrm{d}\nu)$.

The above data processing requires that the peak at $\alpha$ has no overlap with the peak at 0 in the time domain. This condition is met only when the value of $\alpha$ is larger than the temporal width $\Delta\tau$ of $S(\tau)$. Assuming $S(\tau)$ is of a Gaussian shape, $\Delta\tau$ is expressed as $\Delta\tau = 4 \ln 2/\pi\Delta\nu$ with $\Delta\nu$ being the spectral width of $S(\tau)$. Hence, the minimum distance is defined as $L_{\min} = c_0\Delta\tau/2N = 2 \ln 2c_0/\pi\Delta\nu N$. For instance, $L_{\min}$ becomes ~5 μm when $\Delta\nu$ is 80 THz. Another restriction is the non-ambiguity range $L_{\mathrm{NAR}}$, which is given by the frequency resolution $p$ of the spectrometer in accordance with the Nyquist sampling theory, i.e. $\alpha < 1/2p$ as illustrated in figure 7.8. Thus, $L_{\mathrm{NAR}}$ is given as $L_{\mathrm{NAR}} = c_0/4Np$. If all the comb modes are sampled individually with an ideal spectrometer, the resolution $p$ equals the pulse repetition

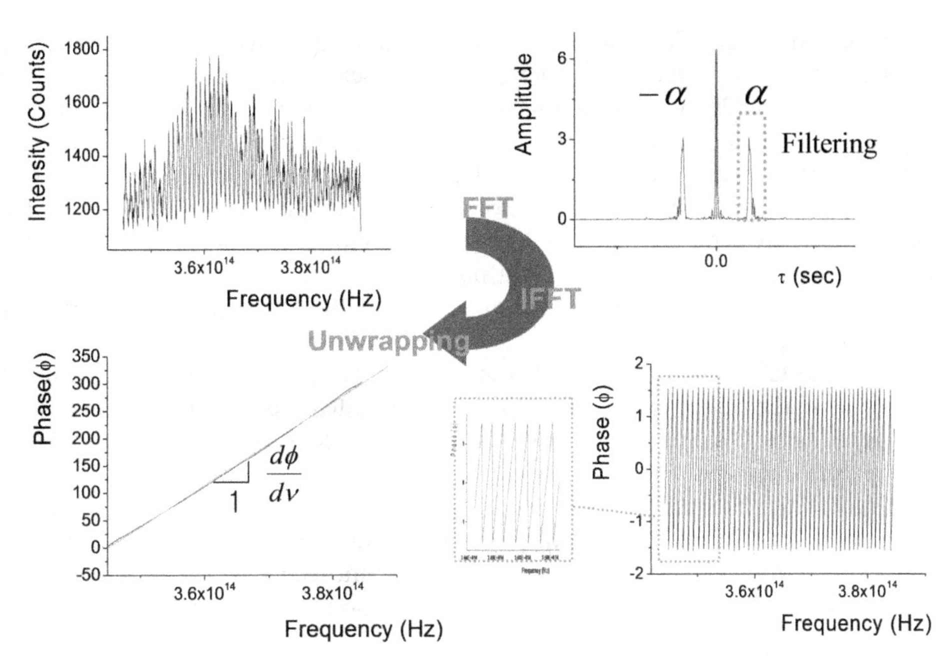


**Figure 7.7.** Data processing for determination of the target distance $L$.

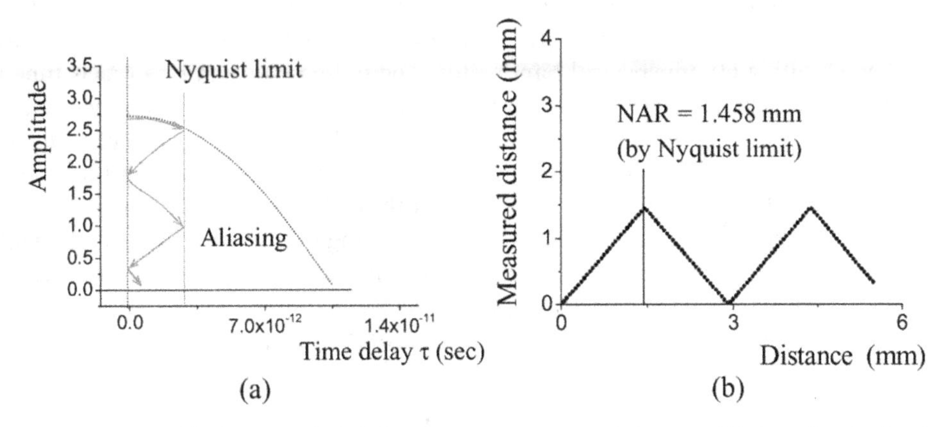


**Figure 7.8.** Non-ambiguity range limited by the spectrometer resolution. (a) Aliasing effect in the time domain. (b) Wrapped measurement result. (Adapted from reference [23] with permission.)

rate of 75 MHz, extending $L_{\text{NAR}}$ to 2.0 m. In reality, the resolution $p$ is much larger, restraining $L_{\text{NAR}}$ to a few millimeters.

## 7.4 Time-of-flight measurement by pulse-to-pulse cross-correlation

The meter, as the basic unit of length, is currently defined as the path length traveled by light in vacuum during a 1/299 792 458 s [28]. This international system (SI) definition suggests that the time-of-flight of light pulses is a direct measure to determine a distance. This method in fact has long been used for long distance measurements such as geodetic survey, satellite laser ranging (SLR) and airborne absolute altimetry [29, 30]. Despite its wide uses, the time-of-flight of light pulses is not a high precision means since its measurement accuracy is ultimately limited by the photo-detectors and associated radio-frequency electronics available today. For instance, the state-of-the-art timing resolution of light pulses is of the order of a picosecond, equivalent to a few hundreds of micrometers in distance [31, 32]. As a consequence, employing ultrashort light pulses in the conventional scheme of time-of-flight measurement leads to no improvement in the measurement accuracy as illustrated in figure 7.9.

In order to take full advantage of using ultrashort light pulses for time-of-flight measurement, the scheme of pulse timing has to be enhanced by making the most of the temporal as well as spectral characteristics of mode-locked lasers. One example is to adopt the balanced optical cross-correlation technique that was in fact made available to stabilize the jitter of femtosecond pulses with the aid of a second harmonic crystal [33]. With its adoption in the time-of-flight measurement, the pulse-to-pulse interval can be quantified to a sub-femtosecond level, allowing a long distance to be measured with a nanometer resolution as demonstrated in several attempts [34–36].

Figure 7.10 shows the hardware system configure to implement the time-of-flight measurement using ultrashort pulses by adopting the balanced optical cross-correlation technique [34, 35]. The interferometer itself is a Michelson type. The light source is an Er-doped fiber laser emitting 150 fs pulses of a 1560 nm central wavelength with a 60 nm spectral bandwidth. The pulse repetition rate ($f_r$) is tuned within a range of ±200 kHz about a nominal value of 100 MHz. The average optical power is 240 mW. The balanced cross-correlation (BXCOR) signal is locked to its zero-crossing point by adjusting the pulse repetition rate by controlling the cavity length of the light source with a 6 kHz bandwidth. A dual servo mechanism is used for control of the cavity length by combining a stepping motor for coarse adjustment with a piezoelectric actuator for fine adjustment. The pulse repetition rate is measured using a frequency counter phase locked to the rubidium atomic clock with a $10^{-12}$ stability. The target distance $L$ is determined as $L = c_0 m/(2f_r n_g)$, in

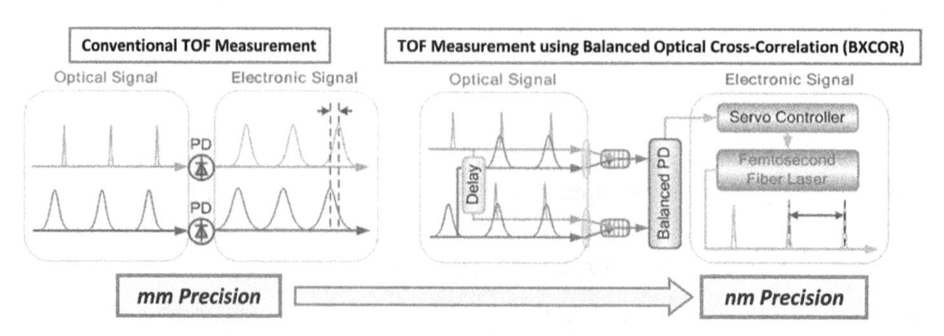


**Figure 7.9.** Time-of-flight measurement using light pulses. (a) Conventional scheme. (b) Balanced optical cross-sorrelation scheme. PD: photo-detector.

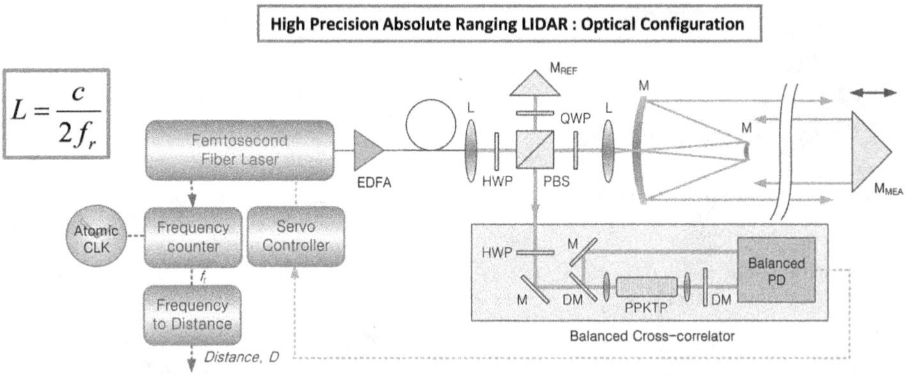

**Figure 7.10.** System configuration for the time-of-flight measurement (above) and its balanced cross-correlation signal (below). PPKTP: periodically poled $KTiOPO_4$ crystal, EDFA: erbium-doped fiber amplifier, PD: photo-detector, M: mirror, L: lens, DM: dichroic mirror, HWP: half-wave plate, QWP: quarter-wave plate.

which $c_0$ is the speed of light in vacuum, $n_g$ is the group refractive index, $f_r$ is the pulse repetition rate and $m$ is an integer multiple.

Figure 7.11 shows a measurement result over a long distance of ~700 m. The measured distance reading over a period of 80 s yields a ~60 μm fluctuation, which is attributable to the building vibration as well as the environmental change of the refractive index. The measurement resolution was evaluated by modulating the target mirror with a sinusoidal modulation of 10 Hz frequency and 1.6 μm amplitude. The measured distance reading in the time domain as well as the frequency domain clearly verifies the induced modulation with a sub-micrometer resolution.

For evaluating an achievable measurement precision, the BXCOR signal is used to estimate the ultimate measurement precision. The BXCOR signal yields the DC voltage signal being linearly proportional to the temporal offset between two pulses. So, a fluctuation of the measured distance can be estimated by reading the BXCOR signal regardless of the variation of the actual target distance. In other words, the timing jitter observed in the BXCOR signal can be directly converted into the measurement fluctuation of the target distance. The BXCOR signal near the zero point of the temporal offset is used to estimate the achievable measurement precision. The ultimate measurement precision turns out to be 8.7 nm (1.1 nm) at 10 ms (1 s) averaging time (figure 7.12).

## 7.5 Time-of-flight measurement by dual-comb interferometry

Dual-comb interferometry, also called multi-heterodyne interferometry, employs a pair of combs having slightly different repetition rates of $f_r$ and $f_r + \Delta f_r$ with $\Delta f_r$ being a small offset [37–40]. One comb is produced from the probe laser and the other comb from the local oscillator (LO) laser. Figure 7.13 illustrates the basic principle of time-of-flight distance measurement by means of dual-comb interferometry.

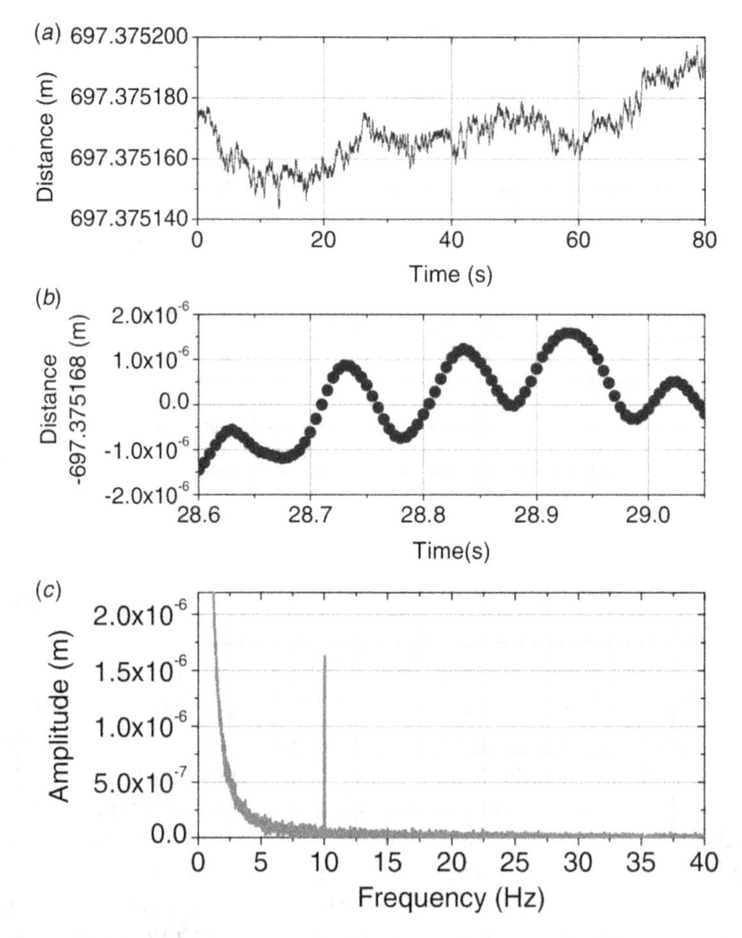


**Figure 7.11.** Long distance meausurement results. (a) Measured distance of ~700 m over a time interval of 80 s. (b) Magnified view of the measured distance with a sinusoidal distance modulation. (c) Fourier-transformed data of the measured distance. (Adapted from reference [34].) 

The output pulse train of the probe laser is split into two trains; one is reflected from the reference mirror and the other from the target mirror with a time delay of $\tau_d$. The probe laser, after recombining the two separated pulse trains, is interfered with the local oscillator laser, of which the cross-correlation interferogram repeats itself at a constant time period of $\Delta T$ ($= 1/\Delta f_r$). This remarkable feature of dual-comb interferometry allows the pulse-to-pulse interference, occurring in a μs real time scale to be monitored in a ns effective time scale with a down-conversion factor of $f_r/\Delta f_r$. Figure 7.14(a) shows how an actual optical system configured to implement the distance measurement by means of dual-comb interferometry [39]. The signal laser is an Er-doped fiber laser emitting 100 fs pulses at a 100 MZ repetition rate ($f_r$). The local oscillator laser is similar to the signal laser except its repetition rate has a 5 kHz offset ($\Delta f_r$). The measured data shows the cross-correlation interferogram

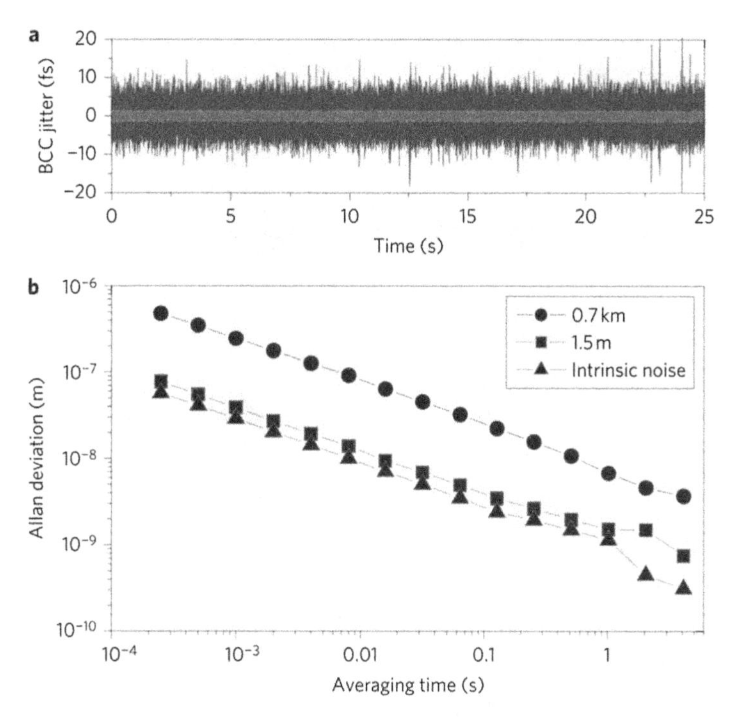


**Figure 7.12.** Estimation of achievable measurement precision. (a) Timing jitter singal over 25 s. (b) Allan deviation. (Adapted from reference [34] with permission.)

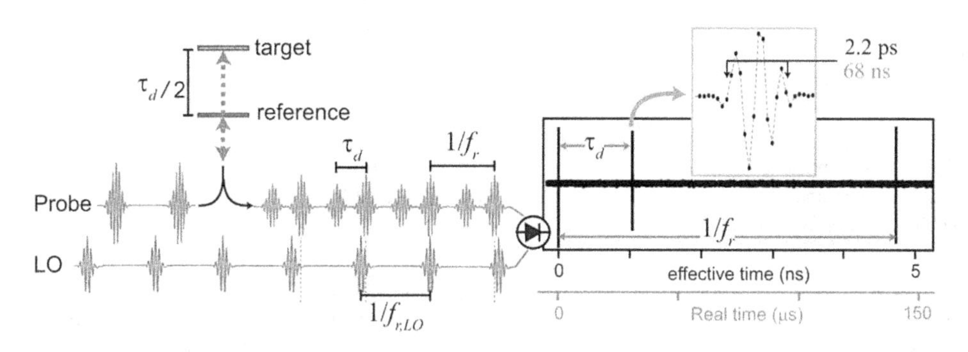


**Figure 7.13.** Basic principle of the dual-comb interferometry. (Adapted from reference [38] with permission.)

repeating itself every 200 μs in the real-time scale. The effective time scale is up-converted by a the scale factor of 20 000 ($f_r/\Delta f_r$). The non-ambiguity range is given as 1.5 m (= $c/2f_r$). A dead zone is found within the non-ambiguity range where the target pulse is overlapped with the reference pulse. The temporal range of the dead zone scales with the pulse duration, so shorter pulses are more desirable to reduce it. As illustrated in figure 7.14(b), the dead zone can inherently be eliminated by incorporating a polarization-separation scheme to make the reference and target pulses be mutually orthogonal.

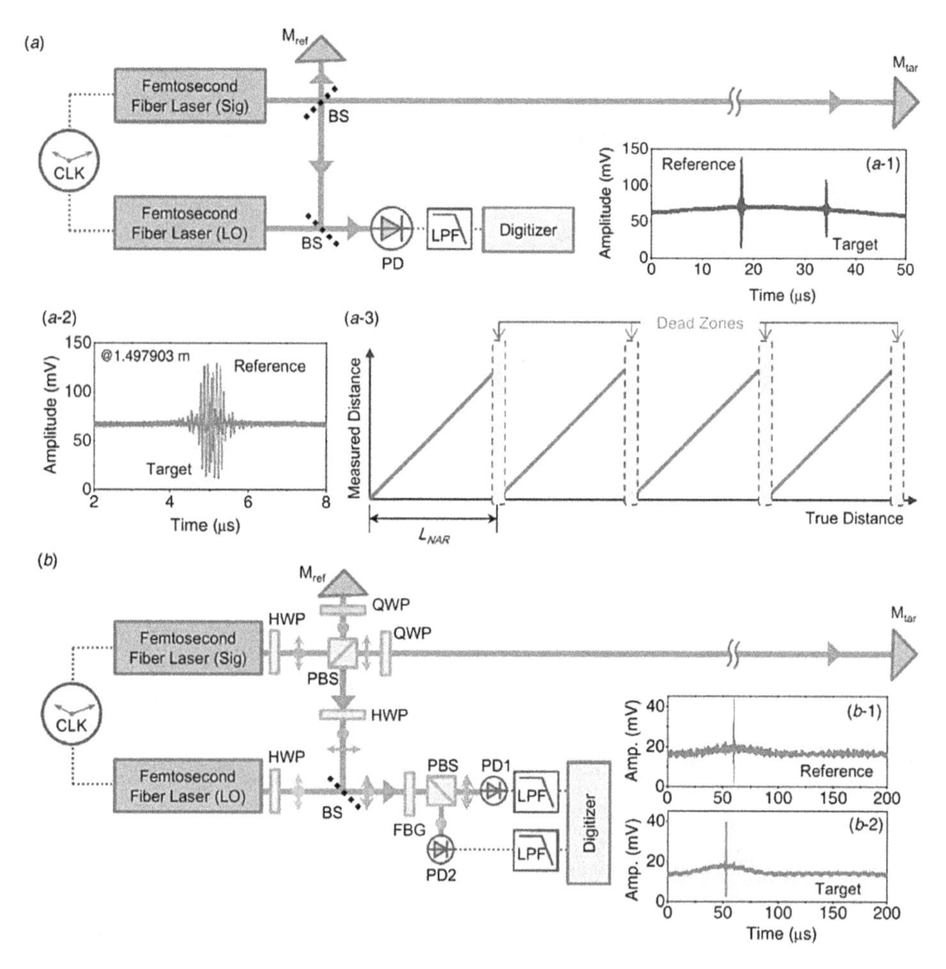


**Figure 7.14.** Optical configuration of dual-comb interferometry. (a) Basic configuration of the dual-comb interferometer and (b) dead-zone free dual-comb interferometer. (Adapted from reference [39].) 

Figure 7.15 shows how the interferograms obtained from the dual-comb interferometer of figure 7.14(b) are processed to determine the target distance $L$. Each interferogram is digitized at a sampling rate of 200 μs in the real time domain and then Fourier-transformed to the frequency domain in consideration of the scale factor of $f_r/\Delta f_r$ between the real time scale and the effective time scale. Then the relative phase difference $\phi(\nu)$ of the target pulse spectrum with respect to the reference pulse spectrum is expressed as $\phi(\nu) = 4\pi nL\nu/c_0$; $c_0$ is the speed of light in vacuum and $n$ is the group refractive index of air, $L$ is the target distance, and $\nu$ is the optical frequency. Finally, the target distance is calculated as $L = (c_0/4\pi n) \times (\mathrm{d}\phi(\nu)/\mathrm{d}\nu)$ in which the phase slope $\mathrm{d}\phi(\nu)/\mathrm{d}\nu$ is obtained by least-squares fitting of the data of $\phi(\nu)$.

Figure 7.16 shows a test result obtained to evaluate the measurement performance of the dual-comb interferometer of figure 7.14(b). First, the measurement

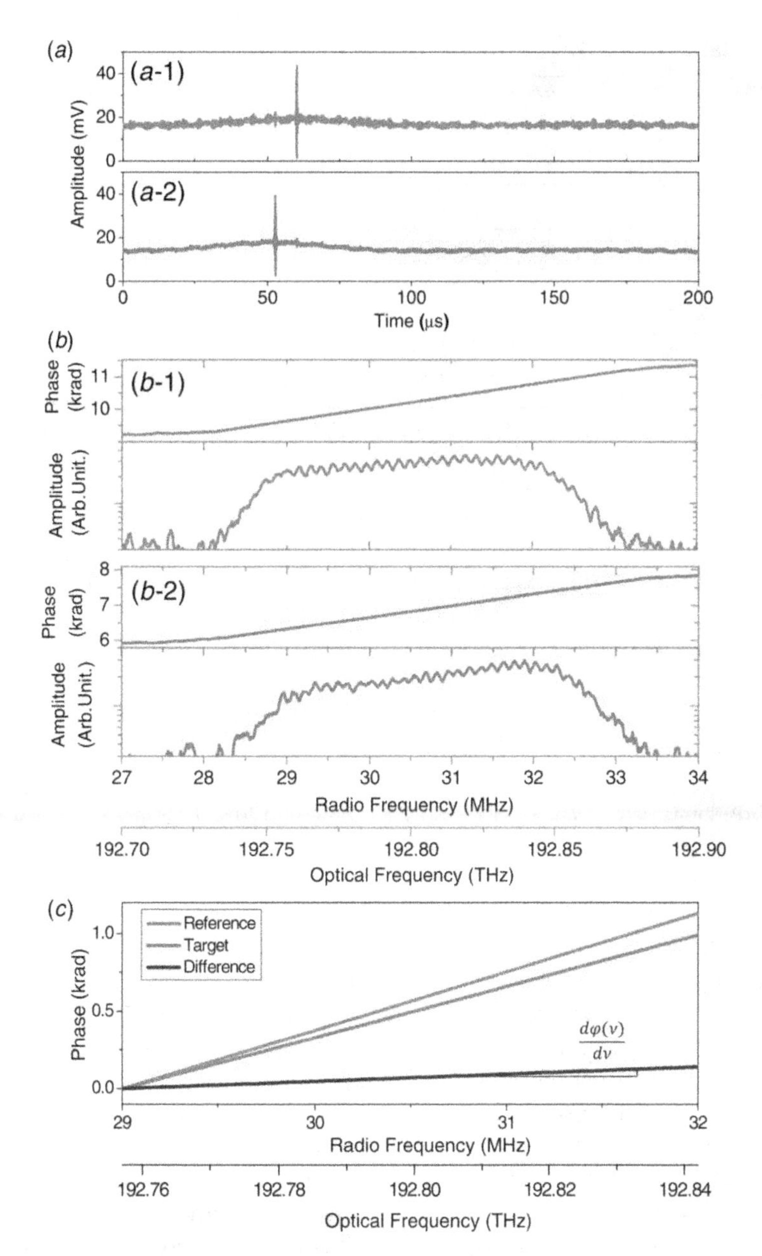


**Figure 7.15.** Signal processing to determine the absolute value of the target distance $L$. (a) Measured interference signal of the reference pulse (gray) and measurement pulse (pink). (b) Calculated spectral amplitude and phase. (c) Least-square fitted first-order slope of the spectral phase distribution. (Adapted from reference [39].) 

linearity is calculated in comparison with a conventional NeHe laser interferometer with a resolution of ~1 nm for a target distance being increased with 1 μm steps from 66 mm using a piezoelectric actuator. The measured distance at each position was averaged during 1 s. The linearity error is found to be 126 nm (358 nm) in terms of

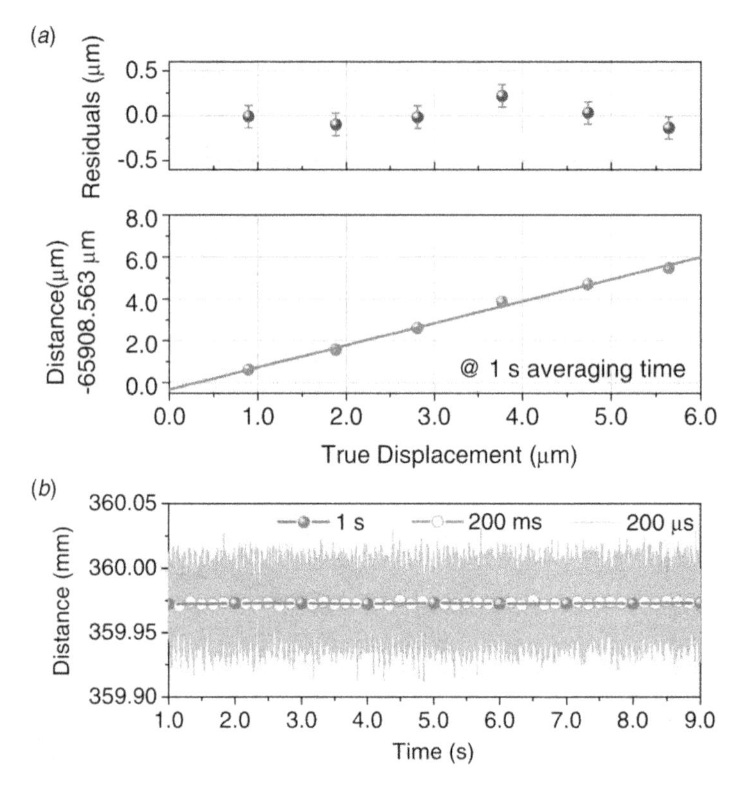


**Figure 7.16.** Performance evaluation. (a) Linearity test result. (b) Measurement stability. (Adapted from reference [39].) 

standard deviation (peak-to-valley). Second, the measurement stability is obtained for a target distance of 360 mm as 15.2 μm, 1.09 μm and 254 nm at 200 μs, 200 ms and 1 s averaging time, respectively, in terms of standard deviation. It is important to note that the measurement stability has a strong dependency on the averaging time, which is reckoned mainly attributable to the pulse jitters within the probe laser and the local oscillator laser.

The non-ambiguity range (NAR) of the dual-comb interferometer of figure 7.14(b) is limited to 1.5 m by the pulse repetition rate, i.e. $\Lambda_{NAR} = c/2f_r$. For a distance longer than $\Lambda_{NAR}$, a synthetic wavelength needs to be introduced by shifting the pulse repetition rate of the probe laser by an amount of $\delta_r$. Then, $\Lambda_{NAR}$ is slightly changed to $\Lambda'_{NAR} = c/2(f_r + \delta_r)$ as illustrated in figure 7.17. By combining $\Lambda_{NAR}$ with $\Lambda'_{NAR}$, a second-order synthetic wavelength can be synthesized as $\Lambda_{NAR\_2nd} = \Lambda_{NAR} \times \Lambda'_{NAR}/(\Lambda_{NAR} - \Lambda'_{NAR}) = c/2\delta_r$. With this method, in principle, the maximum measurable distance can be extended significantly, but its effectiveness is practically constrained by the stability of $f_r$, $\Delta f_r$ and pulse jitters.

Figure 7.18 shows an optical setup configured to utilize the dual-comb-based time-of-flight measurement for detection of multiple targets. The signal laser and the local oscillator laser are Er-doped fiber femtosecond lasers emitting 90 fs duration pulses with 10 mW optical power. The pulse repetition rate is 100 MHz for the signal

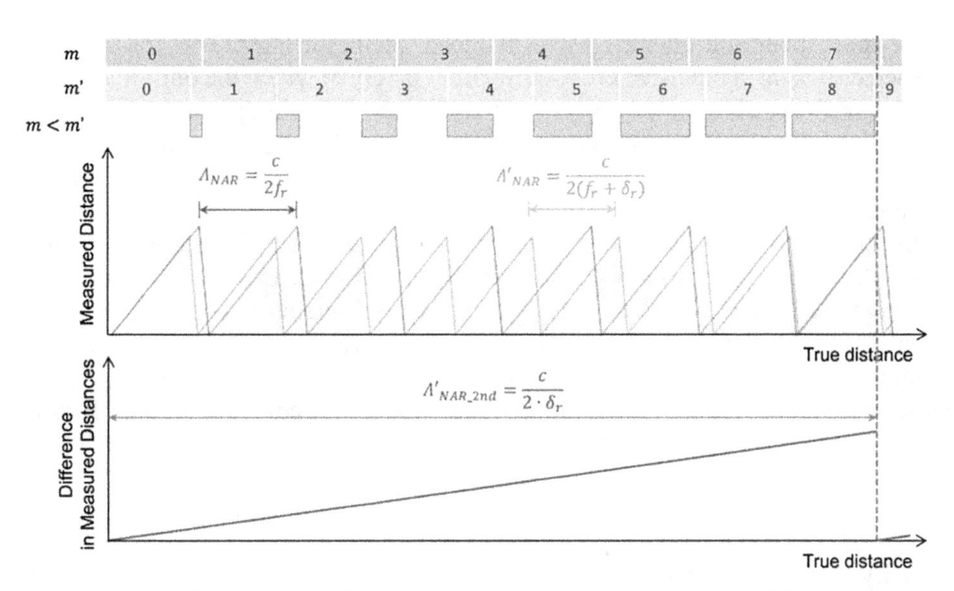


**Figure 7.17.** Extension of the non-ambiguity range. (Adapted from ref [39].) 

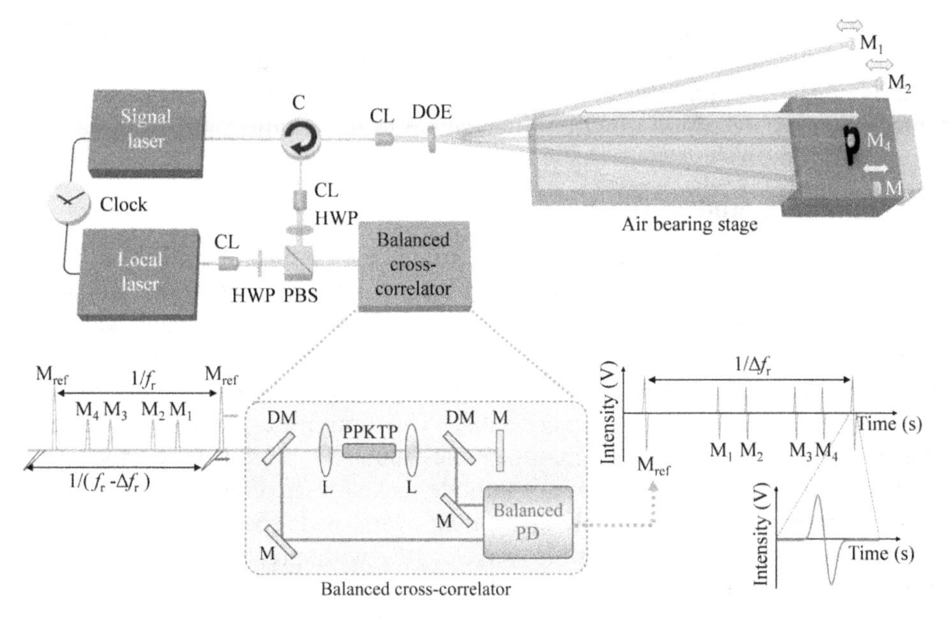


**Figure 7.18.** Optical configuration of dual-comb-based time-of-flight measurement. (Adapted from reference [41] with permission.)

laser and 100.02 MHz for the local oscillator. A diffractive optical element (DOE) is used to split the signal laser towards four target mirrors simultaneously. The reference zero-datum point of distance measurement is set at the partially coated fiber surface located before the DOE. The reflected beams are combined into a single-mode fiber and interfered with the local oscillator laser. The pulse locations are detected using a balanced cross-correlator (BCC), of which the output appears as an S-shaped electric signal for every pulse. The zero-crossing point of the S-shaped signal indicates the corresponding target location. This principle enables multiple targets to be simultaneously dealt with, realizing multi-degrees of freedom sensing of the target in 3D space.

In the time domain, the BCC signal reflected from the reference mirror surface leads other BCC signals arriving from the target mirrors. Each target distance $d$ is determined as $d_i = c\Delta T_i/(2N)$; $c$ is the speed of light in vacuum, $\Delta T_i$ the time-of-flight of the pulse between the reference datum and $i$th target mirror in the effective time domain, and $N$ the group refractive index of air. $\Delta T_i$ is measured in the 'real time scale', and converted to the 'effective time scale' by a conversion scale factor of $f_r/\Delta f_r$. To improve the measurement resolution, the BCC signal is cross-correlated with a standardized BCC signal pattern as shown.

Figure 7.19 shows an exemplary measurement in which $\Delta f_r$ is selected to be 2 kHz, so the target distance is updated at a rate of 0.5 ms. The measurement stability traced at a very short target distance of 0.015 m is 0.936 μm at 0.5 ms averaging and it improves to 17 nm at 0.5 s averaging. The measurement linearity evaluated over a 1.0 m travel in comparison to an incremental type displacement interferometer is 279 nm in peak-to-valley—79 nm in standard deviation value—without notable cyclic errors. Three target mirrors can be measured simultaneously, each mirror being modulated with different amplitudes for distinction. The test result confirms no noticeable cross talks between target mirrors even when the three target positions are captured with a high signal-to-noise ratio.

## 7.6 Summary and outlook

Mode-locked lasers provided the potential to make a significant contribution to the progress of dimensional metrology during the last several years by means of laser-based optical interferometry. In comparison to conventional single-wavelength lasers or broad spectral light sources, mode-locked lasers are capable of enhancing the performance of optical interferometry to meet the ever-growing demand on measurement uncertainty and functionality. This optimistic prediction is based on several spectral and temporal characteristics uniquely provided by mode-locked lasers. Firstly, the frequency comb of a mode-locked laser offers a large number of discrete optical frequencies evenly distributed over a wide spectral bandwidth. This permits the generation of multiple wavelengths with precise traceability to the atomic clock, so long distances up to a few meters can be measured in an absolute way without losing the measurement uncertainty of laser-based interferometry. Secondly, ultrashort pulses emitted from a mode-locked laser in the time domain can be utilized to implement time-of-flight measurement with a timing resolution at a sub-femtosecond level, allowing measurement precision in the nanometer regime. Thirdly, a mode-locked laser is able to extend the 3D surface profiling capability of optical interferometry as a new source providing well-controlled temporal coherence together with high spatial coherence. This unique characteristic permits a rough

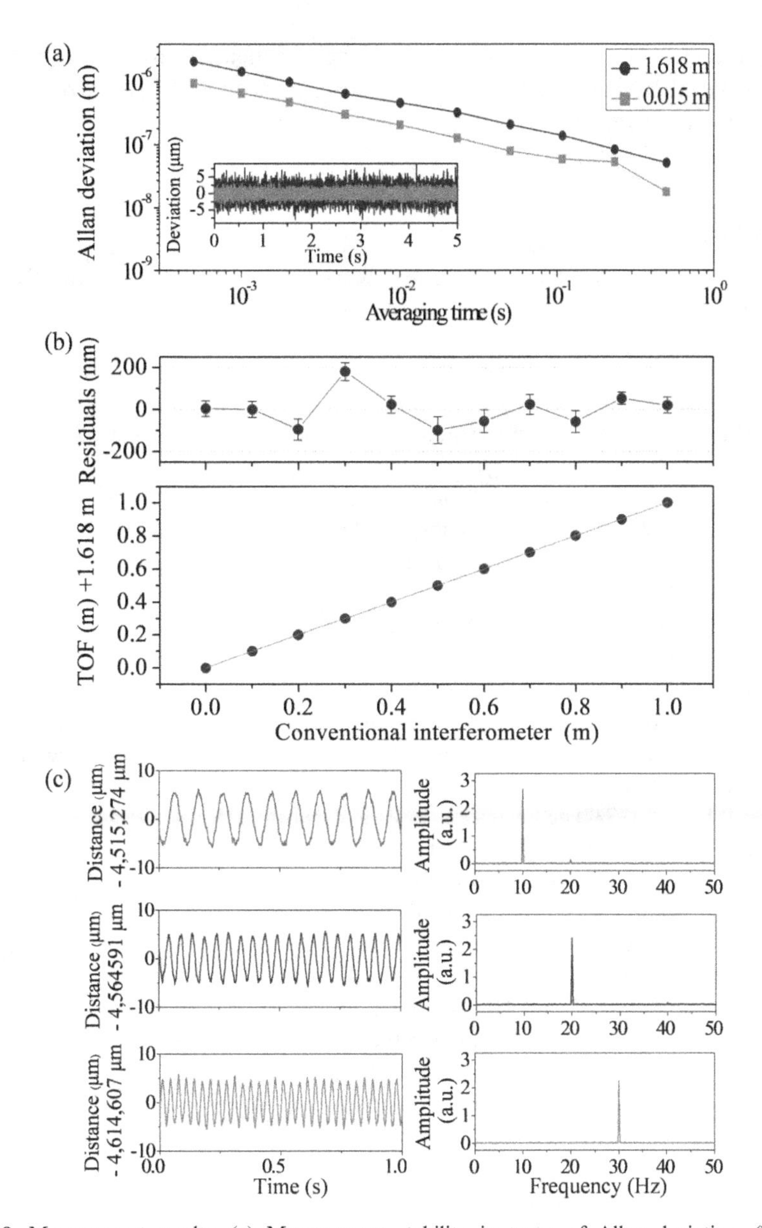


**Figure 7.19.** Measurement results. (a) Measurement stability in terms of Allan deviation, (b) linearity compared with an incremental HeNe laser interferometer and (c) multiple target measurement with different modulation frequency for individual target mirror. (Adapted from reference [41] with permission.)

surface to be profiled with a large field-of-view beyond the capability of white light sources long used to conduct low-coherence optical interferometry.

Despite the many advantages discussed here, not many mode-locked lasers have yet been applied to actual industrial applications of dimensional metrology. The main reason is the cost and complexity of mode-lacked lasers available today. Much effort is being exerted worldwide to make the use of mode-locked lasers more robust to handle at a reasonable price. No doubt in the very near future, mode-locked lasers will be a general tool of dimensional metrology, which will benefit manufacturing industries by effectively responding to the ever-growing demand for measurement precision and functionality.

## Acknowledgments

This work was supported by the National Research Foundation of the Republic of Korea (NRF-2012R1A3A1050386).

## References

[1] Gao W *et al* 2015 Mesurement technologies for precision positioning *CIRP Ann. Manuf. Technol.* **64** 773

[2] Kim S-W 2009 Combs rule *Nat. Photon.* **3** 313–4

[3] Weichert C, Köchert P, Köning R, Flügge J, Andreas B, Kuetgens U and Yacoot A 2012 A heterodyne interferometer with periodic nonlinearities smaller than ±10 pm *Meas. Sci. Technol.* **23** 094005

[4] Bobroff N 1993 Recent advances in displacement measuring interferometry *Meas. Sci. Technol.* **4** 907–26

[5] Demarest F C 1998 High-resolution, high-speed, low data age uncertainty, heterodyne displacement measuring interferometer electronics *Meas. Sci. Technol.* **9** 1024–30

[6] Heydemann P L M 1981 Determination and correction of quadrature fringe measurement errors in interferometers *Appl. Opt.* **20** 3382–4

[7] Bender P L *et al* 1973 The lunar laser ranging experiment *Science* **182** 229–38

[8] Fujima I, Iwasaki S and Seta K 1998 High-resolution distance meter using optical intensity modulation at 28 GHz *Meas. Sci. Technol.* **9** 1049–52

[9] Collins S F, Huang W X, Murphy M M, Grattan K T V and Palmer A W 1993 A simple laser diode ranging scheme using an intensity modulated FMCW approach *Meas. Sci. Technol.* **4** 1437–9

[10] Xiaoli D and Katuo S 1998 High-accuracy absolute distance measurement by means of wavelength scanning heterodyne interferometry *Meas. Sci. Technol.* **9** 1031–5

[11] Meiners-Hagen K, Schödel R, Pollinger F and Abou-Zeid A 2009 Multi-wavelength interferometry for length measurements using diode lasers *Meas. Sci. Rev.* **9** 16–26

[12] Udem T, Holzwarth R and Hänsch T W 2003 Optical frequency metrology *Nature* **416** 233–7

[13] Jin J, Kim Y-J, Kim Y, Kim S-W and Kang C-S 2006 Absolute length calibration of gauge blocks using optical comb of a femtosecond pulse laser *Opt. Express* **14** 5986–5974

[14] Salvadé Y, Schuhler N, Lévêque S and Floch S L 2008 High-accuracy absolute distance measurement using frequency comb referenced multiwavelength source *Appl. Opt.* **47** 2715–20

[15] Hyun S, Kim Y-J, Kim Y, Jin J and Kim S-W 2009 Absolute length measurement with the frequency comb of a femtosecond laser *Meas. Sci. Technol.* **20** 095302

[16] Kim Y, Kim S, Kim Y-J, Hussein H and Kim S-W 2009 Er-doped fiber frequency comb with mHz relative linewidth *Opt. Express* **17** 11972–7

[17] Kim Y, Kim Y-J, Kim S and Ki S-W 2009 Er-doped fiber comb with enhanced fceo S/N ratio using Tm:Ho-doped fiber *Opt. Express* **17** 18606–11

[18] Jang Y-S and Kim S-W 2017 Compensation of the refractive index of air in laser interferometer for distance measurement: A review *Int. J. Precis. Eng. Manuf.* **18** 1881–90
[19] Birch K P and Downs M J 1993 An updated Edlen equation for the refractive index of air *Merologia* **30** 155–62
[20] Ciddor P E 1996 Refractive index of air: new equations for the visible and near infrared *Appl. Opt.* **35** 1566–73
[21] Wang G, Jang Y-S, Hyun S, Chun B J, Kang H J, Yan S, Kim S-W and Kim Y-J 2015 Absolute positioning by multi-wavelength interferometry referenced to the frequency comb of a femtosecond laser *Opt. Express* **23** 9121–9
[22] Jang Y-S, Wang G, Hyun S, Kang H J, Chun B J, Kim Y-J and Kim S-W 2016 Comb-referenced laser distance interferometer for industrial nanotechnology *Sci. Rep.* **6** 31770
[23] Joo K-N and Kim S-W 2006 Absolute distance measurement by dispersive interferometry using a femtosecond pulse laser *Opt. Express* **14** 5954–60
[24] Joo K-N, Kim Y and Kim S-W 2008 Distance measurements by combined method based on a femtosecond pulse laser *Opt. Express* **16** 19799–806
[25] Cui M, Zeitouny M G, Bhattacharya N, van den Berg S A and Urbach H P 2011 Long distance measurement with femtosecond pulses using a dispersive interferometer *Opt. Express* **19** 6549–62
[26] van den Berg S A, Persijn S T, Kok G J P, Zeitouny M G and Bhattacharya N 2012 Many-wavelength interferometry with thousands of lasers for absolute distance measurement *Phys. Rev. Lett.* **108** 183901
[27] Ciddor P E 1999 Refractive index of air. 2. Group index *Appl. Opt.* **38** 1663–7
[28] Giacomo P 1984 News from the BIPM *Metrologia* **20** 25–30
[29] Quinn T J 2003 Practical realization of the definition of the metre, including recommended radiations of other optical frequency standards (2001) *Metrologia* **40** 103–33
[30] Felder R 2005 Practical realization of the definition of the metre, including recommended radiations of other optical frequency standards (2003) *Metrologia* **42** 323–5
[31] Degnan J J 1985 Satellite laser ranging: current status and future prospects *IEEE Trans. Geosci. Remote Sens.* **GE-23** 398–413
[32] Pellegrini S, Buller G S, Smith J M, Wallace A M and Cova S 2000 Laser-based distance measurement using picosecond resolution time-correlated single-photon counting *Meas. Sci. Technol.* **11** 712–6
[33] Kim J, Cox J A, Chen J and Kartner F X 2008 Drift-free femtosecond timing synchronization of remote optical and microwave sources *Nat. Photon.* **2** 733–36
[34] Lee J, Kim Y-J, Lee K, Lee S and Kim S-W 2010 Time-of-flight measurement using femtosecond light pulses *Nat. Photon.* **4** 716–20
[35] Lee J, Lee K, Lee S, Kim S-W and Kim Y-J 2012 High precision laser ranging by time-of-flight measurement of femtosecond pulses *Meas. Sci. Technol.* **23** 065203
[36] Shi H, Song Y, Liang F, Xu L, Hu M and Wang C 2015 Effect of timing jitter on time-of-flight distance measurements using dual femtosecond lasers *Opt. Express* **23** 14057–69
[37] Coddington I, Swann W C, Nenadovic L and Newbury N R 2009 Rapid and precise absolute distance measurements at long range *Nat. Photon.* **3** 351–6
[38] Liu T-A, Newbury N R and Coddington I 2011 Sub-micron absolute distance measurements in sub-milisecond times with dual free-running femtosecond Er fiber-lasers *Opt. Express* **19** 18501–9

[39] Lee J, Han S, Lee K, Bae E, Kim S, Lee S, Kim S-W and Kim Y-J 2013 Absolute distance measurement by dual-comb interferometry with adjustable synthetic wavelength *Meas. Sci. Technol.* **24** 045201

[40] Zhang H, Wei H, Wu X, yang H and Li Y 2014 Reliable non-ambiguity range extension with dual-comb simultaneous operation in absolute distance measurements *Meas. Sci. Technol.* **25** 125201

[41] Han S, Kim Y-J and Kim S-W 2015 Parallel determination of absolute distances to multiple targets by time-of-flight measurement using femtosecond light pulses *Opt. Express* **23** 25874–82

**IOP** Publishing

Modern Interferometry for Length Metrology
Exploring limits and novel techniques
**René Schödel (Editor)**

# Chapter 8

# Absolute distance measurement using frequency scanning interferometry

**Armin Reichold**

## Abstract

In Chapter 8 we discuss a range of interferometric techniques which utilise light sources with a continuously changing frequency. We refer to these techniques as frequency scanning interferometry (FSI).We will focus primarily on the measurement of absolute distances with such techniques and we emphasise the physical effects that limit the performance of FSI techniques throughout the chapter.

Following an introduction into the importance of distance measurements in human endeavours in section 8.1, section 8.2 provides the derivation of a mathematical framework in which to describe the light waves emitted from FSI sources. We try to make this derivation applicable to both the radar and optical parts of the electro-magnetic spectrum.

To gain familiarity with the implicit formulation of the mathematical framework we first use it to derive the well known Doppler shift formula in the case of light being reflected from a moving mirror.

Still in the mathematical core of the chapter we describe the interference of two FSI waves emitted from the same source after they have undergone a time variable path difference. Section 8.2 ends in the description of the interference frequency as a function of the time variable path length difference and frequency tuning parameters in a realistic scenario, typical for large volume metrology applications.

In section 8.3 we introduce the reader to a range of analysis techniques which can be used to determine the absolute optical path length difference in an FSI interferometer from the interference frequency or phase described mathematically in section two. We separate the techniques in section three into those that measure length on a scan by scan or a continuous basis.

doi:10.1088/2053-2563/aadddcch8  

In section 8.4 we briefly describe the hardware elements of FSI systems such as light sources, data acquisition and processing, reference interferometers and frequency references. We explain how they are used and how they limit the capabilities of FSI techniques.

To aid the readability and understanding of this chapter and in particular its mathematical aspects we have included a glossary and a list of mathematical symbols used in this chapter. We have also included a compact list of references that illustrate both the fundamentals and the applications of the ideas discussed.

*'If I knew something about it, I wouldn't lecture on it !'*
*Arnold Johannes Wilhelm Sommerfeld*

## 8.1 Introduction

If you have decided to read this chapter you may well have been introduced to the immensely widespread needs for the measurement of distances that pervade nearly all of humanities activities. Absolute distances as opposed to distance changes measured to date span 20 orders of magnitude, ranging from $10^{-12}$ to $3.8 \times 10^{8}$ m (distance from the Earth to the Moon) or even 25 orders if we consider the distance from the Earth to the Voyager One spaceship of $1.9 \times 10^{13}$ m which is measured by radio signal round trip timing. In the above range we have exclude indirectly inferred cosmic or sub-atomic distances.

If we include the August 2017 gravitational wave observations of a kilo nova[1] [1] and the associated relatively accurate measurement of the luminosity distances the range would have increased to $40^{+8}_{-14}$ Mpc ($1.23 \times 10^{24}$ m) and would span 36 orders of magnitude.

We could also take high energy physics distances into our consideration. If we take a rough estimate of the current 'resolution' of the LHC[2] to be $10^{-28}$ m and use this as the minimal measured distance we would get a range spanning 52 orders of magnitude.

We will however restrict this chapter to those distances that can be practically measured with high accuracy using electro-magnetic radiation as the detection medium.

Together with the range of measurable distances the requirements for measurement accuracy have been growing steadily and even widespread industrial applications now regularly require relative accuracies of $10^{-7}$ in ranges up to many tens or even hundreds of metres. This chapter aims to provide an insight into one of the methods by which absolute distance in these much more modest ranges can be measured interferometrically and with high accuracy. We start with some observations about interferometry in general.

If the interferometric[3] phase of an interferometer is known at multiple wavelengths for the same optical path difference (OPD) then this OPD can be measured

[1] A binary neutron star merger.
[2] Large hadron collider.
[3] We will introduce the difference between the optical and interferometric phase shortly in section 8.2.

in absolute terms with respect to these wavelengths and not only incrementally as would be the case for a single wavelength.

The more wavelengths we use the better our knowledge of the OPD in both ambiguity range and accuracy. If the wavelengths are present in our interferometer at the same time their associated interference patterns need to be separated for example spectroscopically by a dispersive element such as a grating. In the simultaneous presence of a continuum of wavelengths [2] such techniques are referred to as hyper-spectral interferometry and if several narrowly defined wavelength are used we call it multi-wavelengths interferometry.

This chapter describes interferometric techniques for the measurement of absolute distance[4] using frequency or wavelength scanning light sources. In frequency scanning techniques the wavelengths passing through the interferometer are separated in time and we use one or several continuous ranges of frequencies. One may think of FSI in this sense as 'time dispersed' hyper-spectral interferometry.

Although in FSI techniques there is only one frequency emitted at any given time we will see in section 8.2.4 that there are nevertheless always at least two frequencies detected which will provide some similarities with multi-wavelength interferometry.

Such scanning techniques are commonly referred to as frequency scanning interferometry (FSI) but are also known as frequency or wavelength chirped or modulated interferometry or frequency modulated continuous wave (FMCW) lidar if the frequency range is in the optical or GHz range.

We will focus our discussion to some extend on techniques that digitally record and process the interferometric signals. Although it is possible to measure distances with FSI techniques using exclusively analogue technology as is often the case in continuous wave radar applications the digital signal analysis enables more accurate measurements and can suppress systematic errors. Our discussions of the physical principles will of course apply to frequency scanning interference in general.

FSI techniques exist across a large range of the electro-magnetic spectrum. Successful implementations are dependent on the availability of continuously frequency scanning sources. Useful sources exist with significant continuous tuning ranges (up to 32% of centre frequency) and large tuning speeds in the range between 50 and 200 GHz [3–5]. In this wavelength range they are known as frequency modulated continuous wave radar (FMCW radar).

Most common are, however, sources at several optical frequencies with the best ranges in the near IR region around 780 and 1550 nm. We will dedicate section 8.4.1 to an overview of such sources.

Following the intentions expressed in the title of this book a specific purpose of this chapter is to enable the reader to rigorously understand the limitations of

[4] We will often use the word distance to indicate the measurement output of an FSI technique although only the optical path difference can strictly be measured.

FSI techniques, explore how these limitations depend on hardware, parameters or analysis algorithms and ultimately improve existing or develop new FSI techniques. Given the aims and the length of this chapter we do not intend to present a comprehensive review of all FSI techniques to date. The focus will instead be on a necessarily limited set of examples that highlight specific limitations.

The content of this chapter should be largely understandable by a second or third year undergraduate in physics or engineering having completed an advanced course in physical optics. The mathematical methods use are also typical of this level. In our derivations we will proceed in small steps and comment on intermediate results where this is useful to improve understanding or highlight limitations.

To motivate the interest into FSI techniques we will briefly introduce their generic capabilities in this section. We then present a generic mathematical description of frequency swept optical waves in section 8.2 and from it derive analytical expressions for the intensity, interferometric phase and interferometric frequency[5] of a two beam, frequency scanning interferometer. We follow this in section 8.3 with examples of FSI analysis techniques of increasing complexity pointing out their limitations. Finally we end with a brief discussion of the hardware of FSI interferometers in section 8.4.

The primary interest into FSI techniques stems from their ability to measure absolute distances as opposed to distance changes which are measured in conventional interferometers. In modern implementations [6] absolute distances measured with FSI can be traced to the SI metre and measured at very high repetition rates reaching tens of megahertz. Modern FSI implementations can offer high absolute accuracies [6] approaching $10^{-7}$ and even better displacement accuracies all at distances up to many tens of metres. Due to these abilities FSI techniques have found their way into a wide range of applications in large scale dimensional metrology that monitor absolute geometries from large scale industrial metrology via large telescope [7] to particle accelerators [8] to name just a few.

Another attractive feature of FSI techniques is their ability to recover the distance measurements after the beam has been interrupted without having to bring the OPD back to an *a priori* known value.

## 8.2 Physical description of FSI interferometers

In this section we aim to develop physical and mathematical frameworks in which to understand the properties of FSI interference signals and from these derive limitations for the applicability of analysis techniques and validity of simplified interference models. The final goal will be to understand the intensity of a frequency

[5] Section 8.3 explains why we distinguish strictly between optical and interferometric phase and frequency.

scanning interferometer as a function of time, OPD, target speed and tuning parameters. The treatment that follows applies to all electro-magnetic waves form the radar and the optical range.

### 8.2.1 Assumptions

We will limit our discussion by the list of assumptions below which are typical for applications in metrology. Although these assumptions are normally silently implied we wish to point them out here to better understand the limitations of FSI techniques.

*Linear optics:*

The intensities of our light waves are small enough to warrant the use of linear optics in the propagation medium. If we measure in vacuum, air, or other gases this assumption is practically impossible to break due to the power density limitations of our lasers.

*Optically passive media:*

The medium is not optically active in the sense that it does not significantly influence the polarisation state of light. This assumption is good for gaseous media but we have to take care of elements in the interferometer hardware which can become optically active such as mechanically stressed beam splitters.

*Refractive index:*

There are several assumptions about the refractive index distribution of the medium in which we measure. These are all needed to establish a relationship between the OPD and the geometric length of the measurement arm of our interferometer. These relationships should also hold as a function of time during the frequency scan. For a full treatment of interferometric measurements that include compensation for a measured refractive index we refer to chapter 5.

1. The true refractive index distribution $n_{\text{true}}(\vec{x}, t)$ varies along the measured distance. Therefore the path taken by the wave is not necessarily a straight line. We wish to consider only those index distributions which, when averaged across the measurement arm into an effective single index $n(t) = \langle n_{\text{true}}(\vec{x}) \rangle_L$ will still describe the optical path length taken by the wave when averaged in the same way. This assumption limits refractive index profiles to random spatial fluctuations. It excludes systematically varying distributions which produce systematically bent paths (mirages). If this assumption holds the optical paths are straight lines on average and hence linearly related to the geometric length.
2. The effective refractive index $n(t)$ does not vary systematically with time. It can be time averaged into a constant $n = \langle n(t) \rangle_{\tau_{\text{med}}}$ where $\tau_{\text{med}}$ is the smallest time constant over which the averaged index does not vary

any more. We assume that $\tau_{med} \ll \tau_{scan}$ where $\tau_{scan}$ is the duration of the frequency scan. This avoids situations in which the optical path systematically changes across the entire scan although the geometric length stays fixed. Any rapid time variations of $n(t)$ will therefore appear as variations of the measured distance. We will therefore absorb the time derivative $\frac{\partial n(t)}{\partial t}$ into the time variations of the optical path length $L$. This acknowledges the fact that interferometric techniques alone cannot distinguish small and rapid changes in refractive index from changes in the geometrical path length. As such the variable we measure is the optical path length $L$ and not the geometric distance. To recover the latter we need to independently determine $n(t)$. This book devotes chapter 5 to interferometric refractive index compensation methods and we point the reader to the following reference [9].

3. The effective refractive index $n(t)$ has negligible chromatic dispersion across the tuning range of our light source and is hence not a function of wavelength. In mathematical terms we assume that: $\frac{dn(t)}{d\lambda}$= constant. Therefore the refractive index is not an implied function of time during wavelength scans. This assumption needs to be carefully validated for gases across the wavelengths used because dispersion typically falls steeply with wavelength in our wavelength range and eases the use of longer wavelength sources. At 1550 nm standard dry air has a dispersion of $dn/d\lambda \approx -8.3 \times 10^{-7}\,\mu\text{m}^{-1}$ but at 250 nm this already rises to $-2.7 \times 10^{-4}\,\mu\text{m}^{-1}$. Organic gases commonly present in air such as $CO_2$ have a much larger dispersion even at long wavelength ($\lambda \approx -7.6 \times 10^{-6}\,\mu\text{m}^{-1}$ @ 1550 nm). The variation of refractive indices in the visible and near IR is usually simple and well modelled [10] and can hence be corrected if the gas composition, pressure and temperature are known. We recommend [11] as a useful on-line source for refractive index evaluation.

*Wave-front shape:*

The laser light used for interferometry in this section is modelled in the form of collimated beams the central parts of which can be described approximately as plane waves, the amplitudes of which only have a weak quadratic dependence on the distance travelled. We will therefore neglect this distance dependence. The arguments we make in this section are largely independent of this assumption but our modelling will not include rigorous determination of systematic errors arising from wave-front curvature.

*Scalar waves:*

We will describe light waves using scalar amplitudes which hide any dependence on the polarisation state of the wave. This happens without loss of generality as we assumed an optically passive medium.

### 8.2.2 Mathematical description of a frequency scanning wave

In this sub-section we develop a general mathematical description of an optical wave with time variable frequency. In sub-section 8.2.3 we will use this to derive the Doppler shift on a moving mirror and in sub-section 8.2.4 we will derive the intensity of a frequency scanning interferometer. We follow and extend the methods shown by [12–17] and will limit our discussion to straight line optical paths (see above assumptions on refractive index) and hence a one-dimensional wave formulation using the notation below.

- $x$ and $t$ indicate the spatial and temporal coordinates.
- $L$ indicates an optical path length $L = n\overline{x_i x_j}$ where $x_i$ and $x_j$ are the co-ordinates of two points and $\overline{x_i x_j}$ is the geometrical distance between them.
- $\phi(t, x)$ is the optical phase of the wave at position $x$ and time $t$. It is to be distinguished from the interferometric phase which we denote with $\Phi(t, x)$.
- $A(\phi(t, x))$ indicates a scalar light amplitude as function of its optical phase $\phi(t, x)$.
- $\mathcal{A}(t, x)$ indicates the instantaneous magnitude of a scalar wave. Its time and spatial dependencies are assumed to be very slow compared to those of $A(\phi(t, x))$.
- $u$[6] indicates the speed of propagation of the phase of the wave in the measurement medium (i.e. the phase speed) with $u = c/n$ where $c$ is the speed of light in vacuum and $n$ is the phase refractive index.
- $\omega(t, x)$ or $\nu(t, x)$ indicate the optical circular or regular frequency of the wave. Where $\omega_e(t)$ is the frequency at emission and $\omega^{(n)}$ indicates the $n^{\text{th}}$ order Taylor expansion of the circular frequency with time.
- $k(t, x)$ is the wave number of the wave where $k = 2\pi / \lambda$ and $\lambda$ indicates the wavelength.
- $I(t, x)$ indicates the intensity of a wave. If it is measured with a detection system of response time $\tau_{det}$ at a fixed point $x$ in space where several waves coincide then $I = \left\langle \left(\sum_{i=1}^{n} A_i(t, x)\right)^2 \right\rangle_{\tau_{det}}$ where the $A_i(t, x)$ are the amplitudes that interfere at the detection point.

The 'usual' formulation for a monochromatic plane wave such as

$$A(t, x) = \mathcal{A}(t, x)\cos(\omega \cdot (t - x/c))$$

is not appropriate any more as it only captures linear dependencies on $t$ and $x$ and omits all cross terms between $t$ and $x$. It is our purpose to find a more general expression for $A(t, x)$.

We may still describe optical waves by real, periodic, trigonometric functions and we choose a cosine function to do so. The scalar light amplitude of a planar optical wave as a function of its optical phase is then given in the most general but not very helpful form below:

[6] We have chosen the symbol $u$ rather than $v$ to avoid confusion with the symbol for optical frequency $\nu$.

$$A(t, x) = \mathcal{A}(t, x)\cos(\phi(t, x)) \tag{8.1}$$

In the above equation the factor $\mathcal{A}(x, t)$ depends on time because the source may vary its intensity with time or the propagation medium may vary in extinction with time. The dependence on $x$ reminds us that real waves are never properly planar and always fall off in some weak quadratic form with propagation distance. Because these dependencies on $x$ and $t$ are very slow we will ignore them in the following.

If our wave has time variable frequency it is no longer strictly periodic in time or space as a monochromatic wave would be. For a monochromatic wave the frequency and wave number characterise the temporal and spatial frequencies across all time and space. We now extend this concept and define the instantaneous angular frequency $\omega(t)$ and instantaneous wave number $k(t)$ as a function of $t$. The instantaneous frequency is defined as the rate of change of phase with respect to time $t$ when the position is fixed. Analogously the wave number is defined as the rate of change of phase with distance $x$ when the time is fixed. We express these definitions and their inverses mathematically via partial differentials of the phase $\phi$ or integrals of the frequency or wave number as follows:

$$\omega(t, x = \text{const}) = \frac{\partial\phi(t, x)}{\partial t} \tag{8.2}$$

$$k(t = \text{const}, x) = \frac{\partial\phi(t, x)}{\partial x} \tag{8.3}$$

$$\phi(t, x_0) = \int_{t_0}^{t} \omega(t', x_0)dt' + \phi(t_0, x_0) \tag{8.4}$$

$$\phi(t_0, x) = \int_{x_0}^{x} k(t_0, x')dx' + \phi(t_0, x_0) \tag{8.5}$$

Equation (8.4) allows us to compute the phase at a given location from the time integral of the instantaneous frequency at the same location. We usually don't know the frequency as a function of time and space but we may know or rather define it at the source. To find the phase at a location a propagation distance $x$ away from that at which the frequency was defined, i.e. to propagate the phase in space, we need one more equation which we give in the form of equation (8.6) below.

$$\phi(t, x) = \phi(t - x/u, 0) = \phi(t - nx/c, 0) = \phi(t - L/c, 0) \tag{8.6}$$

Equation (8.6) assumes that the wave propagates a distance $x$ at speed $u = c/n$ where $c$ is the speed of light in vacuum and $n$ is the refractive index of the propagation medium averaged over time, wavelength and optical path as indicated in sub-section 8.2.1. It applies to freely propagating waves and states that the phase of the wave after a propagation distance $x$ is equal to the phase prior to propagation but at an earlier time $t' = t-x/u$ and that the time shift corresponds to the time it took the wave to propagate the distance $x$.

We now assume that we can model the time varying instantaneous frequency of our wave at the point of its emission ($x = 0$) through a converging Taylor expansion up to order $n$ around an arbitrary origin of time. The Taylor expansion is most suitable for tuning curves that are approximately linear or quadratic which we will focus on in this chapter. We want to emphasise that the theory we present here is applicable to arbitrary modulation functions $\omega(t)$. It may only be necessary to model the interesting fraction of the tuning curve via a Taylor expansion or several expansions could be used for periodic 'forward–backward' scans. If we want to represent smoothly periodic modulation patterns it would be more appropriate to use a Fourier expansion. We will use the expansion only to second order because this is the first non-linear order and higher orders would be cumbersome to note explicitly and don't show fundamentally new aspects.

Rather than quoting all non-linear coefficients in the expansion of the tuning curve $\omega(t)$ it is common to use a gross measure of tuning non-linearity $\alpha_{\text{lin}}$ which we define in equation (8.7).

$$\alpha_{\text{lin}} = \left(\frac{\delta\omega}{\Delta\omega}\right)_{t_f} = \frac{\omega(t_f) - \omega_{\text{lin}}(t_f)}{\omega_{\text{lin}}(t_f) - \omega_{\text{lin}}(t_i)} \tag{8.7}$$

$\alpha_{\text{lin}}$ is the ratio of the deviation of the actual frequency from the frequency expected for a purely linear scan divided by the total change of frequency expected from a linear scan. We evaluate the non-linearity at the end of the scan because that is the time at which it will be largest if we only consider a single power of non-linear contributions (as we do here).

The upper plot of figure 8.2 shows an example of a quadratic scan with 30% tuning non-linearity. We note that for a $\alpha_{\text{lin}} = +10\%$ the tuning rate at the end of the scan already increases by 60%.

$$\omega_e^{(n)}(t, x = 0) = \sum_{i=0}^{n} \left.\frac{d^i\omega}{dt^i}\right|_{t=0} \frac{t^i}{i!} \tag{8.8}$$

$$\omega_e^{(2)}(t, x = 0) = \omega_0 + \dot{\omega}_0 t + \frac{1}{2}\ddot{\omega}_0 t^2 \tag{8.9}$$

where $\omega_0 = \omega(t = 0)$, $\dot{\omega}_0 = \left(\frac{d\omega}{dt}\right)_{t=0}$ and $\ddot{\omega}_0 = \left(\frac{d^2\omega}{dt^2}\right)_{t=0}$

From equations (8.4) and (8.9) we can now find the phase of the wave as a function of time at the point of emission $\phi_e^{(2)}(t, 0)$ and the phase after it has travelled a distance $x$, $\phi^{(2)}(t, x)$

$$\phi_e^{(2)}(t, 0) = \omega_0 t + \frac{1}{2}\dot{\omega}_0 t^2 + \frac{1}{6}\ddot{\omega}_0 t^3 \tag{8.10}$$

$$\phi^{(2)}(t, x) = \omega_0\left(t - \frac{x}{u}\right) + \frac{1}{2}\dot{\omega}_0\left(t - \frac{x}{u}\right)^2 + \frac{1}{6}\ddot{\omega}_0\left(t - \frac{x}{u}\right)^3 \tag{8.11}$$

### 8.2.3 Doppler shift from reflection on a moving mirror

We now find the instantaneous optical frequency $\omega_m(t, x)$ of a wave that is detected at position $x$ and time $t$ as shown in figure 8.1. Before its detection at time $t$ and fixed position $x$ the wave was reflected by a non-recoiling[7] mirror at an earlier time $\tau < t$. At the time of reflection the mirror was at position $x_m(\tau)$ and was moving with speed $u_m(\tau)$. The wave frequency at the location of emission $x_e = 0$ is $\omega_e(t,0)$ which is given as a function of time by equation (8.9).

We will ignore any fixed phase change that might occur at the reflection. It is irrelevant to our understanding of the reflected wave and would only complicate notation.

We assume that the only time dependent position in figure 8.1 is the position of the mirror $x_m(\tau)$. In figure 8.1 the detection point $x$ happens to be at $x = 0$[8] but we will keep $x$ as a symbol in what follows to maintain generality but remember that $x$ is assumed constant in time.

We note that the motion path of the mirror $[x_m(\tau), u_m(\tau)]$ is arbitrary as long as the speed stays below the speed of light in the propagation medium. Equation (8.11) defines the emitted wave by defining its phase for all time and space in the rest frame of its source. This wave propagates freely (remains unchanged) until it hits the mirror at $(\tau, x_m(\tau))$.

After the reflection the wave propagates again freely to the detection point. The wave at the detection point can therefore be implicitly represented by the emitted wave by substituting $x = x_m(\tau)$ and $t = \tau$ in $\phi(x, t)$ (equation (8.11)).

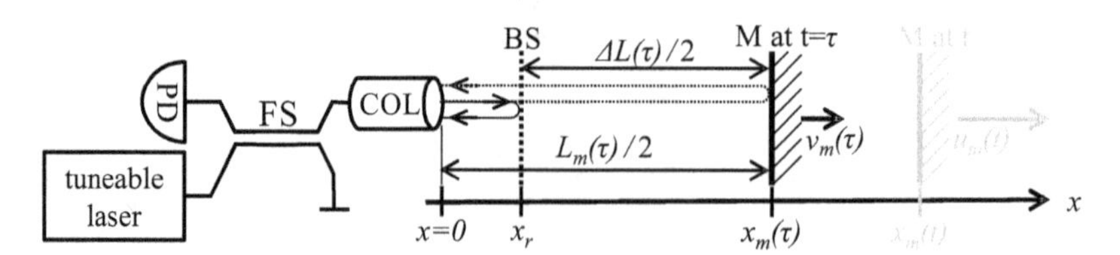


**Figure 8.1.** Basic layout of a fibre coupled Fizeau style interferometer. PD = photo detector, FS = fibre splitter, BS = beam splitter, M = mirror, COL = collimator, $\Delta L$ = optical path difference, $x_m(\tau)$ and $u_m(\tau)$ position and speed of mirror at time of reflection $\tau$. $u_m$ is shown in the positive direction, increasing $x$ and $\Delta L$. Shown in grey are the position $x_m(t)$ and speed $u_m(t)$ of the mirror at time of detection $t$.

[7] The position $x_m$ and speed $u_m$ of the mirror are unaltered by the fact that the wave is reflected from it and as such are independent externally determined variables.

[8] Because we chose the simple form of a Fizeau style interferometer.

This may naively appear to merely evaluate the wave at the mirror position and not at the detection position but the wave travels from the mirror to the detection point without changing its form. In other words, if we knew the reflected wave at the mirror we would also know it at the detection point since it propagates freely in-between.

Given the fact that $\tau$ and $x_m(\tau)$ will be implicitly defined in terms of $x$ and $t$ (see below) we are in fact finding the phase of a wave that would have reached position $x_m(\tau)$ at time $\tau$ given that it must later reach position $x$ at time $t$. The wave that satisfies these conditions will in fact be the wave we are looking for.

The above substitution will make the measured wave a function of three variables $x$, $t$ and $\tau$. The time and position of detection are defined to be $t$ and $x$ and the time of reflection is implied to be $\tau$ if the wave has to be at position $x$ at time $t$. We show this implicit form in equation (8.12).

$$\begin{aligned} \phi_m(t, x, \tau) &= \phi_e(t = \tau, x = x_m(\tau)) \\ &= \omega_0\left(\tau - \frac{x_m(\tau)}{u}\right) + \frac{\dot{\omega}_0}{2}\left(\tau - \frac{x_m(\tau)}{u}\right)^2 + \frac{\ddot{\omega}_0}{6}\left(\tau - \frac{x_m(\tau)}{u}\right)^3 \end{aligned} \tag{8.12}$$

Equation (8.12) is implicit because $\tau$ and $x_m(\tau)$ are implicitly dependent on $x$ and $t$. For simplicity we have dropped the superscript (2) indicating a second order expansion from the left hand side of equation 8.12. We now find the time of reflection $\tau$:

$$\tau = t - \frac{x_m(\tau) - x}{u} \tag{8.13}$$

$$t = \tau + \frac{x_m(\tau) - x}{u} \tag{8.14}$$

This is again an implicit equation. We cannot find $\tau$ analytically for an arbitrary mirror motion path $x_m(\tau)$. We can however find the partial differential $\frac{\partial t}{\partial \tau}$ which we will use later to determine the frequency of the reflected wave. To find $\frac{\partial t}{\partial \tau}$ we remember that $\frac{\partial x_m(\tau)}{\partial \tau} = u_m(\tau)$. To simplify notation we define the ratio of the mirror speed to the speed of the wave as $\beta(\tau) = \frac{u_m(\tau)}{u}$. We have to remember that for the partial differentiation with respect to $\tau$ the independent variable $x$ is considered a constant.

$$\begin{aligned} \frac{\partial t}{\partial \tau} &= 1 + \frac{1}{u}\frac{\partial(x_m(\tau) - x)}{\partial \tau} \\ &= 1 + \frac{u_m \tau}{u} \\ &= 1 + \beta(\tau) \end{aligned} \tag{8.15}$$

Now we are ready to calculate $\phi_m(t, x, \tau)$ We insert $\tau$ from equation (8.13) into equation (8.12) and find

$$\begin{aligned}\phi_m(t, x, \tau) = \omega_0\left(t - \frac{x}{u} - 2\frac{x_m(\tau)}{u}\right)\\ + \frac{\dot{\omega}_0}{2}\left(t - \frac{x}{u} - 2\frac{x_m(\tau)}{u}\right)^2\\ + \frac{\ddot{\omega}_0}{6}\left(t - \frac{x}{u} - 2\frac{x_m(\tau)}{u}\right)^3\end{aligned} \tag{8.16}$$

Equation (8.16) above represents the phase of the measured wave after reflection. In sub-section 8.2.4 we will use it to compute the intensity of an FSI interference signal as a function of time. For this purpose it will be useful to represent it in terms of the optical path length $L_m(\tau)$ travelled by the wave. For notational simplicity we have expressed $L$ a function of $\tau$ rather than $x_m(\tau)$. This will hide the dependence of the phase on the detection position $x$ inside the optical path $L_m(\tau)$ which the wave travelled between the emission and the detection points. If we assume, as shown in figure 8.1, that the measurement happens at the constant position $x = 0$ then $L_m(\tau) = 2n\, x_m(\tau)$ giving:

$$\begin{aligned}\phi_m(t, L_m(\tau)) = \omega_0\left(t - \frac{L_m(\tau)}{c}\right)\\ + \frac{\dot{\omega}_0}{2}\left(t - \frac{L_m(\tau)}{c}\right)^2\\ + \frac{\ddot{\omega}_0}{6}\left(t - \frac{L_m(\tau)}{c}\right)^3\end{aligned} \tag{8.17}$$

As the final step in this sub-section we will compute the instantaneous optical frequency of our detected wave $\omega_m(t, x, \tau)$ from equation (8.16) using the definition of frequency in equation (8.2), remembering to use the chain rule when differentiating $\frac{\partial x_m(\tau)}{\partial t}$. This will define the Doppler shift of this wave.

$$\begin{aligned}\omega_m^{(2)}(t, x, \tau) &= \frac{\partial \phi_m^{(2)}(t, x)}{\partial t}\\ &= \omega_0\left(1 - \frac{2}{u}\frac{\partial x_m(\tau)}{\partial \tau}\frac{\partial \tau}{\partial t}\right)\\ &+ \dot{\omega}_0\left(1 - \frac{2}{u}\frac{\partial x_m(\tau)}{\partial \tau}\frac{\partial \tau}{\partial t}\right)\left(t - \frac{x}{u} - 2\frac{x_m(\tau)}{u}\right)\\ &+ \frac{\ddot{\omega}_0}{2}\left(1 - \frac{2}{u}\frac{\partial x_m(\tau)}{\partial \tau}\frac{\partial \tau}{\partial t}\right)\left(t - \frac{x}{u} - 2\frac{x_m(\tau)}{u}\right)^2\end{aligned} \tag{8.18}$$

We denote the first bracket on each line of equation (8.18) by $Z(\tau)$ which can be simplified by using equations (8.15) and remembering that $\frac{dx_m(\tau)}{d\tau} = u_m(\tau)$.

$$\begin{aligned} Z(\tau) &= \left(1 - \frac{2}{u}\frac{\partial x_m(\tau)}{\partial \tau}\frac{\partial \tau}{\partial t}\right) \\ &= \left(1 - \frac{u_m(\tau)}{u}\frac{1}{1+\beta(\tau)}\right) \\ &= \frac{1-\beta(\tau)}{1+\beta(\tau)} \end{aligned} \tag{8.19}$$

This gives us the final result for the Doppler shifted frequency

$$\begin{aligned} \omega_m^{(2)}(t, x, \tau) &= Z(\tau)\left(\omega_0 + \dot{\omega}_0\left(t - \frac{x}{u} - 2\frac{x_m(\tau)}{u}\right) + \frac{\ddot{\omega}_0}{2}\left(t - \frac{x}{u} - 2\frac{x_m(\tau)}{u}\right)^2\right) \\ \omega_m^{(2)}(t, x, \tau) &= Z(\tau)\omega_e(t - \frac{x}{u} - 2\frac{x_m(\tau)}{u}) \end{aligned} \tag{8.20}$$

Equation (8.20) tells us that the frequency of the measured wave at time $t$ and position $x$ is strictly proportional to the frequency of the emitted wave either at an earlier time $t' = t - 2x_m(\tau)/u$ or a previous position $x' = x - 2x_m(\tau)$. The factor of proportionality is $Z(\tau)$. It applies to an arbitrarily modulated wave and not just to the second order expansion shown.

Most importantly we have to recognise that the Doppler shift is not a property of the wave but a property of the relative motion between the source, mirror and detector.

We recognise the factor $Z(\tau)$ multiplying the emitted frequency as the square of the well known [18] relativistic, longitudinal Doppler shift factor describing the frequency shift observed when a source is moving with fractional speed $\beta$ away from an observer.

The square is expected because in our case the reflector 'observes' the source as moving away from it before the reflection and then the detector 'observes' the reflector move away at the same speed after the reflection. This result holds for an arbitrary frequency scan and an arbitrary mirror trajectory. We note that $Z$ does not depend on the acceleration of the mirror at the time of reflection. Equation (8.20) is however still an implicit equation in the sense that $\tau$ is still a function of $t$ and $x$. If the emitted frequency is fixed at say $\omega_0$ and the optical paths involved are short this problem can easily be glossed over in the sense that the observed frequency at time $t$ is equal to the emitted frequency $\omega_0$ multiplied by $Z(\tau)$. Because the speed of light is so large $\tau$ is very close to $t$ even if path differences are in the km range. We therefore commonly approximate the observed frequency as a function of time to be a measure of the target speed at that same time.

If, however, the emitted frequency is time dependent with a significant rate of change and we are interested in real time determination of the moving targets position we need to bootstrap the above equation by stopping the target at a known time $t'$ for a period longer than $2x_m(t')/u$ and measure this constant $x_m(t')$. After this we may allow the target to move again and we can reconstruct the target motion by measuring $\omega_m(t, x, \tau)$ but we will receive an answer only after a small delay of at least $2x_m(t')/u$.

The rate of change $\dot{\omega}_0$ of the emitted frequency now provides a new technological time scale, other than the natural $x_m(\tau)/u$ to our problem. The larger $\dot{\omega}_0$ the more important the difference between $t$ and $\tau$. To emphasise this we have shown the mirror position both at the time of reflection $\tau$ and the time of measurement $t$ in figure 8.1.

### 8.2.4 Mathematical description of frequency scanning interference

We now have all ingredients necessary to derive the mathematical form of the detected interference intensity $I_{\text{det}}$ in a two beam, frequency scanning interferometer, as depicted in figure 8.1.

We label the two detected wave amplitudes $A_r$ (the reference wave) and $A_m$ (the measurement wave). $A_r$ travels from the collimator to the stationary beam splitter and back and $A_m$ travels from the collimator to the moving mirror and back.

To simplify the following algebra we can set the optical path travelled by the reference wave to zero ($L_r = 0$). This is equivalent to setting the origin of time to the time at which the reference wave has been reflected at the beam splitter or to moving the beam splitter back to the collimator. This will not reduce the generality of our results as they will ultimately only depend on the difference $\Delta L$ between the optical paths of $A_r$ and $A_m$ which is made equal to $L_m$ by the above definitions.

As indicated earlier we will ignore the weak dependence of the magnitudes of the waves on the optical path length but we keep the slow dependence on time that arises from intensity variations of the source.

Before we can proceed we need to clarify the method we will use to compute the intensity of a wave. We wish to present a derivation which applies to both low and high frequency waves. Low frequency waves can be detected by amplitude (voltage) detectors such as antennas whereas waves at optical frequencies can only be detected by power detectors such a photo-diodes. A theory suitable for voltage detection describes the interaction between a real, classical wave and a classical amplitude detector and the time resolution of the detector decides which frequency components will be resolved.

A power detection would be modelled by the interaction of a complex quantum mechanical wave describing the probability density and hence the rate of arriving photons with a classical detector. The appropriate theory would automatically lose frequency information at the optical frequency.

We will use the voltage detection theory even though we wish to describe both low and high frequency (radar and optical) FSI systems. After applying the appropriate time averaging step the low frequency parts of the results will be identical for both methods but the power detection method would lose the optical frequency parts which could be resolvable for low optical frequencies.

The reference and measurement wave amplitudes we wish to interfere are:

$$A_r(t,0) = \mathcal{A}_r(t,0)\cos(\phi_r^{(2)}(t,0))$$
$$A_m(t,\Delta L) = \mathcal{A}_m(t,\Delta L)\cos(\phi_m^{(2)}(t,\Delta L(\tau)))$$

The corresponding reference and measurement phases are:

$$\phi_r^{(2)}(t,0) = \omega_0 t + \frac{\dot{\omega}_0}{2}t^2 + \frac{\ddot{\omega}}{6}t^3$$
$$\phi_m^{(2)}(t,\Delta L(\tau)) = \omega_0\left(t - \frac{\Delta L(\tau)}{c}\right) + \frac{\dot{\omega}_0}{2}\left(t - \frac{\Delta L(\tau)}{c}\right)^2 + \frac{\ddot{\omega}}{6}\left(t - \frac{\Delta L(\tau)}{c}\right)^3$$

The intensities of the reference and measurement waves on their own would be:

$$I_r(t,0) = \frac{1}{2}\langle \mathcal{A}_r^2(t,0)\rangle_{\tau_{\mathrm{det}}}$$
$$I_m(t,\Delta L) = \frac{1}{2}\langle \mathcal{A}_m^2(t,\Delta L)\rangle_{\tau_{\mathrm{det}}}$$

where the factors of 1/2 arise from the time averages of the $\cos^2(\phi_r)$ and $\cos^2(\phi_m)$ terms which oscillate at optical frequencies and which we assume to be too fast to be resolved. The remaining time averages over the slowly varying magnitude functions are to be taken over the response time of the detector $\tau_{\mathrm{det}}$.

The interference between the reference and the measurement wave will lead to the following detected intensity:

$$I_{\mathrm{det}}(t,\Delta L) = \langle A_r^2(t,0) + A_m^2(t,\Delta L) + 2A_r(t,0)A_m(t,\Delta L)\rangle_{\tau_{\mathrm{det}}}$$

which can be re-written in terms of an average intensity $I_0(t,\Delta L)$ which only varies slowly with time and the interference intensity $I_{\mathrm{int}}(t,\Delta L)$ which carries the interesting time dependence:

$$I_{\mathrm{det}}(t,\Delta L) = I_0(t) + I_{\mathrm{int}}(t,\Delta L)$$

The average and interference intensities relate to the intensities of the individual reference and measurement wave intensities as follows:

$$\begin{aligned} I_0(t,\Delta L) &= I_r(t) + I_m(t,\Delta L) \\ I_{\mathrm{int}}(t,\Delta L) &= \sqrt{I_r(t)I_m(t,\Delta L)}\,\langle\cos(\phi_r(t)\phi_m(t,\Delta L))\rangle_{\tau_{\mathrm{det}}} \end{aligned} \tag{8.21}$$

The interesting term in equation (8.21) is of course the cross- or interference-term $I_{\mathrm{int}}$ because the other two terms are constant after the time averaging to within the slow time variation of the laser intensity. We will transform $I_{\mathrm{int}}$ using the trigonometric 'product-to-sum' identity: $\cos(a)\cos(b) = 1/2(\cos(a+b) + \cos(a-b))$

$$\begin{aligned} I_{\mathrm{int}} &= \sqrt{I_r I_m}\,\langle\cos(\phi_r + \phi_m) + \cos(\phi_r - \phi_m)\rangle_{\tau_{\mathrm{det}}} \\ I_{\mathrm{int}} &= \sqrt{I_r I_m}\,\langle\cos(\phi_+) + \cos(\phi_-)\rangle_{\tau_{\mathrm{det}}} \end{aligned} \tag{8.22}$$

We note that the phases $\phi_+$ and $\phi_-$are not optical phases of the detected wave. They represent different spectral components of the optical phase. The lower frequency component $\phi_-$will turn out to be the interference phase.

In order to compute the time averages in equation (8.22) we have to evaluate how fast $\phi_+$ and $\phi_-$vary with time. The computation is very similar to that leading to the frequency of the Doppler shifted wave in equation (8.18)

$$\begin{aligned} \phi_+^{(2)} &= \phi_r^{(2)} + \phi_m^{(2)} \\ \phi_+^{(2)} &= \omega_0\left[t + \left(t - \frac{\Delta L}{c}\right)\right] \\ &+ \frac{\dot{\omega}_0}{2}\left[t^2 + \left(t - \frac{\Delta L}{c}\right)^2\right] \\ &+ \frac{\ddot{\omega}_0}{6}\left[t^3 + \left(t - \frac{\Delta L}{c}\right)^3\right] \end{aligned} \tag{8.23}$$

$$\begin{aligned} \omega_+^{(2)} &= \frac{\partial \phi_+}{\partial t} \\ &= \omega_0[1 + Z] \\ &+ \dot{\omega}_0\left[t + Z\left(t - \frac{\Delta L}{c}\right)\right] \\ &+ \frac{\ddot{\omega}_0}{2}\left[t^2 + Z\left(t - \frac{\Delta L(\tau)}{c}\right)^2\right] \end{aligned} \tag{8.24}$$

To evaluate $\omega_+^{(2)}$ we need to find the technically plausible values for $\dot{\omega}_0$, $\ddot{\omega}_0$, $Z$, $\Delta L$ and $t$ which we give in table 8.1. How these were estimated will be explained in subsection 8.4.1.

**Table 8.1.** Extreme ranges of parameters influencing the frequency of an interference signal.

| Name | Minimum | Typical | Maximum | Units | Explanation |
|---|---|---|---|---|---|
| $\Delta L$ | 0 | 1 | 1000 | m | OPDs |
| $\lvert u_m \rvert$ | 0 | 0.01 | 1000 | m s$^{-1}$ | target speeds |
| $\lvert a \rvert$ | 0 | 0.1 | 10 | m s$^{-2}$ | target accelerations |
| $\lvert Z - 1 \rvert$ | 0 | $6.6 \times 10^{-11}$ | $1 + 6.6 \times 10^{-6}$ | 1 | $Z$ for above speeds |
| $\lvert \dot{\omega}_0 \rvert$ | 0.0008 | 0.08 | 1.6 | Prad s$^{-1}$ | tuning speeds in FSI |
| $t$ | 0.001 | $1 \times 10^3$ | $1 \times 10^5$ | ms | scan durations in FSI |
| $\Delta L / c$ | 0 | $3 \times 10^{-6}$ | $3 \times 10^{-3}$ | ms | measurement arm delays |
| $\frac{\Delta\omega}{\omega_0}$ | $6 \times 10^{-4}$ | $6 \times 10^{-3}$ | 0.32 | 1 | fractional tuning range |

The first term in equation (8.24) is very close to twice $\omega_0$. From table 8.1 we see that the $\Delta L / c$ terms can at most double the value of the terms in the round brackets in equation (8.24) so that the square brackets in the last two lines are still of the same order as $t$ or $t^2$.

The time dependent terms in equation (8.24) are therefore limited by the scan duration which in turn is limited by the total tuning range. The time dependent terms therefore only introduce a fractional change of order $\Delta\omega / \omega_0$ which is at most 13% for lasers with extreme scanning ranges. Consequently the frequency $\omega_+^{(2)}$ is dominated by the first line of equation (8.24) and remains close to twice the original optical frequency. No current optical detection system can resolve twice the optical frequency and the $\cos(\phi_+)$ terms average to zero.

If we were interested in radio waves at this point we may be able to resolve them and the interference phase would turn out to have two components arising from $\phi_+$ and $\phi_-$. We focus here on the optical systems and average the $\phi_+$ terms to zero.

The $\phi_-$ term in equation (8.22) however varies much more slowly

$$\phi_-^{(2)}(t, \Delta L) = \phi_r^{(2)}(t) - \phi_m^{(2)}(t, \Delta L)$$

$$\begin{aligned} \phi_-^{(2)}(t, \Delta L) = \omega_0\left[t - \left(t - \frac{\Delta L}{c}\right)\right] \\ + \frac{\dot{\omega}_0}{2}\left[t^2 - \left(t - \frac{\Delta L}{c}\right)^2\right] \\ + \frac{\ddot{\omega}_0}{6}\left[t^3 - \left(t - \frac{\Delta L}{c}\right)^3\right] \end{aligned} \tag{8.25}$$

$$\omega_-^{(2)}(t, \Delta L) = \frac{\partial\phi_-(t, \Delta L)}{\partial t}$$

$$\begin{aligned} \omega_-^{(2)}(t, \Delta L) = \omega_0[1 - Z] + \\ \dot{\omega}_0\left[(1 - Z)t + Z\frac{\Delta L}{c}\right] \\ + \frac{\ddot{\omega}_0}{2}\left[(1 - Z)t^2 - Z\frac{\Delta L}{c}\left(2t - \frac{\Delta L}{c}\right)\right] \end{aligned} \tag{8.26}$$

We are now ready to identify $\phi_-$from equation (8.25) as the interferometric phase $\Phi$. With this hindsight it becomes easier to define what we actually mean by interferometric phase which we avoided doing so far.

> The interferometric phase $\Phi$ of an optical wave is a property of its time dependent intensity. It is the spectral fraction of the optical phase $\phi$ of this wave which falls within the detectable bandwidth of the system measuring the waves intensity. It is commonly a low frequency component of the optical phase. It can often be identified as the argument of the periodic function describing the measured intensity after averaging over the response time of the detection system.

We want to point out a commonly used simplified expression for the interferometric phase which is most often not properly justified or derived. We re-write equation (8.25) by opening all brackets and re-ordering the terms to find:

$$\begin{aligned}\phi_{-}^{(2)} &= (\omega_0 + \dot{\omega}_0 t + \ddot{\omega}_0 t^2)\frac{\Delta L}{c} + \left(\frac{\Delta L}{c}\right)^2\left[-\frac{\ddot{\omega}_0}{2}t + \frac{\dot{\omega}_0}{2} + \frac{\ddot{\omega}_0}{6}\frac{\Delta L}{c}\right] \\ &= \omega_e(t)\frac{\Delta L}{c} + \left(\frac{\Delta L}{c}\right)^2\left[-\frac{\ddot{\omega}_0}{2}t + \frac{\dot{\omega}_0}{2} + \frac{\ddot{\omega}_0}{6}\frac{\Delta L}{c}\right]\end{aligned} \tag{8.27}$$

As we will see in section 8.3 we will not be able to extract the absolute value of the interferometric phase but only phase changes. As such any constant or very slowly varying terms in the square brackets are not only extremely small but also irrelevant.

The terms arising from the square brackets vary either explicitly with time or implicitly with via a time variable distance or both. The first and third terms vanish for purely linear scans.

We now assume a strongly non-linear scan with a 50% non-linearity as defined in equation (8.7), a large tuning range of 13% of $\omega_0$ a large OPD of 10 m and a large target speed of 1 m s$^{-1}$ to compute the three contributions to the phase change from the square bracket and its pre-factor under these adverse conditions. They contribute 0.89 rad, 0.018 rad and $2.9 \cdot 10^{-9}$ rad respectively.

This is to be compared with the $\omega_e(t)$ term for the same scan which will change by $5.69 \cdot 10^6$ rad. We can therefore safely neglect the second and third term in the square bracket and maintain an accuracy of $3 \cdot 10^{-9}$. The first term would contribute at the $1.6 \cdot 10^{-7}$ level and for some accuracy requirements or even faster targets this would have to be accounted for.

Neglecting the entire square bracket gives us an approximation of the interferometric phase.

$$\Phi \approx \omega_e(t)\frac{\Delta L}{c} \quad \text{for slow target speed and approx. linear tuning} \tag{8.28}$$

This same expression can also be derived by assuming that the frequency of the reference and measurement wave are identical when they arrive at the detection point. It is therefore sometimes referred to as the 'instantaneous' or 'zero delay' approximation. It only accounts for linear elements from the tuning speed.

Equations (8.26) and (8.25) include the fully relativistic form of the Doppler shift element $Z$. It is useful to look at these expressions for the simplified special case of a fixed frequency where $\dot{\omega}_0$ and $\ddot{\omega}_0$ are both zero. We now have two interpretations of the remaining term $\omega_{-}^{(0)} = \omega_0[1 - Z]$ in equation (8.26). We can think of it either as the beat frequency between the reference wave and the Doppler shifted measurement wave or as the rate of fringes that appear in a regular fixed frequency interferometer with a moving reference arm. These interpretations are in fact equivalent.

We can find another useful simplified form of $\omega_{-}$ by returning to the frequency scanning case but considering a stationary target. We now have $\beta = 0$ and $Z = 1$ which gives the particularly simple form of $\omega_{-}^{(2)}$ $(t, \Delta L = \text{const})$.

$$\omega_-^{(2)}(t, \Delta L = \text{const}) = \dot{\omega}_0 \frac{\Delta L}{c} + \frac{\ddot{\omega}_0}{2} \frac{\Delta L}{c}\left(2t - \frac{\Delta L}{c}\right) \text{for stationary targets} \quad (8.29)$$

For the speeds commonly present in metrological applications of FSI we need not drop all motion terms but we can approximate them to first order in $\beta$. The first order approximations we will make are given below and remain accurate to $2\beta^2 = 1 \cdot 10^{-11}$ for speeds up to 1000 m s$^{-1}$.

$$\begin{aligned} Z &\approx 1 - 2\beta + 2\beta^2 + \mathcal{O}(\beta^3) \approx 1 - 2\beta \\ 1 - Z &\approx 2\beta + 2\beta^2 + \mathcal{O}(\beta^3) \approx 2\beta \end{aligned} \quad (8.30)$$

Inserting these into equation (8.26) yields:

$$\begin{aligned} \omega_-^{(2)}(t, \Delta L) \approx\ & \omega_0 2\beta \\ & + \dot{\omega}_0\left[2\beta t + (1 - 2\beta)\frac{\Delta L}{c}\right] \\ & + \frac{\ddot{\omega}_0}{2}\left[2\beta t^2 - (1 - 2\beta)\frac{\Delta L}{c}\left(2t - \frac{\Delta L}{c}\right)\right] \end{aligned} \quad (8.31)$$

This is an approximate expression for the frequency of an FSI signal with its dependence on the tuning characteristic and the motion of the measurement mirror. Many FSI techniques measure this frequency to determine $\Delta L$. We show the approximate form here because it is easier to see the proportionality of certain terms to target speed.

In figure 8.2 we show how optical and interference frequencies develop over time for a scan that starts at a wavelength of 1600 nm with a tuning rate of 39.2 Trad s$^{-1}$ (initially equivalent to 50 nm s$^{-1}$) and a 30% tuning non-linearity. The target starts at an OPD of $L_0$ of 10 m and moves at one of two constant speeds of $\pm$ 1 m s$^{-1}$.

The top plot shows the optical frequency of the emitted wave $\nu_r$ for a linear and a non-linear scan. We also indicate the frequency deviations from the linear scan $\delta\nu$ and the total linear tuning range $\Delta\nu$.

The bottom plot shows how the interference frequency $f_{\text{int}}$ (solid lines) and the difference between the optical frequencies of the reference and measurement waves, $\Delta\nu = \nu_r - \nu_m$ (dashed lines) for positive (thick red lines) and negative (thin black lines) target speeds.

We note that for the negative target speed the $\Delta\nu$ is negative (phase reduces with time) indicating that the Doppler effect increased $\nu_m$ but $f_{\text{int}}$ remains positive as the frequency scan outweighs this effect and moves the phase forward. We also note that $f_{\text{int}}$ is falling because $L_m$ is decreasing. For negative speeds and positive tuning rates the Doppler effect and the frequency tune work against each other and overall reduce $f_{\text{int}}$ compared to the positive speed case.

To understand the limitations of FSI techniques we have to understand when $f_{\text{int}}^{(2)}$ will exceed $f_{\text{Nyq}}$ (the maximally measurable or Nyquist frequency). We also need to know when it approaches zero because many techniques cannot measure a

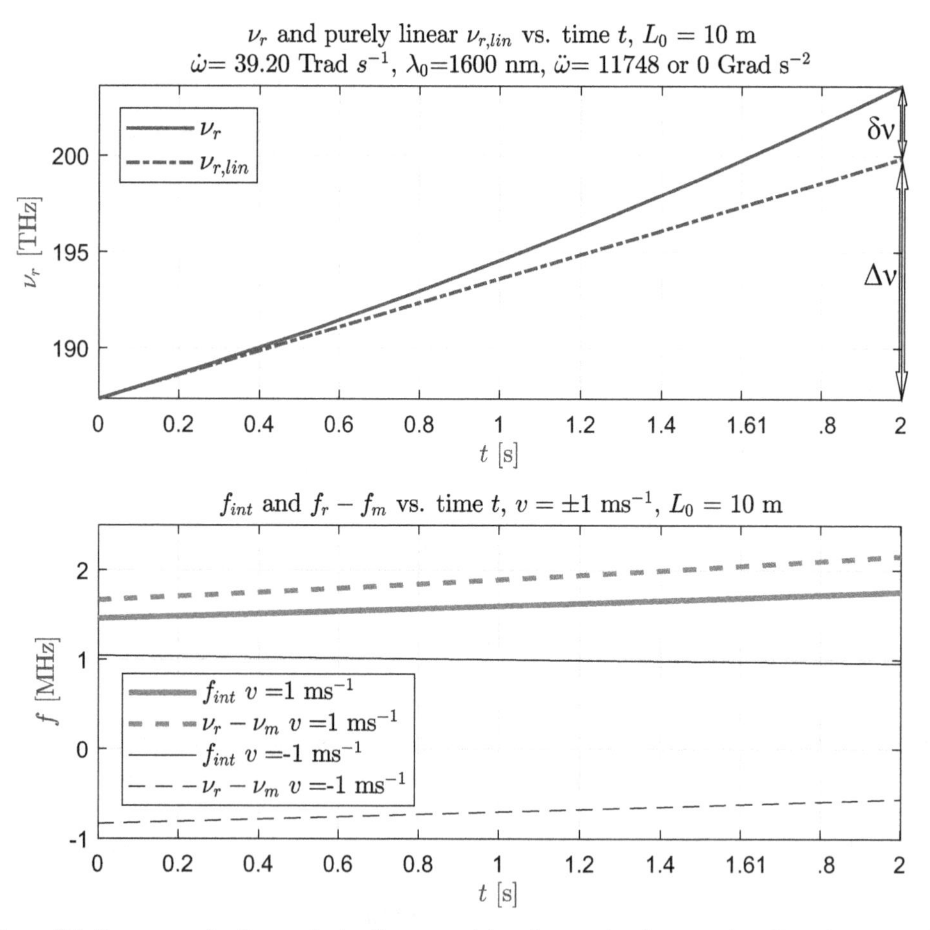


**Figure 8.2.** Frequency of reference (emitted) wave and interference signal versus time. Top: frequency of reference wave at 30% (solid) or 0% (dashed-dot) tuning non-linearity. Bottom: frequency of interference signal (solid) and frequency difference between reference and measurement waves (dash) for positive (thick, red) and negative (thin, black) target speed.

frequency or extract a phase when the frequency drops below the frequency resolution.

Many techniques based on the measurement of a single intensity also cannot measure the sign of a frequency. Figure 8.3 therefore plots the modulus of $f_{\text{int}} = \omega_{\text{int}}^{(2)}/2\pi$ versus pairs of the independent variables $\dot{\omega}_0$, $u_m$ and $\Delta L$.

In figure 8.3 the range of the independent variables has been chosen to represent a typical and currently technically possible range as well as motion speeds possible in metrology problems. For the right column the $\ddot{\omega}_0$ term has been chosen to increase the total tuning excursion at the end of the scan by 10% compared to the purely linear case shown in the left column. Comparing the left column of plots (purely linear tuning) with those on the right (+10% non-linear tuning) we see that $f_{\text{int}}$ always increases in the non-linear case and changes the distance (top row) or the speed (middle row) at which $f_{\text{int}}$ hits zero.

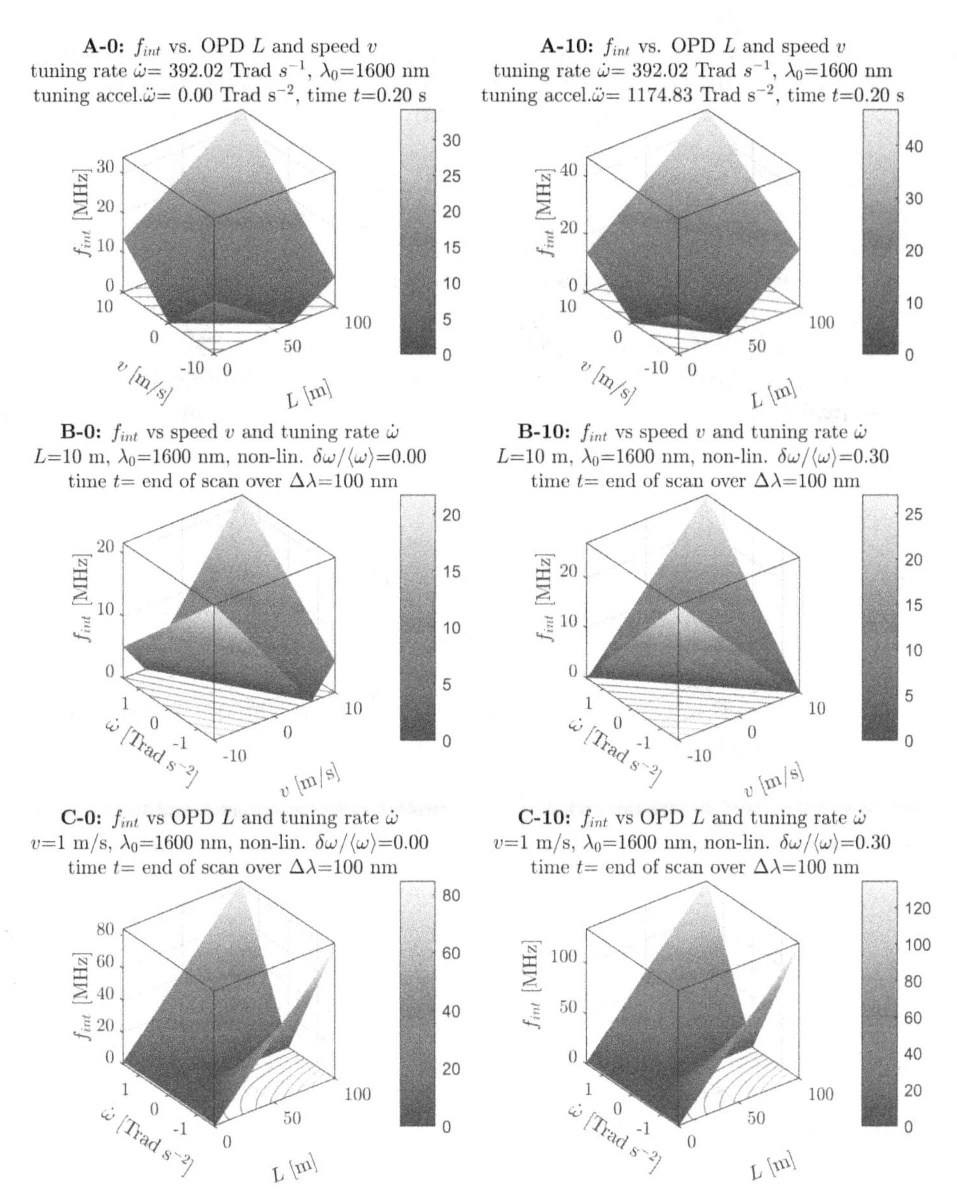


**Figure 8.3.** Absolute frequency of the interference signal. The left column shows a purely linear scan, the right that for a non-linearity of 30%. All times are at the end of the scan. The rows show $f_{\text{int}}$ versus top OPD $L$ and speed $u$; middle $u$ and $\dot{\omega}_0$; bottom $L$ and $\dot{\omega}_0$.

In the top row we can see how $f_{\text{int}}$ changes with speed and OPD. We see that for the given positive initial tuning rate the speed of motion either increases or decreases $f_{\text{int}}$. The sign of this change depends on the sign of the target speed $u_m$. The line of $f_{\text{int}} = 0$ goes through the point ($L = 0$, $u_m = 0$) and lies entirely in the domain of negative speeds. The slight increase of the slope of this line in the $u_m$ versus $L$ plane due to a non-zero $\ddot{\omega}_0$ expresses the increased tuning rate at the end of the scan. In other words, the negative speed necessary to suppress $f_{\mathrm{int}}$ down to zero for a given $L$ is larger in the presence of the additional positive tuning non-linearity.

In the middle row we see how $f_{\mathrm{int}}$ varies with tuning rate and speed at a fixed OPD. As expected $f_{\mathrm{int}}$ achieves the same maximal value when tuning rate and speed are both maximally positive or both maximally negative. This is true for both linear and non-linear scans. The line of $f_{\mathrm{int}} = 0$ goes through the point ($\dot{\omega}_0 = 0$, $u = 0$). Its slope in the $\dot{\omega}_0$ versus $u$ plane reduces as the tuning non-linearity is added showing that the speed necessary to suppress $f_{\mathrm{int}}$ to zero at a given OPD becomes larger as the positive tuning non-linearity is added.

In the bottom row we see how $f_{\mathrm{int}}$ varies with tuning rate and OPD. The left and right plots are identical in shape and only show a scale factor applied to $f_{\mathrm{int}}$ which increases by 60% because this is the increase in instantaneous tuning rate at the end of the scan when the tuning non-linearity is added.

## 8.3 FSI analysis techniques

In this section we will show methods to extract the absolute OPD of a frequency scanning interferometer from the phase or the frequency of its interference signal. Common to many methods is the need to extract the phase of the interferometric signal from its intensity and we will show some methods for this in sub-section 8.3.1.

If the FSI system is to measure distances over an indefinite period it has to tune its laser in a periodic manner because no source can be tuned in wavelength over an indefinite range, nor can any sensor detect waves over an indefinite wavelength range. We refer to a single period of this tuning process $\tau_{\mathrm{scan}}$ as a scan.

We now categorise FSI techniques into two classes based on the relation between the repetition rate of measurements $f_{\mathrm{rep}} = 1/\tau_{\mathrm{rep}}$ and the repetition rate of the periodic tuning process $f_{\mathrm{scan}} = 1/\tau_{\mathrm{scan}}$.

*Shot-by-shot FSI techniques:*

FSI techniques can be categorised as 'shot-by-shot' if they measure a single average distance during the scan. These techniques are often based on the total interferometric phase change during a scan. If wavelength scans are limited to small wavelength ranges 'shot-by-shot' repetition rates can reach the kHz range. Since the accuracy of nearly all FSI techniques benefits from large scanning ranges and scanning speeds are technically limited, common shot-by-shot rates are below the 10 Hz range. We will describe shot-by-shot analysis techniques in sub-section 8.3.2.

*Dynamic FSI techniques:*

FSI techniques can be characterised as dynamic (dFSI) if they measure multiple distances during a scan. We will describe dynamic analysis techniques in sub-section 8.3.3. The rate of distance measured can at most be equal to the rate at which the data acquisition systems (DAQ) can record the intensity of the interferometric signal.

It is often the case that dynamic measurements can only be made in a certain part of the scan which for example excludes extrema of the scan where the tuning speed vanishes. Hence we may be faced with gaps in the time sequence of distance measurements. In some techniques it is possible to stitch shots together without gaps [19] or to alternate between absolute and differential interferometry [19] and so produce a continuous high frequency set of measurements over an extended period covering many scans.

### 8.3.1 Phase extraction methods

There are many ways to extract the interferometric phase of an interference signal from a time series of intensities and we describe a few of them here. As phase extraction is a fundamental element of many techniques and useful in their categorisation we devote a significant sub-section to it. Because the measurable interference intensity depends on the phase only via periodic functions we cannot directly extract the phase from a single intensity measurement. We therefore have to start phase extraction by finding the phase inside $2\pi$-sized intervals of the intensity[9]. We will call this periodic phase the wrapped phase.

The phase as described by our equations in section 8.2 is however a continuous, unlimited variable which can be extracted from a time series of wrapped phases by an unwrapping process which adds or subtracts $2\pi$ to the wrapped phase whenever it leaves its wrapped range. We will not discuss phase unwrapping methods and point the reader to references [12, 20, 21].

The unwrapping process requires us to know how the wrapped phase changes from one to the next intensity measurement. This presents us with the additional problem that the intensity is not only a periodic function of the phase but also symmetric around the locations of its extrema. This means that we cannot obtain the sign of the phase change when the intensity is at an extreme point without assumptions or additional information. There are two principle categories of solutions to this problem.

*Measuring multiple intensities*

In this solution category we have to measure the intensity at several phases with known phase offsets but for the same OPD. We will explain the two most common methods for this below.

The most common sub-class is referred to as polarisation based quadrature detection. It uses orthogonal polarisation states for the reference and measurement waves as shown in figure 8.4. We detect two interference signals as shown. As only horizontally (vertically) polarised light enters the measurement (reference) arm we can selectively delay the reference wave by $\pi/2$ on one of the detectors (PD1).

The two detectors will now measure the interference intensities twice with a phase difference of $\pi/2$ between them making one of the measurements sine and the other cosine like. The phase can now be reconstructed as the arc-tangent of the ratio of both measurements provided that PD1 and PD2 receive the same average intensity which can be arranged if the half-wave plate is appropriately oriented. This method

[9] The ranges are typically either $[-pi, \pi]$ or $[0, 2\pi]$.

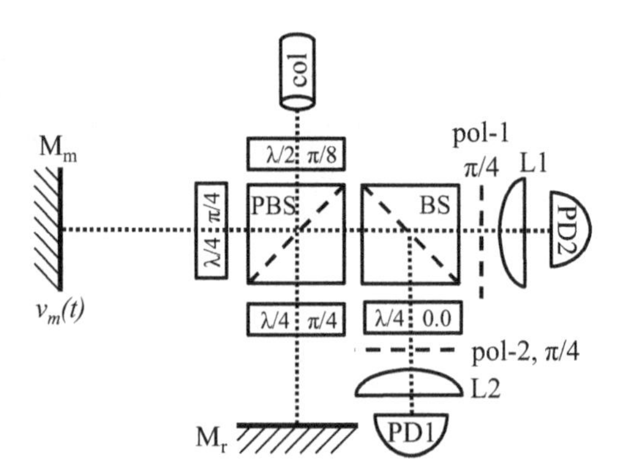


**Figure 8.4.** Polarisation based quadrature readout. PBS = polarising beam splitter transmitting vertical polarisation, BS = beam splitter, $(\lambda/N, \alpha)$ = quarter-wave ($N = 4$) or half-wave ($N = 2$) plate with optic axis at angle $\alpha$ to the vertical. POL = polariser, L = lens, PD = photo detector, COL = collimator emitting vertically polarised wave.

will only work if the propagation medium is optically passive (leaves the polarisation unaffected).

The price we pay is a significantly increased complexity and cost of the front end optics compared to figure 8.1, the need for a polarisation maintaining light delivery system and two detectors and DAQ channels per interferometer.

Polarisation based quadrature however works only approximately because the retarders can only produce a fixed retardance and will only be quarter or half-wave plates at one wavelength of the scan. Since scan ranges are normally small compared to the total wavelength (less than 13% for optical wavelength) the error introduced is similarly small. If we know the wavelength of the source as a function of time and we know the exact retardance we can compensate for this error provided that the OPD of the measurement arm is 'small' and does not change much during the scan. The conditions on the measurement arm length are necessary to justify the assumption that the wavelengths of the two waves traversing the retarders at any given time are approximately the same. In practice this only constrains the product of arm length and tuning speed.

We can improve the accuracy of the polarisation based quadrature method if we replace the fixed retarders can with electro-optic phase retarders and adjust their retardance during the scan. However, the assumption of short measurement path is still necessary.

If a wide beam with an approximately flat transverse phase profile can be obtained back from the measurement arm reflector the above separation of beams by polarisation can be replace by transverse separation in space.

In essence the reference mirror can take the shape of a stepped mirror and the resulting phase-stepped reference wave is interfered with the reference beam on multiple closely spaced detectors. We call this method transverse phase stepping or transverse phase shifting.

Such stepped mirrors can be implemented by spatial light modulators and the multiple detectors can be CCDs or CMOS light light sensors. As SLMs can be actively controlled they can also implement an adjustment of the retardance during the scan. Although these techniques do not require polarisation handling optics, CCD based techniques are limited in readout frequency which in term restricts the useable range of tuning speeds and the detectable range of target speeds because the readout frequency must always be sufficient to measure the interference frequency. SLMs are also a significant cost factor given that one is needed per interferometer.

Another method to obtain measurements at multiple interferometric phases shifts the position of the reference mirror. We refer to this method as temporal phase stepping or temporal phase shifting. For temporal phase stepping to work we have to move the mirror faster than the reference arm will change. In its simplest form the reference mirror would move by $\lambda/4$ giving us a phase change of $\lambda/2$ and we can apply the arc-tangent algorithm from the polarisation based quadrature method to the two consecutive phase-stepped intensity values.

The advantages or temporal phase stepping over transverse or polarisation phase stepping are that we are more or less guaranteed to get the same 'average' intensity underlying both intensity values, we don't need to maintain polarisation in the interferometer and we only need a single detector and DAQ channel.

We will however be relatively sensitive to errors in the mirror's movement. The method becomes more robust against such errors if we use more phase steps. As an example we point to the Carree algorithm which uses four steps of approximately 120° [20].

If we simply move the reference mirror continuously along a periodic trajectory and sample the intensity rapidly we can get many points on the phase axis and fit a sinusoid to the resulting intensity versus phase-shift axis to determine the phase of the interference signal [20]. This becomes very robust against slow changes of the trajectory.

The primary disadvantage of temporal phase stepping methods is that they place a limit on the target speed which must be much slower than the stepping speed of the reference arm. The detection must also happen synchronously with the phase stepping which severely limits the detection rate.

*Assuming monotonous phase changes*

In the following we will show how to extract the phase of the interference signal for an entire frequency scan from one set of the interferometers intensities alone. We will make use of a Hilbert Transform for this purpose which in essence can generate a 90° phase shifted copy of a real signal under certain circumstances described below. Our signal will of course be the detected interference intensity. We look back at equation (8.21) and notice that our interference intensity has the general form.

$$I_{\text{det}}(t, \Delta L) = I_0(t) + b(t)\cos(\Phi(t, \Delta L))$$

Where $I_0(t) = I_r + I_m(t, \Delta L)$ and $b(t) = \sqrt{I_r I_m(t, \Delta L)}$ and we summarise the rapidly time variable part of the signal into $s(t, \Delta L) = b(t)\cos(\Phi(t, \Delta L))$ We now make some assumptions about the frequency spectrum of both $I_0(t)$ and $s(t)$ taken across the entire scan. We obtain the frequency spectrum of $I(t)$ via its Fourier transform $\mathcal{F}[I(t)]$.

1. The frequency spectrum $\mathcal{F}[s(t)]$ lies entirely at frequencies above some threshold $\omega_{hp}$ and does not overlap with the spectrum $\mathcal{F}[I_0(t)]$ which lies entirely below $\omega_{hp}$. We can therefore obtain $s(t)$ separately by high-pass filtering $I_{\text{det}}(t, \Delta L)$ with spectral filter function $W_{hp}(\omega)$.
2. The spectrum $\mathcal{F}[s(t)]$ does not extend to zero frequency.

Under these assumptions we may obtain the phase from a time series of intensities $I(t)$ [12]. The Hilbert transform is a well understood and frequently used method in signal processing. The Hilbert transform of a real function $f(t)$ is defined via a consecutive pair of forward and backward Fourier transforms as follows:

$$\mathcal{H}[f(t)] := \mathcal{F}^{-1}[\mathcal{F}[f(t)]H(\omega)],$$

and the Hilbert window function $H(\omega)$ is defined as

$$H(\omega) := -i\ \text{sgn}(\omega) \tag{8.32}$$

where $\text{sgn}(\omega)$ is the sign function. It is defined as $\text{sgn}(0) = 0$, $\text{sgn}(\omega) = +1$ if $\omega > 0$ and $\text{sgn}(\omega) = -1$ if $\omega < 0$.

The wrapped phase $\Phi(t, \Delta L)$ is then found using a four quadrant arctan function as follows

$$\Phi(t, \Delta L) = \arctan\left(\frac{\text{Im}(\hat{s}(t))}{\text{Re}(\hat{s}(t))}\right)$$

$$\hat{s}(t) = \mathcal{F}^{-1}[\mathcal{F}[I(t)W_{hp}(\omega)](1 + iH(\omega))]$$

The advantage of the Hilbert transform method is the fact that no additional hardware is needed to generate additional measurements of the interference signal at different phases. The disadvantage is the limitation on the interference frequency which in turn limits the acceptable target speed for a given tuning speed or vice versa.

### 8.3.2 Shot-by-shot FSI analyses

The variables from which to obtain a single distance measurement over a given scan are:

*Interferometric phase change* $\Delta\Phi$:

If we assume a stationary target, $\Delta\Phi$ will be strictly proportional to the OPD via $\Delta\Phi = \Delta\omega\frac{L}{c}$, as we will show later. If we allow the OPD to change during the scan $\Delta\Phi$ will turn out to be a function of the average OPD $L_0$, the change in OPD $\delta L$, the total frequency change $\Delta\omega$ as well as the scan's centre frequency $\omega_0$. We will therefore have to determine $\Delta\omega$, $\omega_0$ and $\delta L$ in order to find $L_0$.

Before we discuss how to do this below we want to emphasise that, although $\Delta\Phi$ is the difference between the initial and final interferometric phase we nevertheless have to continuously measure the phase throughout the scan if we scan further than $\pi$ in phase. This is due to the periodicity problem we mentioned earlier and the fact that we need to unwrap the phase across the scan.

*Interferometric frequency $\omega_{\text{int}}$:*

For stationary targets and linear scans $\omega_{\text{int}}$ is strictly proportional to $\frac{L}{c}$ and the constant of proportionality is $\dot{\omega}_0$ as equation (8.29) shows. In this simple linear scan and stationary target case we can simply take the Fourier transform of the interference intensity and will observe a peak $\dot{\omega}_0 \frac{L}{c}$. Measuring $\dot{\omega}_0$ against a frequency standard (see below) will yield $L$. Details of how to cope with non-linear scans with stationary targets follow later and for moving targets we will need to use two scanning lasers as described in section 8.3.3.

*Optical frequency measurement during the scan*

Since methods using $\omega_{\text{int}}$ or $\Delta\Phi$ both need to measure some aspects of the optical frequency $\omega_0$, $\Delta\omega$ or $\dot{\omega}_0$ we continue with a brief description of optical frequency measurement during a scan. There are three principle methods to measure $\Delta\omega$ and $\omega_0$ during a scan.

We can measure by comparison to one of three types of frequency reference. Such a reference could be a suitably calibrated and stable, passive, man-made artefact such as an RF or optical cavity or etalon or reference interferometer (i.e. a low finesse cavity).

Alternatively it could be a 'natural' frequency standard such as a gas absorption spectrum or last, but certainly not least, it could be a frequency comb that is referenced to a natural RF standard.

In the case of a passive man-made artefact its calibration tells us the amount of phase advance the artefact produces per unit frequency change and hence our measurement of $\Delta\omega$ turns out to be the measurement of the phase increment of the reference instrument. The advantage of using a man-made artefact is the fact that it can produce many spectral features if it has a suitably small free spectral range and we can therefore determine the phase increments at many points in the scan. This advantage also allows us to extract $\dot{\omega}_0$ from the same measurement. The primary disadvantage is that we are using a secondary standard (an artefact) which we have to maintain calibrated.

In the case of a 'natural' standard all we need to measure $\Delta\omega$ are two reference features, ideally close to the start and end of the scan. We could then plot the absorption spectrum across the scan versus time and determine, via at least two fits to absorption features, the exact time in the scan when the laser frequency coincided with the known absorption frequencies of the reference features.

The main advantage of natural standards is the fact that they are not man-made. Nature itself defines the absorption frequencies via the laws of physics. 'All' we have to do is to control the secondary properties that can influence these features such as pressure (pressure shift), temperature or composition of a gas cell. To achieve high accuracies we also have to know the probing conditions under which we extract the absorption spectrum. We could for example compensate for the power shift. The power shift is the change of absorption centre with probe beam power which can be of order kHz per mW. We may also wish to control electric and magnetic fields in the reference.

One disadvantage of using natural standards is that we only know the frequencies in the few places where the reference has 'features'. We need to make assumptions about the linearity of the scan to estimate the frequency between the features. In section 8.3.3 we will show an example that combines the use of an artefact and a gas cell to measure both $\Delta\omega$ and $\omega_0$ and even extract the optical frequency at each time in the scan.

Currently the most accurate measurement of frequency would come from the comparison to a stabilised frequency comb. This combines the advantages of both methods as high performance combs can feature millions of spectral peaks over very large wavelength ranges and they can be known with unmatched accuracy. Since we are only interested in frequency changes we do not even have to measure the carrier envelope offset frequency[10] and can use combs with less than octave spanning spectra.

The laser light from our scanning laser can be combined with the light from the comb and passed onto a fast photo detector. The detector's electrical signals are sent to a band-pass filter and then on to an RF power detector. The resulting signal is a series of beat patterns between the frequency of the scanning laser and the many frequencies of the comb. Each time the scanning laser's frequency passes by the frequency of a comb mode the beat frequency will enter the range of the band-pass filter twice providing two signals symmetrically around the time at which the laser and comb mode were at the same frequency. If we wish to improve the beat detection we may chose to combine the scanning laser light only with a sub-set of the combs teeth which have been spectrally pre-selected for example via a grating that can be adjusted to 'track' the scanning laser's frequency as described in [22].

The exact determination of rapidly varying optical frequencies by spectroscopic methods is the subject of significant complexity that cannot be covered in depth in this chapter. We point the reader instead to a recent book on spectroscopy [23].

Now that we are able to measure optical frequency we can turn to the extraction of $L$ on a shot-by-shot basis.

*Extracting OPD $L$ from total phase advance $\Delta\Phi$ in a single scan*

In the very simplest case we would use a second interferometer of fixed, yet unknown, length and measure length ratio's between the two interferometers. We will first develop an equation describing the total phase advance for the slightly simplified case of a purely linear scan and a constant target speed. We will use a fully symmetric scan in which the scan variables $\omega_0$, $\dot{\omega}_0$ and $\ddot{\omega}_0$ are defined in the middle of the scan at $t = 0$.

[10] This frequency offset determines the absolute frequency of comb features in addition to their frequency spacing which is given by the comb's repetition rate.

- The time of the scan will extend from $t_i = -\frac{\Delta t}{2}$ to $t_f = +\frac{\Delta t}{2}$.
- The frequency will linearly change from $\omega_i = \omega_0 - \frac{\Delta\omega}{2}$ to $\omega_f = \omega_0 + \frac{\Delta\omega}{2}$.
- It will follow the linear form $\omega(t) = \omega_0 + \dot{\omega}t$.
- This requires the scan to have a total duration of $\Delta t = \frac{\Delta\omega}{\dot{\omega}_0}$
- The OPD changes linearly from $L_i = L_0 - \frac{\delta L}{2}$ to $L_f = L_0 + \frac{\delta L}{2}$.
- The equation of motion of the target is $L(t) = L_0 + 2u_m t$ where $u_m$ is the target speed.

We recall equation (8.25) for the interferometric phase of the interference signal in a first order (linear) scan, algebraically simplified below

$$\Phi^{(1)}(t, L) = \omega_0\left[\frac{L}{c}\right] + \frac{\dot{\omega}_0}{2}\left[2t\frac{L}{c} - \left(\frac{L}{c}\right)^2\right] \tag{8.33}$$

From the above we can compute the total interferometric phase advance during the scan

$$\begin{aligned}\Delta\Phi &= \Phi(t_f, L_f) - \Phi(t_i, L_i)\\ &= +\,\omega_0\frac{L_f}{c} + \frac{\dot{\omega}_0}{2}\left[2t_f\frac{L_f}{c} - \left(\frac{L_f}{c}\right)^2\right]\\ &\quad -\omega_0\frac{L_i}{c} - \frac{\dot{\omega}_0}{2}\left[2t_i\frac{L_i}{c} - \left(\frac{L_i}{c}\right)^2\right]\end{aligned} \tag{8.34}$$

If we insert the expressions for $t_i$, $t_f$, $L_i$, $L_f$, $\Delta t$ into equation (8.34) we find

$$\Delta\Phi = \omega_0\frac{\delta L}{c} + \Delta\omega\frac{L_0}{c} - \dot{\omega}_0\frac{\delta L}{2}\frac{L_0}{2} \tag{8.35}$$

If we further substitute $\delta L = 2u_m\Delta t = 2u_m\frac{\Delta\omega}{\dot{\omega}_0}$ into the second occurrence of $\delta L$ in equation (8.35) we find:

$$\Delta\Phi = \omega_0\frac{\delta L}{c} + \Delta\omega\frac{L_0}{c}\left(1 - 2\frac{u_m}{c}\right) \tag{8.36}$$

Equation (8.36) can now be interpreted. For a scan with stationary target we find:

$$L_0 = c\frac{\Delta\Phi}{\Delta\omega} \quad \text{if} \quad L = \text{const} \tag{8.37}$$

If the target moves slowly by a total of $\delta L$ and we do not account for this we introduce a fractional error $\Omega = \frac{\sigma_L}{L_0}$

$$\Omega = \frac{\sigma_L}{L_0} = \delta L\frac{\omega_0}{\Delta\omega} \tag{8.38}$$

The factor $\Omega = \frac{\omega_0}{\Delta\omega}$ is much larger than one for all currently existing tunable sources. At optical wavelengths $\Omega_{opt} \geqslant 35$. At radar wavelengths $\Omega_{rad} \geqslant 3$. This generates a high sensitivity to unaccounted target drift which is the biggest problem for FSI techniques using only a single scanning source.

*Extracting $\Delta L$ from total phase advance in a double scan*

If we use equation (8.36) for two scans of opposite scan direction but equal scan speed (i.e. opposite sign of $\dot{\omega}_0$ and hence of $\Delta\omega$), from the difference of the two phase advances we find:

$$\Delta\Phi^{up} - \Delta\Phi^{down} = 2\Delta\omega\frac{L_0}{c}\left(1 - 2\frac{u_m}{c}\right) \tag{8.39}$$

This allows us to compute the average distance during the scan as follows:

$$L_0 = c\frac{\Delta\Phi^{up} - \Delta\Phi^{down}}{2\Delta\omega\left(1 - 2\frac{u_m}{c}\right)} \tag{8.40}$$

Assuming speeds typical of metrology of $\mathcal{O}(1\ \mathrm{cm\,s}^{-1})$ the factor $2\frac{u_m}{c} \approx 6.6 \cdot 10^{-11}$ can be ignored for most accuracy requirements. The accuracy of error cancellation in this dual scan method relies on the fact that the scans have opposite scan direction but identical and known total phase advances and centre frequencies.

An ingenious method developed by a team at NPL not only guarantees the accurate identity of the two opposing scans by using degenerative four wave mixing [24, 25] but also does so without requiring a second frequency scanning source and the associated problems of scan synchronisation.

The degenerative four wave mixing takes place efficiently in a Semiconductor Optical Amplifier (SOA) which is illuminated by a first fixed frequency laser of frequency $\omega_1$ and the frequency scanning laser with frequency range $\omega_2(t) \in [\bar{\omega}_2 - \Delta\omega/2; \bar{\omega}_2 + \Delta\omega/2]$. In this notation $\bar{\omega}$ indicates the average frequency and $\Delta\omega$ is the scan range as usual. The output of the SOA contains, in addition to the input spectrum, the frequency spectrum defined by $\omega_3(t) = 2\omega_1 - \omega_2(t) = \omega_1 - (\omega_2(t) - \omega_1)$ which is shown in figure 8.5.

The frequencies $\omega_1$ and $\omega_2(t)$ were chosen in such a way as to generate $\omega_3(t)$ and $\omega_2(t)$ close together in the C-band of the telecommunications wavelength range yet sufficiently widely separated to allow separate detection of both signals via DWDM[11] filters. Proximity of the frequencies has the advantage that both signals can be transported in the same optical fibre yet they can still be cheaply separated.

A disadvantage arises from the fact that the two centre frequencies of the scans $\bar{\omega}_2$ and $\bar{\omega}_3$ are not exactly identical and the cancellation of the $\delta L$ term in equation (8.36) is imperfect. Nevertheless a very strong suppression of the motion error factor $\Omega$ from equation (8.38) from 35 down to unity was reported.

[11] Dense wavelength division multiplexing.

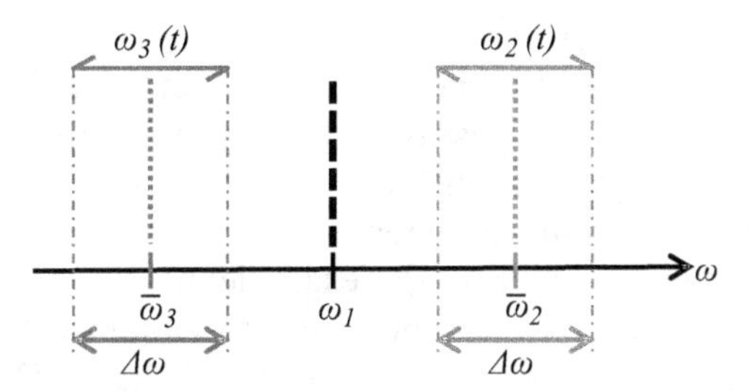


**Figure 8.5.** Frequency spectra on the input ($\omega_2(t)$, blue and $\omega_1$, black) and output ($\omega_3(t)$, red) of a degenerative four wave mixer used to create scans of identical $\Delta\omega$.

To further improve on the motion tolerance the centre frequencies would have to be made very similar and signal separation would have to happen based on polarisation or an amplitude modulation scheme with different modulation frequencies for the two sources.

An alternative approach to achieve the same cancellation uses two independent sources that tune around the same centre frequency yet with different tuning rates. The undesired $\delta L$ term still cancels in the difference of phase increments but the remaining pre-factor in equation (8.40) now contains $\Delta\omega_1 + \Delta\omega_2$ which must both be measured during the scan.

*Extracting* $\Delta$L *from* $\omega_{int}$

If we extend these techniques from their linear form used so far to non-linear scans but still assume a stationary target we will see that the peak in the interference frequency spectrum mentioned earlier is smeared out because $\dot{\omega}_0$ and hence the instantaneous position of the peak varies across the scan. We can recover from this by re-sampling the intensity at time intervals corresponding to a fixed frequency increment. Such increments can be provided by the periods in the output of an etalon or reference interferometer of stable OPD illuminated with the same light. The reference has to be of stable OPD during the scan. The method of re-sampling has the advantage that the abscissa of our data series is still equi-spaced and we can still use computationally efficient digital Fourier transforms (DFT). Alternatively we could assign each original measurement of intensity an abscissa co-ordinate that corresponds to the phase advance of the reference interferometer. Now the data points are no longer equally spaced and we cannot use DFTs. Instead we could use a Lomb-Periodigram which has been used in [26] for this purpose. It is also described as a numerical technique on page 686 in [27] and as C++ code in [28].

If our target moves we can no longer distinguish frequency changes from Doppler shifts from those arising due to frequency scanning of our laser. In this case we point to the methods described in sub-section 8.3.2. The methods for drift compensation in techniques using the total phase advance apply analogously to methods using the interferometric frequency. In the following sub-section we explain how the above methods can be applied to multiple simultaneous measurements of OPDs in a single interferometer.

*Multiple target measurements*

Because the instantaneous interference frequency $\omega_{\text{int}}$ of an FSI interferometer is proportional to the OPD it is possible to measure multiple OPDs in a single interferometer, with a single readout channel, under certain conditions. If these multiple OPDs lead to different interference frequencies they are distinguishable in the same signal channel. This will for example be the case if the OPDs are significantly different (mean $\langle\omega_{\text{int}}\rangle$ proportional to $L$) and if the frequency scan is approximately linear (narrow $\omega_{\text{int}}$ distributions) and the OPDs are approximately constant (Doppler shifts of $\omega_{\text{int}}$ don't make their distributions overlap). In such a case the interference signals from the different OPDs may be separated using band-pass filters.

Two interference frequency distributions are separable if they are separated by more than the frequency resolution:

$$\begin{aligned}\Delta\omega_{\text{int}} &= \min_{\Delta t}(\omega_{\text{int},\,1}) - \max_{\Delta t}(\omega_{\text{int},\,2})\\ \Delta\omega_{\text{int}} &\geqslant \frac{2\pi}{\Delta t}\end{aligned} \tag{8.41}$$

In the above we have expressed the fact that in order to measure any frequency one must always measure the signal over a minimum time $\Delta t$ to achieve a frequency resolution of $\delta\omega \geqslant 2\pi/\Delta t$.

The first obstacle to multi-target measurements is the target motion which can shift the interference frequency via Doppler shifts. At a given time $t$ the interference frequency $\omega_{\text{int},1}(t)$ of the longer OPD $L_1 > L_2$ may be reduced by target motion speed $u_1$ if $u_1$ is in the direction opposing the phase generated by the tune[12]. At the same time target speed $u_2$ of the shorter OPD $L_2$ may have the opposite sign and hence increase the interference frequency $\omega_{\text{int},2}(t)$ of the shorter interferometer. In this case $\omega_{\text{int},1}(t)$ may become indistinguishable (within the frequency resolution) from $\omega_{\text{int},2}(t)$. This problem is irreducible because the frequencies are unresolvable at the same time and hence not separable.

The second problem arises from tuning non-linearities. If the instantaneous tuning speed $\dot{\omega}(t)$ varies during the frequency measurement interval $\Delta t$ then the smallest tuning rate in this period $\dot{\omega}_{\min}$ may create an interference frequency with the longer OPD $L_1$ which becomes indistinguishable (within the given frequency resolution) from the interference frequency generated by the maximum tuning rate $\dot{\omega}_{\max}$ and the shorter OPD $L_2$. This problem can be reduced by optimising $\Delta t$ for the expected instantaneous tuning acceleration $\ddot{\omega}(t)$. The bigger $\ddot{\omega}(t)$, the smaller $\Delta t$ should be until decreasing frequency resolution starts to dominate the issue and the problem becomes unresolvable.

We may use equation (8.31) to estimate the degree to which any two OPDs have to differ to allow separation during measurement time $\Delta t$ in a purely linear scan. For reasons of simplicity we will ignore the second order tuning terms proportional to $\ddot{\omega}_0$ and make the following simplifying assumptions:

[12] If $\dot{\omega}(t) > 0$ then $u < 0$ will reduce the interference frequency.

**Table 8.2.** Contributions in metres to minimum separable OPD difference $\Delta L_{\text{sep}}$ (penultimate column) from equation (8.42) and theoretical limit of OPD resolution $\sigma_L/L = 2\pi/L\Delta\omega$ (final column). The parameters for this were: mean OPD $<L> = 10$ m; starting frequency $\omega_0 = 1.22 \cdot 10^{15}$ rad s$^{-1}$ (1550 nm); tuning rate $\dot{\omega}_0 = 1.57 \cdot 10^{15}$ rad s$^{-2}$ (2000 nm s$^{-1}$); scan duration $\Delta t = 0.1$ s; tuning range $1.57 \cdot 10^{14}$ rad (200 nm); maximal target speed $u_m = 1$ cm s$^{-1}$.

| $4u_m\frac{\omega_0}{\dot{\omega}_0}$ | $4u_m\Delta t$ | $4u_m\frac{\langle L\rangle}{c}$ | $\frac{2\pi c}{\dot{\omega}_0\Delta t} = \frac{2\pi c}{\Delta\omega}$ | $\Delta L_{\text{sep}}$ | $\sigma_L/L$ |
|---|---|---|---|---|---|
| $+3.10 \cdot 10^{-2}$ | $+4.02 \cdot 10^{-3}$ | $-1.33 \cdot 10^{-9}$ | $+1.20 \cdot 10^{-6}$ | $+3.50 \cdot 10^{-2}$ | $4 \cdot 10^{-15}$ |

*Frequency scan linearity*:

We assume the tune to be purely linear with $\dot{\omega}(t) = \dot{\omega}_0 > 0$ and may hence ignore problems of the second kind described above.

*OPD difference*:

We assume that the two OPDs $L_1 > L_2$ have mean $\langle L\rangle$ and differ by $\Delta L = L_1 - L_2$.

*OPD rate of change*:

We finally assume that the OPDs change with constant and opposite sign speeds $u_1 = -u_m$ and $u_2 = +u_m$ during $\Delta t$.

In this case we find the following condition for the minimum difference of OPDs $\Delta L_{\text{sep}}$ at which interference frequencies are separable

$$\Delta L_{\text{sep}} \geqslant 4u_m\left[\frac{\omega_0}{\dot{\omega}_0} + \left(\Delta t - \frac{\langle L\rangle}{c}\right)\right] + \frac{2\pi c}{\dot{\omega}_0\Delta t} \tag{8.42}$$

We evaluate equation (8.42) numerically for a very fast linear frequency scan of long average OPD[13] and list the individual contributions in table 8.2.

Positive (negative) contributions to $\Delta L_{\text{sep}}$ make separation harder (easier). The first term dominates and is positive even though this scan has a very high tuning speed and a long tuning range. The first term is proportional to the ratio at which motion and tuning generate phase. The bigger the tuning rate the easier to separate the signals.

The second term is also positive. It is the total change of the OPD over the scan duration $\Delta t$. Here $\Delta t$ is also the time available for 'detrimental' motion[14] to occur. The smaller the motion speed $u$ the easier separation.

The third term is the only negative term but it is of negligible size. It expresses the fact that longer interferometers are negligibly easier to separate because the light takes a longer time to pass through the measurement arm and hence produces a bigger beat frequency with the light from the shorter reference arm.

The fourth term stems from the frequency resolution limit which is also quite small for the long integration time of 0.1 s.

[13] Fast scans of long average OPD are best at resolving small OPD differences in the same signal.

[14] Remember that we have set up the target speeds such that the OPDs become more similar and hence make separation harder. This term would have the opposite sign if target speeds were such as to increase OPD differences.

The final column shows the theoretical distance resolution in this scan if it was limited only by the Fourier transform resolution limit. This resolution cannot be reached in practice for several reasons. To achieve this would require knowledge of the frequency axis at any time during the scan to approximately 0.8 Hz and the ability to sample the intensity with a timing accuracy of 2 fs. The real distance resolution limit will be dictated by the ability to measure the signal frequency of the sinusoidal intensities versus laser frequency in the presence of noise affecting both intensity and laser frequency. The Cramer–Rao bound [29] describes this limit. For the best current FSI techniques the achievable limit is of the order of $\sigma_L/L \approx 10^{-7}$ and dominated by refractive index uncertainties.

A fuller analysis of the separation from equations (8.31) or (8.26) is best done numerically by plotting the interference frequency differences of two interferometers of differing lengths and motion speeds against the desired parameters. To do this we have to remember to evaluate this difference at the point in the scan at which it is smallest. If we follow the assignment of signs of targets speeds and tuning rate used above and we choose a negative tuning acceleration the smallest difference will always occur at the end of the scan.

The power of simultaneously measuring multiple distances is nicely demonstrated by Hughes *et al* in [30]. The team describes a self-calibrating large volume co-ordinate measurement system which combines multi-target FSI with diffractive multi-beam steering using SLMs.

*Discontinuous tuning ranges*

In all methods discussed so far we have assumed that the frequency changes continuously during a scan. A particular and very common type of discontinuity is a longitudinal mode hop in the laser cavity. A tunable laser can in principle support all frequencies inside its gain bandwidth. A cavity of adjustable length and sufficient finesse surrounding or integral to the gain medium is used to select a single frequency from this range. At each cavity length multiple longitudinal modes can be supported but only one is desired. In addition to adapting its cavity length to the desired output frequency an additional, often individually controlled[15] frequency selection mechanism, commonly a rotating grating, can be used to suppress all but one desired longitudinal mode of the tuning cavity. It is often difficult to achieve this mode selection if the tuning process is very rapid. In these cases the gain of the laser may rapidly move from the desired mode to a competing mode that was not sufficiently suppressed and the frequency will hop discontinuously.

Although such mode hops maintain the sinusoidal relation of the reference interferometer's phase with the frequency of the wave, they break the unwrapping process of the phase because they create a gap of unknown size in the frequency axis and hence disable the reconstruction of the frequency during the scan. Let us assume that two continuous and mode hop free tuning ranges are separated by a rapid and potentially large tuning period during which mode hops have occurred.

[15] Distributed feedback (DFB) lasers utilise a periodic modulation of the gain medium (i.e. fibre Bragg grating or periodic grating inside the diode) to select modes. These are not individually controllable.

If the size of the rapid, discontinuous tune shift can be measured externally to an accuracy much lower than $2\pi$ in units of the interferometric phase of the reference interferometer it can still be possible to 'stitch together' the two continuous regions surrounding the discontinuous gap. The phase of a reference interferometer of fixed length is measured as a sinusoidal function of frequency in the first and second continuous tuning regions. The approximate frequency shift during the rapid scan is measured unambiguously by some other device, for example a set of vernier etalons or a wave metre. The phase from the first continuous scan is then extrapolated to the start of the second continuous scan using the external frequency shift measurement. If the accuracy of this extrapolation is sufficient to predict the reference interferometer phase at the start of the second continuous scan to better than half of the period, the phases in the second periods are shifted to coincide with the predicted phases. In this way knowledge of the frequency differences between the first and the second continuous region remains unaffected compared to a single continuous scan across the full range. Examples of the application to high energy physics detector alignment and a detailed description of such a method are shown in [20, 31, 32].

### 8.3.3 Dynamic FSI analyses

We now turn to FSI methods that can measure multiple distances in each scan. We refer to these as dynamic FSI (dFSI). In principle the shot-by-shot methods described above can be applied to subsections of a scan. They do however lose significant accuracy if less frequency advance is used per shot as shown for example in equation (8.38).

Fourier transforms used in the methods based on $\omega_{\text{int}}$ also reduce in resolution when the time over which they integrate shrinks. We will therefore focus on methods that maintain an accuracy of length measurement typical of the full scan range but obtain multiple measurements during the scan.

The method we describe below was first published in [6] and is an extension of [12]. It has made its way into a commercial metrology instrument [33] and is also described in a patent [34]. The hardware arrangement of this set-up is shown in figure 8.6. We describe dFSI here in a generalised form to ease understanding.

In dFSI two external cavity diode lasers (L1 and L2) scan simultaneously with opposite tuning directions and different tuning speeds across overlapping wavelength ranges in the vicinity of 1550 nm. Both scans last for the same time and hence the slower laser covers a smaller tuning range. The lasers send their light into three sub-systems:

*Frequency measurement system*:

This consists of a gas absorption cell (GC) with many, sharp, well characterised absorption features at frequencies $\omega_{a,n}$ in the scan range. Suitable cells for this frequency range may use low pressure (1–100 Torr) HCN, $C_2H_2$, $CO_3$ in different isotope compositions or gas mixtures depending on the chosen scanning ranges. The lower the pressure the smaller the dominant pressure broadening.

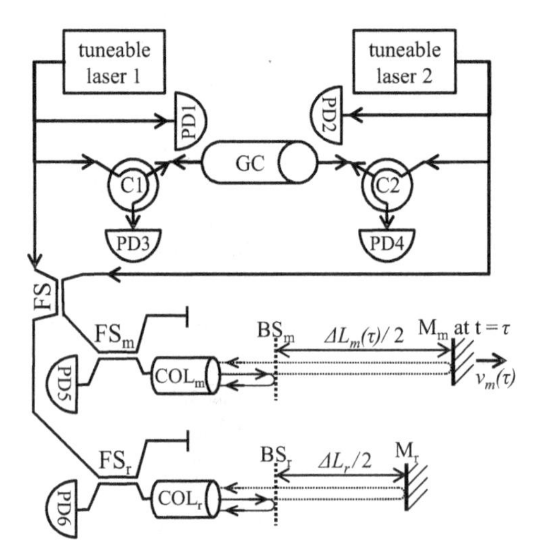


**Figure 8.6.** Hardware set-up for the dFSI technique. The symbols have the same meaning as in figure 8.1. C1 and C2 are circulators.

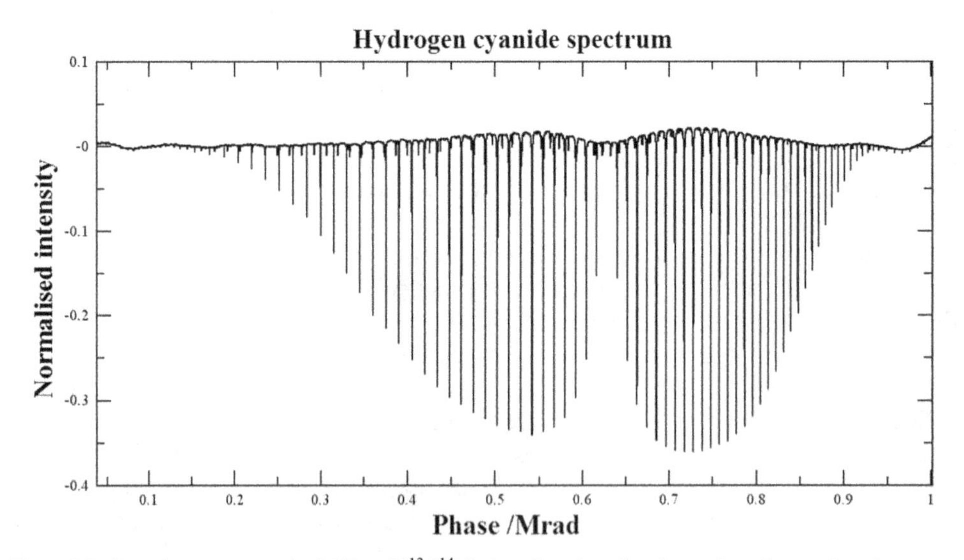


**Figure 8.7.** Intensity spectrum of a 25 Torr $H^{13}C^{14}N$ plotted against the phase of a reference interferometer.

Figure 8.7 shows the absorption spectrum of an HCN cell measured in real time during a fast dFSI scan.

We adjust the laser scan ranges in such a way that a prominent absorption peak is approximately at the centre of both scans. We refer to this peak as the principle absorption peak and its frequency is $\omega_{a,p}$.

The lasers send a fraction of their light in opposite directions through the same gas volume. The fibre circulators C1 and C2 send the light from each laser via the gas cell onto its own photo detector (PD3 and PD4). Directly after each laser's fibre output a small fraction of the light is split of onto a photo detector (PD1 and PD2) to monitor each laser's power. The four splitters in this section are implemented as planar light circuits. They are not separately labelled.

*Reference interferometer*:

The majority of the light from both lasers is combined in the fused taper fibre splitter FS which sends a fraction of the combined light into another fused taper fibre splitter $FS_r$ which then passes 50% into the reference interferometer and dumps the remaining 50%.

The reference interferometer's construction is such that it is dimensionally stable over the short time scales of the scans. It therefore has to be vibrationally and thermally isolated and suitably constructed. In this system the reference interferometers are entirely made of fibre. In sub-section 8.4.3 we discuss how to correct for the dispersion of the fibre which introduces an apparent change in OPD during the scan.

The light returning from the reference interferometers passes back through $FS_r$ and 50% is guided onto photo detector PD6. Using a fibre splitter in the position of $FS_r$ only utilises 25% of the available light. A fibre circulator instead of $FS_r$ would use 100% of the light but could internally introduce more multiple reflections which interfere with the analysis of the interference signals. It would also be significantly more expensive.

*Measurement interferometer*:

The measurement interferometer is allowed to change its OPD during the scans within limits discussed later. It is illuminated and readout in the same way as the reference interferometer. It is the most inexpensive of the sub-systems and multiple measurement interferometers can be served by a single reference and frequency measurement system. The number of measurements is practically limited either by the available laser power (which can often be amplified to Watt scales) or by the capacity of the DAQ system.

All photo detector readings are digitised at discrete times $t_i$ at a high sampling frequency. The data is digitally recorded and analysed off-line on a personal computer. Many of the computations for this analysis are time intensive and should preferably be performed using parallel computing resources such as GPUs.

For the theoretical description of this technique we use the approximate expression for the interferometric phase given in equation (8.28) which we copy below. In this 'instantaneous' approximation the time $t$ at which an intensity is recorded and the time $\tau$ when the measurement wave has hit the measurement mirror are approximately the same.

$$\Phi \approx \omega_e(t)\frac{\Delta L}{c} \quad \text{for slow target speed and approx. linear tuning} \tag{8.28}$$

We will denote phases of the reference and measurement interferometer with a subscript $r$ and $m$ respectively. Subscripts 1 and 2 indicate phases arising from laser 1 and 2. Both lasers illuminate both interferometers simultaneously.

Because both lasers scan with different speeds the corresponding intensity patterns they generate have different signal frequencies $f_1$ and $f_2$ and can therefore be separated from the measured intensity using band-pass filters. To achieve this the scans need to have approximately constant and sufficiently different tuning rates or else their interference frequency spectra will overlap. This is one of the main limitations of this technique. It requires high degrees of control over the laser scanning process.

Once the signals from the two lasers have been normalised by the power measurements and separated by band-pass filters the corresponding phases can be obtained using the Hilbert Transform method described on page 8.3.1.

We will define some times during the scan as follows:

$\mathbf{t}_i$ = start of the scan (initial time).

$\mathbf{t}_f$ = end of the scan (final time).

$\mathbf{t}_{p1}$ = time when $\omega_1$ of laser 1 coincides with centre of principal absorption peak.

$\mathbf{t}_{p2}$ = time when $\omega_2$ of laser 2 coincides with centre of principal absorption peak.

$\mathbf{t}_c \approx$ time near the centre of the scan.

$\mathbf{t}_j$ = $j^{\text{th}}$ discrete time at which all intensities were recorded.

$\mathbf{t}$ = arbitrary time during the scan (any one of $t_j$ or any interpolated value).

In the following the indices $i$, $p$ or $c$ appearing instead of the '?' on expressions of the form $\Delta_?\Phi(t)$[16] indicate that unwrapping started at times $t_i$, $t_p$ or $t_c$ and finished at time $t$. We first unwrap all phases from $t_i$ to $t$ producing four phase differences for all times $t$.

$$\Delta_i\Phi_{r,\,1}(t) = \frac{\Delta L_r}{c}(\omega_1(t) - \omega_1(t_i)) \tag{8.44}$$

$$\Delta_i\Phi_{r,\,2}(t) = \frac{\Delta L_r}{c}(\omega_2(t) - \omega_2(t_i)) \tag{8.45}$$

$$\Delta_i\Phi_{m,\,1}(t) = \frac{\Delta L_m(t)}{c}\omega_1(t) - \frac{\Delta L_m(t_i)}{c}\omega_1(t_i) \tag{8.46}$$

$$\Delta_i\Phi_{m,\,2}(t) = \frac{\Delta L_m(t)}{c}\omega_2(t) - \frac{\Delta L_m(t_i)}{c}\omega_2(t_i) \tag{8.47}$$

Next we will determine the phases at which each laser reaches each of the absorption features in its tuning range. We plot the absorption spectrum versus the unwrapped reference interferometer phase for each laser as shown in figure 8.7 and

[16] The ? can take any of the values $i$, $p$ or $c$.

fit functions describing the shape of the absorption peaks to each of the peaks. These functions could be approximated by Gaussian functions if we fit only small sections around the centre of each peak. For higher accuracy we could use Voigt functions[17]. If we fit Voigt functions we can also determine the Lorentzian and Gaussian width from which we can determine the cell pressure and temperature which in turn can be used to correct the absorption frequencies for pressure shifts.

The unwrapped phases of the reference interferometers at the peaks $\Delta_i\Phi_{r,\,1}(t_{a,n})$[18] are then plotted against the known absorption peak frequencies $\omega_{a,n}$. The resultant linear relation is fitted and the slope of this fit is $\Delta L_r/c$. This fit can be done for each laser and we obtain two independent measurements of $\Delta L_r$ which we average to get the best estimate for $\Delta L_r$

Now we look at the unwrapped reference interferometer phases of laser 1 at time $t_{p,1}$ and of laser 2 at time $t_{p,2}$. At these times both lasers are at the frequency of the principal absorption peak $\omega_{a,p}$. We also look at the measurement interferometer phases unwrapped from $t_i$ to $t_c$.

$$\Delta_i\Phi_{r,1}(t_{p,1}) = \frac{\Delta L_r}{c}\big(\omega_{a,p} - \omega_1(t_i)\big) \tag{8.48}$$

$$\Delta_i\Phi_{r,2}(t_{p,2}) = \frac{\Delta L_r}{c}\big(\omega_{a,p} - \omega_2(t_i)\big) \tag{8.49}$$

$$\Delta_i\Phi_{m,\,1}(t_c) = \frac{\Delta L_m(t_c)}{c}\omega_1(t_c) - \frac{\Delta L_m(t_i)}{c}\omega_1(t_i) \tag{8.50}$$

$$\Delta_i\Phi_{m,2}(t_c) = \frac{\Delta L_m(t_c)}{c}\omega_2(t_c) - \frac{\Delta L_m(t_i)}{c}\omega_2(t_i) \tag{8.51}$$

Subtracting equations (8.48) from (8.44), (8.49) from (8.45), (8.50) from (8.46) and (8.51) from (8.47) yields unwrapped phase difference w.r.t. times $t_{p,1}$, $t_{p,2}$ and $t_c$ as shown below.

$$\Delta_p\Phi_{r,1}(t) = \frac{\Delta L_r}{c}\big(\omega_1(t) - \omega_{a,p}\big) \tag{8.52}$$

$$\Delta_p\Phi_{r,2}(t) = \frac{\Delta L_r}{c}\big(\omega_2(t) - \omega_{a,p}\big) \tag{8.53}$$

$$\Delta_c\Phi_{m,1}(t) = \frac{\Delta L_m(t)}{c}\omega_1(t) - \frac{\Delta L_m(t_c)}{c}\omega_1(t_c) \tag{8.54}$$

$$\Delta_c\Phi_{m,2}(t) = \frac{\Delta L_m(t)}{c}\omega_2(t) - \frac{\Delta L_m(t_c)}{c}\omega_2(t_c) \tag{8.55}$$

[17] Convolutions of Gaussian with Lorentzian functions describing Doppler and pressure broadened natural line shapes.

[18] These fitted phases and corresponding times are interpolations between those measured at discrete times $t_j$.

As we already know $\Delta L_r$, we can use equations (8.52) and (8.53) to determine the laser frequencies $\omega_1(t)$ and $\omega_2(t)$ at all times including the interpolated time $t_c$. This leaves only $\Delta L_m(t)$ and $\Delta L_m(t_c)$ as unknowns in equations (8.54) and (8.55).

We find $\Delta L_m(t_c)$ by dividing equation (8.54) by (8.55) and isolating $\Delta L_m(t_c)$ to find:

$$\Delta L_m(t_c) = \frac{\omega_2(t)\Delta_c\Phi_{m,1}(t_c) - \omega_1(t)\Delta_c\Phi_{m,2}(t_c)}{\omega_2(t_c)\omega_1(t) - \omega_1(t_c)\omega_2(t)} \tag{8.56}$$

Equation (8.56) can be evaluated for all $t = t_j$ where $t \neq t_c$. The solutions increase in numerical stability the further $t$ is from $t_c$. Since $t_c$ is approximately in the centre of the scans we evaluate equation (8.56) for all $t$ that are further away from $t_c$ than a chosen threshold. We then average all values for $\Delta L_m(t_c)$.

Now that $\Delta L_m(t_c)$ is known we can use equations (8.54) and (8.55) to find two independent solutions for $\Delta L(t_j)$ which we can then average.

*Dynamic FSI corrections*

In summary the dFSI technique measures the absolute length of a temporary reference interferometer for each shot using a gas cell as frequency reference. The reference interferometer only has to be stable for the duration of the scan and the gas cell provides the long term reference. Because both lasers scan through the principle absorption feature the phases of the measurement interferometer of both laser's can be related to each other which ultimately allows us to measure the length.

The absolute accuracy of dFSI relies on accurately known absorption frequency differences in the gas cell and the measurement of the laser frequency against these features. The fits of the absorption peaks against reference interferometer phase may suffer from systematic errors at a given peak position because peaks can overlap with smaller side peaks or the laser intensity normalisation may be imperfect or chromatic effects in the cell read-out produce other variations. Similar effects can exist if the gas cell pressure is not well known and the pressure shifts cannot be accurately calculated.

If these systematic errors are constant and a given peak is always shifted systematically by a fixed amount then these systematics can be compensated or corrected by a 'quasi-calibration' process.

The straight-line fits of known peak frequency versus reference interferometer phase would suffer a fixed shift in slope. This shift can be corrected if the system measures a distance which is calibrated by other means. The ratio of the measured and the known distance can be then used as a correction factor which can compensate for all systematics that result in a fixed change of slope.

The performance of a dFSI system corrected in such a way is well documented in [6]. The system, including refractive index calculations from a single temperature and pressure sensor achieves a relative Gaussian distance uncertainty of $1.24 \cdot 10^{-7}$ over distances from 1 to 20 m. The systematic uncertainties from refractive index corrections and absorption feature systematics are approximately equal.

## 8.4 FSI hardware technology

In this section we briefly describe the main alternatives for the key elements of FSI systems. We also explain the technological developments that have and are still driving recent progress in FSI techniques and we show where new technologies can influence the future development of FSI techniques. It is not possible to list in any useful level of detail the possible hardware components or even just their manufacturers. We therefore have to limit ourselves to the principle features of technologies and their relevance to FSI techniques.

### 8.4.1 FSI sources

At the heart of each FSI technique lies a tunable light or microwave source. The range of useful light sources which are dominated by lasers of all kinds has exploded in the last 20 years. One of the drivers behind this development has been the telecommunications and related industry which have invested billions of dollars, euros, pounds, yen, etc into the development of fibre-optics and advanced lasers, many of which are tunable. The development of their production technology has made them extremely cost efficient which is a requirement for their practical application in FSI systems. Cost efficient, primarily diode based tunable lasers of high coherence length—which interferometry practitioners are often most interested in—are now readily available across a wide range of wavelengths. Small mode hop free tuning ranges[19] of order 1 nm are available from diode lasers at a few discrete short wavelengths from 375 nm to 633 nm. Long tuning ranges often reaching 100–200 nm are available from 915 to 1070 nm and from 1260 to 1770 nm.

It should be noted that pulsed, electronically tuned Ti:sapphire lasers have also been used to generate pulsed FSI systems [35] with extremely wide tuning ranges exceeding 100 nm which were used for surface profile measurements. Although the pulsed nature of this technique seems very different from the continuous wave approach described here the basic principles of the distance measurement are identical.

Most of the long tuning range lasers are external cavity diode lasers (ECDL) the fastest of which can reach tuning speeds up to 2000 nm $s^{-1}$ and tuning ranges up to 200 nm. Short term coherence lengths of such lasers vary from a few 10 to several 1000 m. Unamplified laser powers commonly reach a few mW but can reach the 100 mW range. In the range around 1000 nm, 1500 nm and 2000 nm ytterbium, erbium and thulium doped fibre amplifiers offer extensions in power up to the one Watt range.

Although these higher powers lend themselves somewhat to frequency doubling, the doubling efficiencies of CW sources (per length of doubling material) remain relatively low around 0.6% $W^{-1}$ $cm^{-1}$. Doubled lasers have therefore not yet found their way into FSI systems. Because the doubling efficiency has to be optimised for a narrow range of wavelengths through phase matching or quasi phase matching this also restricts the frequency range of the doubled source.

[19] See section 8.3.2 how to use non-continuous tuning ranges.

It should also be noted that extremely high tuning speeds over small wavelength ranges are available from electrically modulated distributed feedback laser diodes. Such DFB lasers can reach tuning speeds of several 10 000 nm $s^{-1}$.

In the radar range of frequencies from 50 to 250 GHz transceivers built as bipolar, SiGe monolithic integrated microwave circuits offer extremely wide bandwidth (up to 32.8% of centre frequency). These are combined with extremely fast tuning rates that allow periodic tuning across the entire range at kHz repetition rates corresponding to tuning speeds of $\dot{\omega} \geqslant 3.0 \cdot 10^{14}$ rad $s^{-1}$.

Such radar systems have been used [3–5] to implement FSI systems with 290 nm distance repeatability and sub-micron accuracy.

### 8.4.2 FSI DAQ and processing

FSI interferometer signals can reach extremely high interference frequencies if the interferometers are long and the tuning rates and/or motion speeds of the measured distances are high. The interference frequencies shown in graph B-0 of figure 8.3 demonstrate that fast scans in 10 m OPD interferometers can exceed signal frequencies of 20 MHz even in completely linear scans. The total data volumes of FSI scans are also necessarily large because the scans have to cover large frequency ranges to achieve accuracy and large frequency excursions produce large total phase advances which have to be sampled with at least a few points per $2\pi$ of phase advance. From the simple form for the length of a stationary short interferometer as a function of its total phase advance in equation (8.37) we can see that the total phase generated is:

$$\Delta\Phi = L_0 \frac{\Delta\omega}{c}$$

If we consider a scan range of $7.85 \cdot 10^{13}$ rad (100 nm) starting at $1.22 \cdot 10^{15}$ rad s (1550 nm) we will generate a total phase advance of $2.62 \cdot 10^{6}$ rad. If we sample with four points per fringe and we acquire two bytes of data per measured intensity we generate nearly 21 MByte of data per interferometer. As we need at least data from a measurement and a reference interferometer (or six lines in the case of dFSI) we will generate a total of 42 MByte (126 MByte for dFSI) per scan. Such a scan may only take 50 ms which means that we produce data rates of 840 MByte $s^{-1}$ (2.52 GByte $s^{-1}$ for dFSI). These numbers explain why high performance DAQ and consecutive fast processing can play an important role in modern FSI techniques.

This is why the second technological driver behind FSI techniques is the increasing ability of modern DAQ systems to detect and digitise optical signals at ever increasing bandwidths which can now reach tens of Gigahertz with 8 bit resolution. As acquisition speed falls to approximately 100 MHz, digitisation depths of 16 bits are common and effective resolutions of 12 bit are achieved off the shelf.

DAQ development has again been driven largely by the telecommunications market in its quest for faster optical data transmission. The necessary optical detection elements are usually photo-diodes which can cover the entire optical and near infra-red wavelength region. Today's fastest diodes are often directly fibre coupled and can reach bandwidths of tens of GHz at responsivities approaching one Ampere per Watt of optical input power.

Together with the high digitisation rates comes the necessity to rapidly store and process the acquired data. Data transfer speeds have evolved together with the DAQ systems and the currently prevailing generations of data transport protocols all utilise multiple serial connections and feature aggregate bandwidth in the range around 100 Gbit $s^{-1}$. While the telecommunication industry has focussed on the development of the large form factor ATCA (advanced telecommunication computation architecture) the last 15 years have seen the rise of the uTCA or $\mu$TCA (Micro Telecommunication Computation Architecture) standard which is aimed at smaller scale applications and highly suitable for advanced FSI DAQ systems [36].

The range of commercially available uTCA system components now covers all that is needed to build a high performance DAQ system for metrology applications [37].

The processing needs of FSI techniques can also be significant. The dFSI technique performs at least two Fourier transforms for each of its six intensity measurement series per scan. If processing times of the order of a second per scan are desired the use of parallel computing resources becomes mandatory. Modern PC based GPUs with fast PCIe bus access (16 lanes of PCIe 2.0 or larger) can handle such data throughput and processing speeds. The data transfer rates typical of dFSI techniques cannot reliably be performed on single second time scales using standard networking protocols (TCP/IP) over ethernet connections. For continuous real time data transport from the DAQ system to the computation platform eight lanes of PCIe connectivity are needed or the computing platform needs to move directly into the DAQ system as is often the case in uTCA DAQ crate solutions.

### 8.4.3 FSI reference interferometers

The purpose of an FSI reference interferometer is to measure either just the total frequency advance or the frequency as a function of time during the scan. If the reference interferometer also incorporates the absolute distance scale of the system it has to be stable long term. If, as is the case in dFSI methods, the reference interferometer only measures the frequency during the scan it only has to be stable during a scan. Many methods for constructing such interferometers exist. Most of them run into problems with stability or compactness when the interferometers OPD has to be large. As an ideal reference interferometer should have an OPD similar to that of the measurement interferometer[20] and since the measurement interferometer can vary widely in length it may be necessary to use multiple reference interferometers.

Research FSI systems such as the one described in [32] which rely solely on large reference interferometers as their primary scale have been built and operated for decades. Their large and costly, evacuated constructions are based on thermally ultra stable materials such as invar and are hardly practicable for industrial applications.

[20] If reference and measurement interferometers have similar OPDs their phases will vary with similar rate and they represent similar limitations on the tuning range and DAQ speed.

Other constructions utilise highly stable glass or ceramic cavities in folded linear or circular geometries. Common to the above is that they avoid any dispersion in the reference interferometer because without special measures this would look as if the interferometer is changing length or, if we assume it is stable, it would imprint its length change onto the length measurement it helps to perform.

If a method for dealing with dispersion in the reference interferometer can be used then thermally and vibrationally isolated fibre based references would offer extremely compact and cost efficient solutions. Such a fibre is used as the reference interferometer in a dFSI experiment [6]. The dispersion correction method used for this will soon be described in full in [38]. The method is briefly summarised below.

The non-linear part of the dispersion curve[21] of standard single mode telecommunication fibre does not vary significantly with temperature even though the total length of the fibre may change significantly with temperature. The reference interferometer in a dFSI system only has to be stable over second time scales because its absolute length is re-measured automatically at each scan. If we pre-determine the dispersion of such a fibre during a calibration experiment in terms of the additional phase advance produced compared to a non-dispersive interferometer in a way that is independent of the length of the non-dispersive interferometer we can later subtract this additional phase.

During such a calibration experiment we use a single scanning laser to determine the unwrapped interferometric phase of the fibre interferometer (referred to as 'the fibre') and the non-dispersive reference interferometer (referred to as 'the reference') and the absorption spectrum of a suitable gas across the widest possible frequency range. We set both interferometric phases to zero at an absorption feature near the centre of the scan and plot the interferometric phase of 'the reference' versus that of 'the fibre'. We now fit a straight line to this plot and subtract it from the phases of 'the reference'. The resulting data set is referred to as the fibre's calibration curve. It measures the additional non-linear phase (the phase residual) generated by 'the fibre' as a function of 'the fibre's' interferometric phase. Each time this fibre is used and its phase is unwrapped and set to zero at the same absorption feature the phase residual stored in the calibration curve can simply be subtracted from the phase of 'the fibre'. The resulting corrected interferometric phase represents that of a dispersion free interferometer irrespective of the temperature of the fibre.

### 8.4.4 FSI frequency references

*'Never measure anything but frequency!'*
*Arthur Schawlow to Theodor Hänsch,* [39]

We have shown in sub-section 8.3.3 that knowledge of the frequency differences between multiple points on a laser scan can be efficiently exploited to provide absolute length scales to FSI measurements. We now need to look for systems which

[21] With the dispersion curve we mean the group refractive index versus wavelength.

contain accurately known frequency differences in the wavelength range of the light source used for FSI. We cannot hope to provide a comprehensive list of suitable references but we will categorise them and provide examples of references that cover a wide range of frequencies.

The currently most performant frequency references in the optical and near infrared regions are frequency combs which have been used for interferometry work in multi-frequency [40, 41] or FSI methods [42]. It would exceed the scope of this chapter to provide an overview of the different technologies used for frequency combs here and we point the reader to a reference book [43], an interesting introductory presentation about frequency combs by Scott Diddams here [44] and an associated video [45].

Frequency combs combine several excellent qualities for an FSI frequency reference system. Combs can have a large number of densely spaced features. Several 100 000 'teeth' can be generated by some combs with tooth separations (repetition rates) ranging from tens of MHz up to a THz. They also feature accurately defined frequency separation between the features (sub-Hz knowledge is possible) and some combs provide absolute tooth frequencies to similar levels of accuracy. The cost of such high quality combs is currently prohibitively high if their use in a commercially viable metrology system is intended. They do, however, play a significant role in research systems for many types of interferometry.

In May 2018 a collaboration of several groups led by Daryl T Spencer from NIST, Boulder, published results obtained with a frequency comb in the 1550 nm region constructed entirely in integrated photonics technologies [46] capable of linking an RF reference source via comb stages of 22GHz and 1 THz repetition rate and achieved a synthesised frequency stability of $7.7 \cdot 10^{-15}$ across a 4 THz section around 1550 nm with a 1 Hz absolute frequency uncertainty. These developments could be the start of a miniaturisation and dramatic cost reduction in high performance frequency combs that would allow them to be used in metrology instruments. While the integrated photonics technology has the potential to significantly reduce the cost of combs, the miniaturisation it brings also drives up the repetition frequencies which is detrimental to its use as an FSI frequency reference as fewer features are available across the scan-spectrum.

There are also much less expensive frequency comb technologies with reduced performance but nevertheless promising applicability to distance metrology such as electro optical combs used in [47] for multi-frequency absolute interferometry.

The second important category of frequency references are gas absorption cells. Gas cells are many orders of magnitude cheaper than frequency combs and hence play a larger role for practical FSI systems. The quality of the gas as a frequency reference depends primarily on the line width and the 'useful' interaction cross-section between the gas atoms or molecules and the interferometers light. The useful fraction of the cross-section is the one leading to the desired absorption process rather than other processes such as scattering.

Secondly, we care about the sensitivity of the line centre and width to external parameters such as pressure, temperature, probing laser intensity, interaction time with the probing laser, electric or magnetic fields, isotopic purity, other gas admixtures, etc.

Tertiary aspects are the number of available features in the relevant wavelength range, availability of isotope pure gases and saturability of the absorption line.

The desire to have many transition of very similar energies (narrowly spaced features) suggest the use of series of low energy excitations. Such narrowly spaced excitation series with many series members can be realised by molecular rotations and their overtones with vibrations. It is important to use small molecules if the excitations are to be in the optical range of wavelength. Large molecules with large moments of inertia will have too large a moment of inertia and hence excitation energies which are too low. They can also have too many interacting degrees of freedom so that the combined spectrum can be extremely complex to understand. We will list a few molecular excitations of small molecules below.

Other narrowly spaced excitation series, albeit with smaller numbers of features can be the hyperfine structure series of electronic excitations. The number of possible hyperfine multiplets is vast and each only covers a small frequency range. We will therefore not attempt to list them here.

As a rule of thumb we may assume that the lower the excitation energy of an excited state the more easily it will be disturbed by external thermally driven collisions. The rotational and vibrational excitations are however often significantly more sensitive to thermally driven collisions than electronic excitations which may happen deep inside an atom where they are well protected by the surrounding electrons. Rotation and vibration states therefore usually suffer more collisional (pressure) broadening than electronic excitations and it is desirable to use them in low pressure gas cells to reduce the pressure broadening.

All types of absorption features will suffer similarly from Doppler broadening so low temperatures are desirable or even better, forms of Doppler free, saturated spectroscopy should be used. Due to their short life-times rotational excitations often require extremely high light intensities to saturate. Such intensities are difficult to achieve with conventional methods over reasonably long distances (to obtain sufficient absorption contrast).

One method to saturate such gases has used tapered nano-fibres in which the fibre carrying the probe light has been smoothly drawn out to such narrow diameters that the majority of the light energy is highly concentrated into an evanescent wave which exists in a very narrow region on the outside of the fibre where it is able to interact with the chosen gas [48]. This technique is limited in its practical applicability by the fragility of the nano-fibre and in spectral resolution by the extremely short time that molecules spend in the narrow layer of light surrounding the fibre.

Another method to saturate short lived molecular gases utilises hollow core fibres which can concentrate the light energy in the hollow core which can also be filled with gases. It is however difficult to first empty and then fill these cores as the extremely small cross-sections severely limit the diffusion rates into and out of the long cores.

**Table 8.3.** List of useful gases for frequency referencing in the telecommunications wavelength range [49–51] stating the range in which useful features exist.

| Name | Wavelength range |
|---|---|
| $H^{13}C^{14}N$ | 1525–65 |
| $H\,^{12}_{2}\,C_2$ | 1510–45 |
| $^{13}C^{16}O$ | 1595–1635 |
| $^{12}C^{16}O$ | 1560–1600 |

A range of molecular excitations in the particularly interesting telecommunications range of frequencies is given in table 8.3.

Last but not least we would like to mention man-made artefacts as frequency references. If these are to be most useful for FSI applications they can take the form of high finesses, long etalons with short free spectral ranges (FSR) to produce many sharp features.

Because they are to be used across a large frequency range they should not be of solid construction because the FSR would depend on wavelength. If the dispersion is to be 'calibrated out' an additional frequency reference is needed to locate the resulting calibration curve on the frequency axis as described in section 8.4.3.

The etalon should therefore be evacuated or filled with a very low dispersion gas such as helium. Alternatively it could be filled with the same gas that fills the measurement interferometer. In this case the etalon does not really serve as a frequency reference but as a length reference in which the refractive index of the measurement arm is automatically corrected for. For use as a calibrated reference interferometer it is not necessary to construct a high finesse system. A simple Michelson or Fizeau style interferometer would suffice.

Great care has to be taken in their construction to make them almost independent of temperature and pressure changes or other mechanical influences. This will often require the use of expensive low expansion materials and costly production techniques. Such high finesse and long, precision etalons are not cheap. They also need to be carefully and potentially regularly calibrated against traceable references such as a suitable calibrated frequency comb to determine their actual FSR.

## Acknowledgements

Since my 'day jobs' are those of a particle or accelerator physicist and university and college academic this chapter had to be written in my 'free' time and it took—as all such works do—much longer than anticipated. I therefore have to primarily thank my wife Ute for her patience and willingness to give up family time and instead support me in this project through many long weekends.

I would also like to thank Dr Matthew Warden for long discussions on the phone and his reading of the early versions of this chapter which have strongly influenced it and in particular the bibliography in a major way. Other inputs to the bibliography have come from Dr Florian Pollinger who was also instrumental in consolidating my interest in metrology through his welcoming support during my time as a visiting scientist at PTB. The section on frequency standards and my own understanding of this topic have benefited greatly from the friendly and competent advice of Dr Uwe Sterr also at PTB in Braunschweig. Dr John Dale has been the second reader of this chapter and I thank him for his useful comments.

## Symbols and acronyms

$c$ — Phase speed of light in vacuum

$x$ — Independent position variable (one-dimensional)

$x_r$ — Position of the reference reflections

$x_0$, $x_i$, $x_j$ — Specific positions

$\overline{x_i x_j}$ — Geometrical path length between positions $x_i$ and $x_j$

$x_m(\tau)$ — Position of the mirror (in the measurement arm) at time of reflection $\tau$

$\tau$ — time at which the mirror has to be at the position $x_m(\tau)$ so that the reflected wave will reach position $x$ at time $t$

$u_m(\tau)$ — Speed of measurement mirror at time $\tau$, also referred to as target speed

$\beta(\tau)$ — Ratio of measurement mirror speed and speed of light in the medium,

$$\beta(\tau) = \frac{u_m(\tau)}{u}$$

$Z(\tau)$ — Doppler shift factor $Z(\tau) = \frac{1-\beta(\tau)}{1+\beta(\tau)}$; factor by which the frequency of a wave, reflected from a moving mirror at time $\tau$ with speed $u_m(\tau)$ is changed

$t$ — Independent time variable

$t_0$ — Specific moment in time

$t_i$, $t_f$ — Initial and final times of a frequency scan

$t_{p1}$, $t_{p2}$, $t_c$ — Special times during a dynamic dual frequency scan, see section 8.3.3 for definitions

$t_j$ — Discrete times during which intensities are digitised

$n_{\text{true}}(\vec{x}, t)$ — The true refractive index of a wave propagation medium as a function of space $x$ and time $t$

$n(t)$ — Effective refractive index after averaging across entire optical path

$n$ — Time averaged effective refractive index, $n = \langle t \rangle_{\tau_{\text{med}}}$

$\tau_{\text{med}}$ — Minimum averaging time over which $n$ becomes a constant

$\lambda$ — Wavelength of a wave

$\frac{dn(t)}{d\lambda}$ — Wavelength dispersion of the propagation medium

$L$ — Optical path length $L = n$ times the geometrical path length, sometimes also represents the optical path length difference if the reference wave travels through zero path lengths

$L_r$, $L_m$ — Optical path length of reference and measurement wave of interferometer

$L_i$, $L_f$ — Initial and final OPD during a frequency scan

$L_0$ or $\langle L \rangle$ — Average OPD during a frequency scan

$\delta L$ — Amount by which an OPD varies during a frequency scan

$\Delta L$ — Optical path length difference

| | |
|---|---|
| $\Delta L_{\text{sep}}$ | Minimal amount by which two OPDs have to differ to be separable when simultaneously present in a single interferometer |
| $\sigma_L$ | Error in the measurement of an OPD or OPD resolution |
| $A(x, t)$ | Scalar light amplitude as a function of time and space co-ordinates |
| $A_r(x, t)$, $A_m(x, t)$ | Scalar light amplitude of interferometers reference or measurement wave |
| $\mathcal{A}(t, x)$ | Magnitude of a scalar light amplitude |
| $u$ | Propagation speed of the phase of a wave in the propagation medium $u = c/n$ |
| $k(t, x)$ | Wave number of an optical wave, $k = 2\pi/\lambda$ |
| $\omega(t, x)$ or $\nu(t, x)$ | Circular and regular optical frequency of wave at time $t$ and position $x$ |
| $\omega_e(t)$ | Circular frequency of a light wave at the point of emission |
| $\omega_o$ | Circular frequency of a light wave at the start of a scan |
| $\omega^{(n)}(t, x)$ | $n^{\text{th}}$ order Taylor expansion in time of the circular frequency |
| $\omega_{lin}$ | Part of the optical frequency of a wave that varies linearly with time $\omega_{\text{lin}} = \omega^{(1)}(t, x)$ |
| $\omega_m(t, x, \tau)$ | Circular frequency of the measurement wave at time $t$ and position $x$ after reflection at a moving mirror at time $\tau$ |
| $\omega_+$ | Circular frequency with which $\phi_+$ varies; this is not the frequency of a wave but the high frequency component of an interference intensity |
| $\omega_-$ | Circular frequency with which $\phi_-$ varies; this is not the frequency of a wave but the low frequency component of an interference intensity |
| $\omega_1(t)$ | Circular frequency of the first laser in a dual scanning FSI system |
| $\omega_2(t)$ | Circular frequency of the second laser in a dual scanning FSI system |
| $\Delta\omega$ | Total circular frequency advance during a frequency scan |
| $\phi(x, t)$ | Absolute optical phase of a wave |
| $\phi_e(t)$ | Phase of an optical wave at the location of its emission |
| $\phi_r$, $\phi_m$ | Optical phase of the reference and measurement wave of an interferometer |
| $\phi_+$, $\phi_-$ | Sum and difference of phases from reference and measurement waves $\phi_+ = \phi_r + \phi_m$, $\phi_- = \phi_r - \phi_m$; this is not an optical phase but a component of an interference signal phase |
| $\Phi$ | Interferometric phase; see definition in section 8.2.4 |
| $\Phi_{r,1}(t)$, $\Phi_{r,2}(t)$ | Interferometric phase of the reference interferometer illuminated by light from Laser 1 or 2 in a dual FSI scan |
| $\Phi_{m,1}(t)$, $\Phi_{m,2}(t)$ | Interferometric phase of the measurement interferometer illuminated by light from Laser 1 or 2 in a dual FSI scan |
| $\Delta\Phi$ | Total interferometric phase advance during a frequency scan |
| $\Delta\Phi^{\text{up}}$, $\Delta\Phi^{\text{down}}$ | Total interferometric phase advance during a frequency scan from an upward and downward scanning source |
| $\omega_{\text{int}}$, $f_{\text{int}}$ | (Circular) frequency with which the interference intensity varies |
| $\tau_{\text{det}}$ | Response time of an optical detector |
| $I(x, t)$ | Intensity of a light wave at position $x$ and time $t$ |
| $I_r(t, x)$, $I_m(t, x)$ | Intensities of reference and measurement wave in an interferometer |
| $I_{\text{det}}(t, x)$ | Detected total intensity of an interferometer |
| $I_0(t, x)$ | Slowly time varying average part of $I_{\text{det}}(t, x)$ |
| $I_{\text{int}}(t, x)$ | Rapidly varying interference part of $I_{\text{det}}(t, x)$ |
| $\alpha_{\text{lin}}$ | Gross measure of tuning non-linearity, see equation (8.7) |
| $\dot{\omega}(t)$ | Tuning speed of an optical source |
| $\dot{\omega}_0$ | Tuning speed of an optical source at the start of a scan, part of the Taylor expansion $\omega^{(n)}(t, x)$ |

| | |
|---|---|
| $\ddot{\omega}(t)$ | Tuning acceleration of an optical source |
| $\ddot{\omega}_0$ | Tuning acceleration of an optical source at the start of a scan, part of the Taylor expansion $\omega^{(n)}(t, x)$ |
| $\tau_{scan}$ | Duration of a frequency scan |
| $f_{scan}$ | Repetition frequency of periodic frequency scans, $f_{scan} = 1/\tau_{scan}$ |
| $f_{rep}$ | Repetition frequency of a measurement process |
| $i$ | The complex unit number, $i = \sqrt{-1}$ |
| $\mathcal{F}$, $\mathcal{F}^{-1}$ | Fourier and inverse Fourier Transform |
| $\mathcal{H}$ | Hilbert transform |
| Im, Re | Imaginary and real part of a number of function |

## Acronyms

| | |
|---|---|
| ATCA | Advanced telecommunication computing architecture, a DAQ system standard used widely in the telecommunication industry |
| BS | Beam splitter |
| CCD | Charge couple device (image sensor) |
| CMOS | Complementary metal oxide semiconductor, a technique for producing integrated circuits in semi-conductors |
| COL | Collimator |
| crab cavities | Radio frequency cavities used to rotate bunches of particle at the collision points of particle colliders so that the bunches collide head on |
| DFB laser | Distributed feedback laser containing an integrated grating in its gain medium |
| DFT | Digital Fourier transform |
| DAQ | Data acquisition system |
| ECDL | External cavity diode laser |
| FSI | Frequency scanning interferometry |
| FMCW | Frequency modulated continuous wave |
| FS | Fibre splitter |
| FSR | Free spectral range |
| GMT | Giant Magellan Telescope, a planned 30 m class telescope |
| GPU | Graphics processing unit |
| HL–LHC | High luminosity–large hadron collider, the upgrade form of the current large hadron collider |
| IR | Infrared |
| LHC | Large hadron collider |
| M | Mirror |
| $\mu$TCA or uTCA | Micro telecommunications computing architecture complementary to ATCA but aimed at physics applications |
| OPD | Optical path length difference |
| PBS | Polarising beam splitter |
| PCIe | Peripheral component interconnect express is the standard bus protocol for personal computers |
| PD | Photo detector |
| SLM | Spatial light modulator |
| TCP/IP | Transmission control protocol/Internet protocol |

## References

[1] Abbott B P *et al* 2017 Gw170817: Observation of gravitational waves from a binary neutron star inspiral *Phys. Rev. Lett.* **119** 161101

[2] Ruiz P D and Huntley J M 2017 Single-shot areal profilometry using hyperspectral interferometry with a microlens array *Opt. Express* **25** 8801–15

[3] Pohl N, Jaschke T, Scherr S, Ayhan S, Pauli M, Zwick T and Musch T 2013 Radar measurements with micrometer accuracy and nanometer stability using an ultra-wideband 80 ghz radar system *2013 IEEE Topical Conference on Wireless Sensors and Sensor Networks (WiSNet)* pp 31–3

[4] Pohl N, Jaeschke T and Aufinger K 2012 An ultra-wideband 80 ghz fmcw radar system using a sige bipolar transceiver chip stabilized by a fractional-n pll synthesizer *IEEE Trans. Microwave Theory Tech.* **60** 757–65

[5] Jaeschke T, Bredendiek C, Küppers S and Pohl N 2014 High-precision d-band fmcw-radar sensor based on a wideband sige-transceiver mmic *IEEE Trans. Microwave Theory Tech.* **62** 3582–97

[6] Dale J, Hughes B, Lancaster A J, Lewis A J, Reichold A J H and Warden M S 2014 Multi-channel absolute distance measurement system with sub ppm-accuracy and 20 m range using frequency scanning interferometry and gas absorption cells *Opt. Express* **22** 24869–93

[7] Rakich A, Dettmann L, Leveque S and Guisard S 2016 A 3D metrology system for the GMT *Ground-based and Airborne Telescopes VI, SPIE Proceedings* **vol 9906** 990614 p (https://www.google.com/url?sa=t&rct=j&q=&esrc=s&source=web&cd=1&ved=2ahUKEwiJmefWkp3eAhXHK8AKHQBCCAsQFjAAegQIAxAC&url=https%3A%2F%2Fwww.gmto.org%2FSPIE_2016%2FSPIE_2016_9906-112.pdf&usg=AOvVaw2p9YHKI0Veao7z1i95l8qw)

[8] Sosin M, Dijoud T, Durand H M and Rude V 2016 Position monitoring system for hl-lhc crab cavities *Proceedings, 7th International Particle Accelerator Conference (IPAC 2016) (Busan, Korea, May 8-13, 2016)* WEPOR018 (https://www.google.com/url?sa=t&rct=j&q=&esrc=s&source=web&cd=1&ved=2ahUKEwj-scKrk53eAhXrBcAKHfHbDg4QFjAAegQIBxAC&url=http%3A%2F%2Faccelconf.web.cern.ch%2Faccelconf%2Fipac2016%2Fpapers%2Fwepor018.pdf&usg=AOvVaw3zR1mYnEHmp37i9MIcTqxk)

[9] Jang Y-S and Kim S-W 2017 Compensation of the refractive index of air in laser interferometer for distance measurement: A review *Int. J. Precis. Eng. Manuf.* **18** 1881–90

[10] Ciddor P E 1996 Refractive index of air: new equations for the visible and near infrared *Appl. Opt.* **35** 1566–73

[11] Polyanskiy M refractiveindex.info. https://refractiveindex.info/?shelf=other&book=air&page=Ciddor.

[12] Warden M S 2011 Absolute distance metrology using frequency swept lasers *PhD Thesis* (UK: Oxford University) (https://ora.ox.ac.uk/objects/uuid:a20e9c69-e580-48d8-bbce-51d8f4186450#permalinkModal)

[13] Zheng J 2004 Analysis of optical frequency-modulated continuous-wave interference *Appl. Opt.* **43** 4189–98

[14] Zheng J 2005 Continued analysis of optical frequency-modulated continuous-wave interference *Appl. Opt.* **44** 765–69

[15] Zheng J 2006 Coherence analysis of optical frequency-modulated continuous-wave interference *Appl. Opt.* **45** 3681–87

[16] Zheng J 2006 Optical frequency-modulated continuous-wave interferometers *Appl. Opt.* **45** 2723–30

[17] Rakhmanov M 2006 Reflection of light from a moving mirror: derivation of the relativistic Doppler formula without Lorentz transformations ArXiv Physics e-prints. physics/0605100.
[18] Einstein A 1905 Zur elektrodynamik bewegter körper *Ann. Phys.* **322** 891–921
[19] Lancaster A J 2015 Absolute distance interferometry capable of long-term high frequency measurements of fast targets, *PhD Thesis* University of Oxford, Department of Physics (https://ora.ox.ac.uk/objects/uuid:d597a926-ba4c-4b25-b2b7-695faf88979d#permalinkModal)
[20] Coe P A 2001An Investigation of Frequency Scanning Interferometry for the alignment of the ATLAS semiconductor tracker *PhD Thesis* University of Oxford, Department of Physics (https://www.google.com/url?sa=t&rct=j&q=&esrc=s&source=web&cd=1&cad=rja&uact=8&ved=2ahUKEwjAq7uQlJ3eAhVkL8AKHY__AVEQFjAAegQICRAC&url=https%3A%2F%2Fcore.ac.uk%2Fdownload%2Fpdf%2F44189377.pdf&usg=AOvVaw0LXzBFcKGZU6GY6KTu1BjS)
[21] Novák J, Novák P and Mikš A 2008 Multi-step phase-shifting algorithms insensitive to linear phase shift errors *Opt. Commun.* **281** 5302–9
[22] Rohde F, Benkler E, Puppe T, Unterreitmayer R, Zach A and Telle H R 2014 Phase-predictable tuning of single-frequency optical synthesizers *Opt. Lett.* **39** 4080–83
[23] Tranter G E, Lindon J C and Koppenaal D W 2017 *Encyclopedia of Spectroscopy and Spectrometry* 3rd edn (Amsterdam: Elsevier)
[24] Martinez J J, Campbell M A, Warden M S, Hughes E B, Copner N J and Lewis A J 2015 Dual-sweep frequency scanning interferometry using four wave mixing *IEEE Photon. Technol. Lett.* **27** 733–36
[25] Krzczanowicz L and Connelly M 2014 40 gb/s nrz-dqpsk data all-optical wavelength conversion using four wave mixing in soa *25th IET Irish Signals Systems Conference 2014 and 2014 China-Ireland International Conference on Information and Communications Technologies (ISSC 2014/CIICT 2014)* pp 349–51
[26] Green J R 2007 Development of a prototype frequency scanning interferometric absolute distance measurement system for the survey & alignment of the international linear collider *PhD Thesis* University of Oxford, Department of Physics (https://www.google.com/url?sa=t&rct=j&q=&esrc=s&source=web&cd=1&cad=rja&uact=8&ved=2ahUKEwjXibqjlZ3-eAhWKKcAKHUulDq0QFjAAegQICRAC&url=http%3A%2F%2Fwww-pnp.physics.ox.ac.uk%2F~reichold%2FjrgThesis080303.pdf&usg=AOvVaw3oYYK2ojw6XQCB4t2dYdyT)
[27] Press W H, Teukolsky S A, Vetterling W T and Flannery B P 2007 *Numerical Recipes: The Art of Scientific Computing* 3rd Edition (New York, NY, USA: Cambridge University Press)
[28] Press W H, Teukolsky S A, Vetterling W T and Flannery B P 2002 *Numerical Recipes in C+ +* (Cambridge: Cambridge University Press)
[29] Warden M S 2014 Precision of frequency scanning interferometry distance measurements in the presence of noise *Appl. Opt.* **53** 5800–6
[30] Hughes B, Campbell M A, Lewis A J, Lazzarini G M and Kay N 2017 Development of a high-accuracy multi-sensor, multi-target coordinate metrology system using frequency scanning interferometry and multilateration *Proc. SPIE 10332, Videometrics, Range Imaging, and Applications XIV, 1033202 (26 June 2017)* **vol 10332** pp 1–9
[31] Gibson S M, Coe P A, Mitra A, Howell D F and Nickerson R B 2005 Coordinate measurement in 2-d and 3-d geometries using frequency scanning interferometry *Opt. Lasers Eng.* **43** 815–31 Fringe Analysis special interest group (FASIG)

[32] Gibson S M, Coe P A, Dehchar M, Fopma J, Howell D F, Nickerson R B and Viehhauser G 2008 First data from the ATLAS inner detector FSI alignment system *Proc. 10th International Workshop on Accelerator Alignment (IWAA 2008) (KEK Tsukuba, February 11-15, 2008)*
[33] Etalon A G Absolutemultiline$^{TM}$. http://www.etalon-ag.com/en/products/absolute-multiline-technology/.
[34] Warden M and Urner D 2011 Apparatus and method for measuring distance *Patent pub. no.: Wo/2012/022956*
[35] Yamamoto A, Kuo C-C, Sunouchi K, Wada S, Yamaguchi I and Tashiro H 2001 Surface shape measurement by wavelength scanning interferometry using an electronically tuned ti: sapphire laser *Opt. Rev.* **8** 59–63
[36] Larsen R 2017 Microtca.4 / 4.1 hardware standards and software guidelines progress overview, *Proc. Technology and Instrumentation in Practicle Physics'17, TIPP17 (Beijing, 26 May 2017)* (https://www.google.com/url?sa=t&rct=j&q=&esrc=s&source=web&cd=1&ved=2ahUKEwi208nplp3eAhUGVsAKHe2VCEoQFjAAegQIBBAC&url=https%3A%2F%2Findico.ihep.ac.cn%2Fevent%2F6387%2Fsession%2F23%2Fcontribution%2F233%2Fmaterial%2Fslides%2F&usg=AOvVaw0wP9qhbNu2eR9pXlq1g7XQ)
[37] VadaTech Inc. 2017 http://www.vadatech.com/product.php?product=498&catid_prev=0&-catid_now=0. URL: http://www.vadatech.com/product.php?product=498&catid_prev=0&catid_now=0
[38] Dale J, Hughes B, Lancaster A J, Lewis A J, Reichold A J H and Warden M S 2018 An optical fibre as a reference interferometer for frequency scanning interferometry *Opt. Express* in preparation
[39] Hänsch T W 2006 Nobel lecture: Passion for precision *Rev. Mod. Phys.* **78** 1297–309
[40] Minoshima K and Matsumoto H 2000 High-accuracy measurement of 240-m distance in an optical tunnel by use of a compact femtosecond laser *Appl. Opt.* **39** 5512–17
[41] Salvadé Y, Schuhler N, Lévêque S and Le Floch S 2008 High-accuracy absolute distance measurement using frequency comb referenced multiwavelength source *Appl. Opt.* **47** 2715–20
[42] Baumann E, Giorgetta F R, Coddington I, Sinclair L C, Knabe K, Swann W C and Newbury N R 2013 Comb-calibrated frequency-modulated continuous-wave ladar for absolute distance measurements *Opt. Lett.* **38** 2026–28
[43] Ye J and Cundiff S T 2005 *Femtosecond Optical Frequency Comb: Principle, Operation, and Applications* (Norwell, MA, USA: Kluwer Academic Publishers/Springer)
[44] Diddams S 2015 Fundamentals of frequency combs: What they are and how they work *Proc., KISS Workshop: 'Optical Frequency Combs for Space Applications'*
[45] Diddams S 2015 Fundamentals of frequency combs: What they are and how they work, 2015 URL: http://www.youtube.com/watch?v=njtHAxqo7bU.
[46] Spencer D T *et al* 2018 An optical-frequency synthesizer using integrated photonics *Nature* **557** 81–5
[47] Zhao X, Qu X, Zhang F, Zhao Y and Tang G 2018 Absolute distance measurement by multi-heterodyne interferometry using an electro-optic triple comb *Opt. Lett.* **43** 807–10
[48] Takiguchi M, Yoshikawa Y, Yamamoto T, Nakayama K and Kuga T 2011 Saturated absorption spectroscopy of acetylene molecules with an optical nanofiber *Opt. Lett.* **36** 1254–56
[49] Gilbert S L and Swann W C 2001 Acetylene $^{12}C_2H_2$ absorption reference for 1510 nm to 1540 nm wavelength calibrationsrm 2517a, *NIST Special Publication 260-133 2001 Edition, Standard Reference Materials* (https://www.nist.gov/document-10369)

[50] Gilbert S L and Swann W C 2002 Carbon monoxide absorption references for 1560 nm to 1630 nm wavelength calibration srm 2514 ($^{12}C^{16}O$) and srm 2515 ($^{13}C^{16}O$) *NIST Special Publication 260-146 2002 Edition, Standard Reference Materials* (https://www.nist.gov/document-10353)
[51] Gilbert S L and Swann Chih-Ming Wang W C 1998 Hydrogen cyanide $H^{13}C^{14}N$ absorptino reference for 1530 nm–1560 nm wavelength calibration—srm 2519 *NIST Special Publication 260-137 1998 Edition, Standard Reference Materials* (https://www.nist.gov/document-10362)

**IOP** Publishing

Modern Interferometry for Length Metrology
Exploring limits and novel techniques
**René Schödel (Editor)**

# Chapter 9

## Picometre level displacement interferometry

**Birk Andreas and Christoph Weichert**

Optical displacement interferometers are used in many applications, but only a few of them require distance metrology at the picometre level. In addition, the meaning of 'picometre level' is different depending on the moving range of the desired application. Some error sources of displacement interferometers limit their resolving capability while others are length-proportional. Length-proportional errors result in significant contributions to the uncertainty of interferometer systems on top the other error sources, which define their precision. Resolving displacements of 10 pm, which is smaller than the size of a hydrogen atom, is only possible with averaging, spatially by means of the beam diameter and temporally by means of filters during the phase evaluation. In applications with moving ranges above 10 mm and with the need of traceability, length-proportional errors become significant for interferometer systems with a relative uncertainty better than $1 \times 10^{-9}$. The realisation of displacement interferometers with a relative uncertainty better than $1 \times 10^{-9}$ is the focus of this chapter.

Interferometer systems are used in scientific as well as industrial applications. Exemplary applications of displacement interferometers with a relative uncertainty below $1 \times 10^{-9}$ are the calibration of the silicon lattice parameter within a setup with moving range of 10 mm and the calibration of line scales with a length up to 500 mm (slightly more than 18 inches). In many industrial applications with moving ranges between 10 mm and 500 mm, interferometric encoder systems are used. They can provide a resolution better than 5 pm [1], a repeatability better than 0.1 nm for displacement measurements over several hundreds of millimetres [2], periodic nonlinearities below 0.1 nm [3] and a drift stability below 0.1 nm per hour [4]. Additionally, optical encoder systems are nearly insensitive to variations of the laser frequency and the refractive index [5] in the case of an adequate symmetrical design and a small air gap between the encoder head and the grating scale [6]. These facts make them a good choice for picometre level precision displacement measurements and increasingly popular for industrial applications, where they replace displacement interferometers [7]. In some applications interferometric encoder systems cannot replace displacement interferometers, because of the additional effort of their integration into a machine structure fulfilling the extended Abbe's criteria [8] and their lack of traceability. The grating scales of the encoder systems defining the measured length exhibit individual deviations to the metre definition damping down the comparability of measurement results. Indeed, the nonlinear deviations can be corrected using multi-sensor based error separations [9], but the linear scale must be calibrated using optical displacement interferometers [10]. Therefore, the measurement uncertainty of encoder systems is defined among others by the limitations of optical interferometers and the ability to transfer their measurement uncertainty.

Optical interferometers can provide repeatable and reproducible distance measurements at the picometre level. However, this fact does not result in traceable measurements. General overviews of the limitations of interferometer systems and their integration into a specific application are given by Steinmetz [11], Bobroff [12] and Badami and DeGroot [13]. Figure 9.1 shows a rough overview of typical error sources of displacement interferometers. Indeed, the uncertainty of length measurements does not only depend on the interferometer system, but also on its integration in the experiment or the machine structure. Influences like the cosine error, the parallax error or the correction of the mirror topography can outnumber the uncertainty of the interferometer system itself. The data age of the phase values is important where the interferometer system is used for comparison measurements or as feedback in control loops. Therefore, phase meters provide a data age compensation with fixed and variable delays [14]. Assuming a slow variation of the velocity

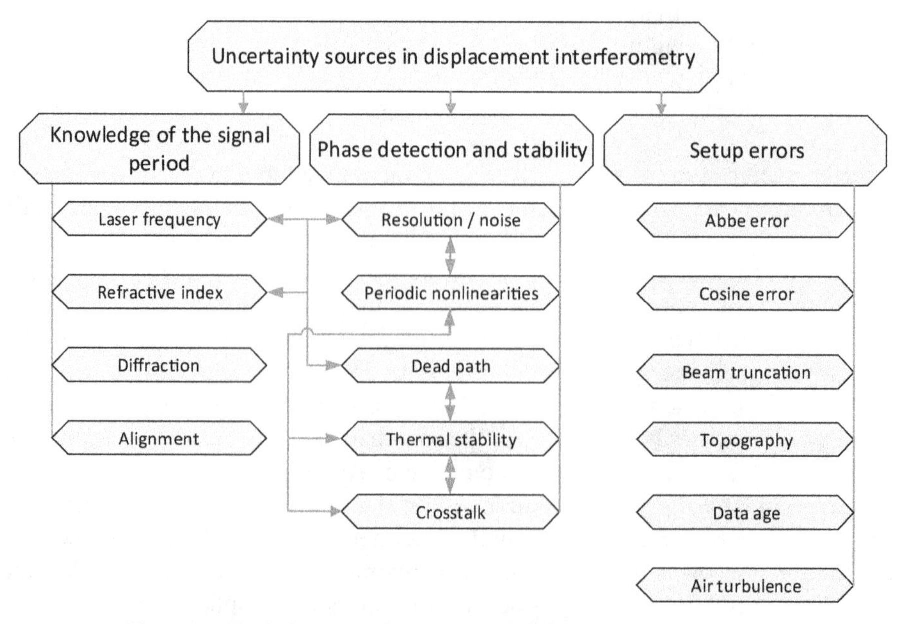


**Figure 9.1.** Typical sources of uncertainty in displacement interferometry.

of the moving mirror in comparison to the acquisition time of the phase meter, a prediction of the position value by means of the current velocity can increase the synchronisation between comparison systems [15]. Picometre level uncertainties can be achieved with the application of a data age compensation [16]. The errors resulting from the integration of the interferometer system in a machine structure are not discussed in detail in this chapter. The focus of attention is on intrinsic errors of interferometer systems limiting their uncertainty. Traceable measurements are essential to allow comparability of scientific results and they are fundamental to the modern production philosophy of interchangeable parts. The uncertainty of displacement measurements is always a combination of their precision and their traceability.

As described in the introduction of the book, the displacement of a mirror can be determined with a displacement interferometer based on measured phase values ($\varphi$) and the signal period. The signal period depends on the frequency of the light ($f$), the refractive index of the ambient medium ($n$) and a correction factor ($\delta$). Consequently, the uncertainty of all four quantities can limit the uncertainty of the interferometer system:

$$
\begin{aligned}
z &= \frac{4\pi f}{c} \cdot n \cdot \delta \cdot \varphi, \\
\frac{u_z}{z} &= \sqrt{\left(\frac{u_f}{f}\right)^2 + \left(\frac{u_n}{n}\right)^2 + \left(\frac{u_\varphi}{\varphi}\right)^2 + \left(\frac{u_\delta}{\delta}\right)^2}.
\end{aligned}
\tag{9.1}
$$

To achieve picometre level displacement measurements, all four contributions are of importance and require particular characteristics of the displacement interferometer system. This chapter is structured in a similar sequence to equation (9.1), examining the frequency stabilisation of the light source, the refractive index determination, errors of the evaluated phase and the correction factor. A clear sub-division of the different error sources in categories is difficult, since many of them are linked to each other as suggested partly in figure 9.1. Three examples of this are the following: electrical or optical cross talk and multi-reflections cause periodic nonlinearities and also influence the stability of the interferometer system, variations of the laser frequency affect the stability when the interferometer exhibits a dead path (initial path length difference of the reference and the measurement beam) and the phase noise can limit the quality of a compensation of periodic nonlinearities. The uncertainty contributions discussed below appear in single frequency (homodyne) and multi-frequency (heterodyne) interferometer systems similarly. The basic principles of homodyne or heterodyne interferometers are explained in chapter 1. In the case of a fundamental difference, it is explicitly emphasised.

The usage of a homodyne or heterodyne interferometer is often just a question of personal preference. Both systems can reach similar levels of uncertainty, but there are differences depending on the targeted application. The resolving capability is defined for both systems by different parameters, to name just two: the shot noise and the laser intensity noise. However, with both systems it was demonstrated that picometre level resolution is achievable, for example for homodyne systems in the references [1, 17] and for heterodyne systems in reference [18]. A crucial factor for choosing the kind of system for a specific application is the expected speed of the moving mirror. Despite additional bandpass limitations of the amplifiers and the phase meter electronics, for heterodyne interferometers the maximal moving speed ($v$) is finally limited by the beat frequency of the system. In the classical case, the Doppler-shifted frequency can be calculated as follows

$$\tilde{f} = f \cdot \left(1 - 2\frac{v}{c}\right),$$
$$\Delta f = \tilde{f} - f = -\frac{2v}{\lambda}. \tag{9.2}$$

For a heterodyne system with a beat frequency of 2.5 MHz using a 633 nm light source the theoretically maximal detectable moving speed would be 0.79 m $s^{-1}$. For applications with higher speeds aiming for sub-nanometre accuracy, like wafer scanners or absolute gravimeters, a larger beat frequency is required. These can end up in comparatively large beat frequencies of several hundreds of MHz by directly using the two orthogonal polarised modes of a frequency stabilised HeNe laser. For homodyne systems, 'only' the bandwidth of the amplifiers and the phase meter electronics limit the maximal moving speed. Independently of the system used, a relativistic correction of the interferometric signals must be considered in the case of moving speeds above 0.1 m $s^{-1}$ to meet the demand of sub-nanometre accuracy [19].

The effort one can save with homodyne systems in applications with large moving speeds has to be repaid in the case of very small speeds. Especially, the usage of the signals of homodyne systems as feedback in a control loop to keep a specific position (zero sensor) can result in additional errors. These errors are caused by the origin of periodic nonlinearities at homodyne interferometers. At these systems periodic nonlinearities cannot be avoided but only corrected using for example ellipse parameters [20]. These ellipse parameters correct for different amplitudes and offsets of the interferometric signals as well as a phase deviation from $\pi/2$, which is connected for example with the intensity of light sources. In the case where the intensity varies, the ellipse parameters have to be adapted to reduce the resulting errors. For this a phase variation is needed, for example introduced by the movement of a mirror or by a frequency variation of the light source in combination with a system design including a dead path. However, the first step to traceable interferometric measurements is knowledge of the laser frequency.

## 9.1 Influences of the laser light source

### 9.1.1 Frequency stabilization—the first step to traceability

Since the redefinition of the unit of length in the International System of Units in 1983 [21] the metre has been defined using the value of the speed of light in a vacuum $c_0$ = 299 792 458 m $s^{-1}$. The International Committee for Weights and Measures (Comité international des poids et mesures, CIPM) recommended methods for the practical realization of the metre and 'that in all cases any necessary corrections be applied to take account of actual conditions such as diffraction, gravitation, or

**Table 9.1.** Commonly used wavelengths for the realisation of the metre in dimensional metrology [26].

| Frequency/THz | Fractional uncertainty | Wavelength/nm | Laser/absorber |
|---|---|---|---|
| 473.6127 | $1.5 \times 10^{-6}$ | 632.9908 | HeNe unstabilised |
| 473.612 353 604 | $2.1 \times 10^{-11}$ | 632.991 212 58 | HeNe/$I_2$ R(127) 11–5, a(16) |
| 563.260 223 513 | $8.9 \times 10^{-12}$ | 532. 245 036 104 | 2f Nd:YAG/$I_2$ R(56) 32–0, a(10) |

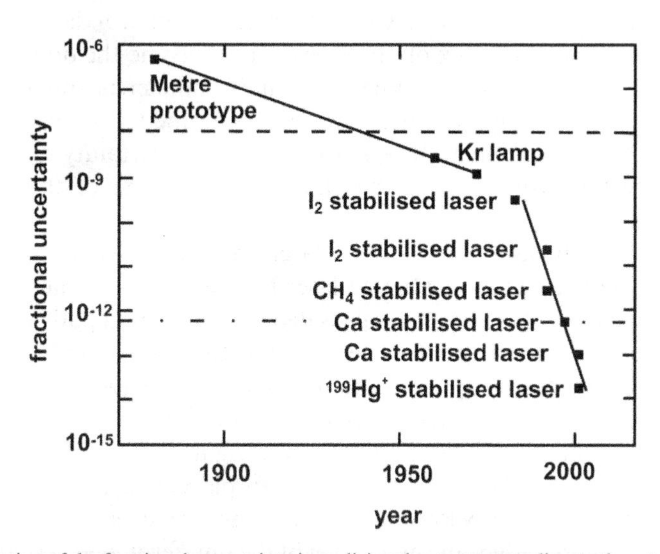


**Figure 9.2.** Evolution of the fractional uncertainty in realizing the metre according to the *mise en pratique* [26].

imperfection in the vacuum' [21]. One recommended method is to realize the metre by means of recommended optical radiations, whose vacuum wavelength, frequency and uncertainty are listed and updated [22–25]. The improvement of the realization of the definition of the metre over the last few decades is illustrated in figure 9.2 [26]. Table 9.1 lists the commonly used wavelength standards to realise the metre in interferometric length measurements. With cost reduction accompanying the growth of optical telecommunication, light sources with a wavelength of about 1310 nm and 1550 nm has become increasingly popular for interferometer systems. Picometre level interferometry is achievable with infrared light sources, since the resolving capability of interferometer systems nowadays is not limited by the ability of the phase meter to determine a fraction of the period.

Due to their utmost commercial importance HeNe lasers are spotlighted in detail. Unstabilised HeNe lasers, operating on the 3s2 $\rightarrow$ 2p4 transition, are included in the list of standard frequencies for the realization of the metre. Their centre frequency mainly depends on the isotopic mixture of neon and the distribution and sweep of modes within the Doppler width of the gain curve. These errors result in a fractional standard uncertainty of $1.5 \times 10^{-6}$ [27]. The uncertainty of the centre line limits the applicability of unstabilised HeNe lasers for picometre level interferometry to a moving range of a few micrometres.

Most of the commercial displacement interferometers are using a gain curve stabilised HeNe laser as light source. Their frequency can be calibrated in comparison with an iodine-stabilised HeNe laser [28] or a frequency-comb [29]. They can provide a relative frequency stability of $2.5 \times 10^{-9}$ over 24 h, without turning the laser head on and off, and a repeatability of less than $2 \times 10^{-9}$ [30]. Therefore, gain curve stabilised HeNe lasers can be used for picometre level displacement interferometry for millimetre moving ranges. In the case when the interferometer is not situated in a vacuum environment, for larger distances the variations of the refractive index of air nevertheless become the dominant source of uncertainty in most applications. When the interferometer is situated in a proper environment and picometre level measurements are aimed at moving ranges above one millimetre, the centre line must be known with an uncertainty smaller than $10^{-9}$, either by a direct measurement or by stabilisation to one of the recommended vacuum wavelengths.

In the case of a measurement of the laser frequency, for example by a beat frequency measurement with a stabilised laser source, the dead path of the interferometer system must be known and compensated for. The initial path length difference between the interfering beams can be determined on the basis of optical design and the manufacturing tolerances or by a variation of the laser frequency. The accuracy of both methods is limited to the micrometre range. Independent of the dead path of the interferometer system, the time-of-flight of the light reduces the contrast of the interferometric signal in the case of a mirror displacement and a laser frequency variation. Besides the necessary knowledge of the laser frequency, a stabilisation of the frequency can therefore reduce the noise level of the interferometer system [31]. A laser frequency stabilisation does not prevent frequency variations with a higher frequency than the control bandwidth, for example introduced by ground vibrations or acoustic waves. Therefore, the installation conditions of the laser source can also limit the precision of interferometric displacement measurements. For picometre level measurements the laser should be placed on a vibration isolated bench, shielded and in a thermally stable environment.

### 9.1.2 The laser as noise source

The linewidth of the laser source reduces the interference contrast and increases the noise of the detected phase in the case of the path length difference between the interfering beams. A frequency variation of the laser equates the time derivation of the phase noise. Following the calculations of Salvade [31, 32], the variance of the measured phase fluctuations over the observation time ($T$) can be roughly approximated according to equation (9.3), based on the power spectral density ($S_{\delta\nu}(f)$) of the light source and the interferometric delay ($\tau = 2\Delta z/c$):

$$\langle \Delta\varphi^2_{\tau,\,T} \rangle = 4\pi^2\tau^2 \int_0^\infty S_{\delta\nu}(f) \left( \frac{\sin(\pi f T)}{\pi f T} \right)^2 \mathrm{d}f. \tag{9.3}$$

Assuming a constant power spectral density represented by the linewidth ($\Delta\nu$) of the laser source divided by $\pi$, the resulting standard deviation can be approximated by

$$\sigma_{\Delta\nu} = \sqrt{\frac{2\pi\tau^2\Delta\nu}{T}}. \tag{9.4}$$

Two examples within this simple approximation for an acquisition time of $T = 20\,\mu s$ are the following: a displacement of $\Delta z = 500\,\text{mm}$ and a linewidth of $\Delta\nu = 1\,\text{kHz}$ (exemplarily for a frequency-doubled Nd:YAG laser) lead to a noise contribution of $\sigma_{\Delta\nu} = 2.4\,\text{pm}$, while a displacement of $\Delta z = 10\,\text{mm}$ and a linewidth of $\Delta\nu = 1\,\text{MHz}$ (exemplarily for a HeNe laser) leads to a noise contribution of $\sigma_{\Delta\nu} = 1\,\text{pm}$. Indeed, a constant power spectral density is a poor approximation of the one of a HeNe laser. But the example illustrates that the influence of the laser linewidth is not negligible for picometre level displacement interferometry and is worth more detailed analysis based on the actual power spectral density of the laser and the actual bandwidth of the phase evaluation (instead of the acquisition time).

A variation of the laser power within the detection bandwidth of the interferometer cannot be distinguished from a phase variation. The intensity noise of the laser affects the measurement signal as well as the reference signal, when it is picked optically. Therefore, their uncertainty contributions resulting from the intensity noise are correlated. The relative intensity noise of a laser depends on the frequency. Figure 9.3 shows a measured intensity noise of a frequency-double Nd:YAG laser with a wavelength of $\lambda = 532\,\text{nm}$. Depending on the chosen beat frequency, the uncertainty contribution of the relative intensity noise (RIN) of the laser can outnumber the influence of the shot noise [33]. For the example shown in figure 9.3, a RIN of $-135$ dB Hz$^{-1}$ results into a noise contribution of 7.1 pm while the contribution of the shot noise is 3.9 pm, assuming an optical power of 60 μW and a sensitivity of the photodiode of 0.28 A W$^{-1}$.

A fundamental requirement for picometre level interferometry is a single-mode light source or a multi-mode one with stabilised frequency offset between the modes [34]. Spurious radiation from other transitions causes nonlinearities and phase shifts

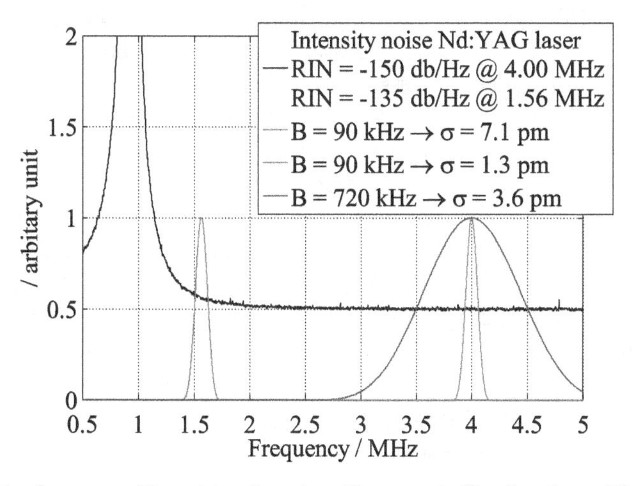


**Figure 9.3.** Example of a measured laser intensity noise with respect to filter functions of the phase meter with different bandwidth (B) and beat frequencies.

depending on their frequency difference and intensity. Therefore, variations of the intensity ratio of the spurious radiation to the desired one affect the stability of the interferometer system. Spurious radiation can occur by running a single-mode laser source with undesired parameters (e.g. a Nd:YAG laser with a smaller pump diode current), at relatively long HeNe laser (>250 mm tube length) [27] or it can be induced by optical feedback re-entering the laser. Therefore, it is preferable to place an optical isolator (e.g. a Faraday isolator) between the laser source and the interferometer optics to cancel out unwanted reflections (especially in the case of plane mirror setups).

## 9.2 Correction of the refractive index

The displacement of the moving mirror is evaluated from the measured phase variation, the vacuum wavelength and the refractive index of the ambient medium. In the case of a stabilised laser source, the uncertainty of the refractive index is often a more significant uncertainty contribution than the uncertainty of the vacuum wavelength. Depending on the application, the beam paths are in air, vacuum or a specific gas, like helium. The influence of the refractive index of air can be reduced using two-colour or multi-colour interferometers, as described in reference [35] and in more depth in chapter 5 'Interferometry in air with refractive index compensation'. Beside this approach, three other methods are common for the determination of the refractive index:

- The modified Edlén formulae [36] can be used to calculate the refractive index of air for visible light based on measured values of the ambient conditions: temperature, pressure, humidity and $CO_2$ concentration. The uncertainty of the ambient parameter measurements dominates the uncertainty of the evaluated refractive index. Assuming uncertainty-free measurements of the ambient conditions the correction of the refractive index would still lead to a relative uncertainty of $8 \times 10^{-10}$ due to the standard deviation of the fitted parameters [37]. In addition, the calculation based on a limited number of factors can be affected by the presence of contaminants, for example due to the presence of hydrocarbons [12].
- A differential refractometer, also called a wavelength tracker, can be used to measure variations in the refractivity. The length-proportional error of a displacement interferometer, which is corrected with this method, is limited by the mentioned factors of the Edlén formulae, since the initial value follows from a parametric evaluation based on the measurement of the environmental parameters. However, a wavelength tracker provides a higher bandwidth compared to the time delay inherent in the measurement of the environmental parameters and therefore can be sufficient in many applications where precision is of primary importance [38].
- The absolute value of the refractive index can be directly determined using a refractometer. An expended uncertainty of $3 \times 10^{-9}$ was achieved by Egan and Stone [39].

In addition to the mentioned limitations, the uncertainty of the refractive index determination can be limited by gradients of the refractive index. These gradients occur due to the decrease in pressure with altitude, due to temperature stratification and due to rapid climatic changes [40]. Due to the spatial separation of the refractive index determination and the interferometer beam, air turbulences cannot be compensated for by the mentioned methods. Under laboratory conditions, without extensive realisation of a laminar air flow, the influence of air turbulences is uncorrelated for two beams with an interspace of more than 12 mm [41]. The error resulting from air turbulences is in the range of several nanometres for a beam path length of around 100 mm and the beams have to be enclosed to reduce the error to the sub-nanometre level [42].

For the realisation of measurements at picometre level the interferometer setup should be placed into an evacuated or specified (like helium) environment. This deterministic approach has been established over decades to increase machine tool accuracy [43]. The extended Edlén formulae from Bönsch [36] is valid for dry air from atmospheric pressure down to 0 Pa [44]. Below the atmospheric pressure, air can be approximated as a mixture of ideal gases and the refractive index of the residual gas can be approximately calculated using equation (9.5) [45]:

$$(n-1) = \sum x_j(n_{0j}-1) \cdot \frac{p}{p_0}\frac{T_0}{T}. \tag{9.5}$$

This equation uses of the additivity of the Lorentz–Lorenz relation and is valid for pressure regimes, where the compressibility factor can be approximated as one. The refractive index of a gas mixture can be calculated in this case with the mole fraction ($x_j$) and the refractive index ($n_{0j}$) of the particular gases under normal conditions ($p_0$, $T_0$). For the mole fraction of dry air, a wavelength of $\lambda = 532$ nm and a temperature of $T = 293\ 14$ K the pressure dependence of air is $\partial n/\partial p = 2.7 \times 10^{-9}/\mathrm{Pa}$. Therefore, a correction of the refractive index becomes negligible for pressure below 0.001 Pa for picometre level displacement interferometry. This degree of evacuation can be reached without the use of ultra-high-vacuum equipment and baking out the vacuum chamber. A vacuum chamber sealed with elastomers combined with a turbo pumping station is sufficient.

Depending on the general requirements of an experiment, a pressure above 0.001 Pa can be preferable, for example to avoid the cost of a turbo pump or to allow for thermal balancing via convection [46]. The heat conduction of a gas is nearly insensitive to the pressure until the mean free path is larger than the dimensions of the chamber [47].

In the case of a medium vacuum (100 Pa to 0.1 Pa [48]), the refractive index must be corrected for picometre level interferometry. The accurate determination of the refractive index of the residual gas is limited mainly by two factors: the uncertainty of the pressure measurement [44] and the presence of contaminants. Vacuum pressure sensors often have a characteristic nonlinear curve and the indicated pressure is a function of the temperature and gas species [49]. Therefore, the pressure gauge must be calibrated to provide traceable values with an appropriate uncertainty [50]. The uncertainty of the sensors depends on the measurement principle and on the pressure value. A relative uncertainty of $10^{-3}$ of the pressure sensor leads to a relative displacement uncertainty of $2.7 \times 10^{-12} \cdot p/\text{Pa}$ at a wavelength of $\lambda = 532$ nm and a temperature of 20 °C. The pressure should be in the range of 1 Pa or the relative measurement uncertainty of the pressure has to be decreased inversely proportional to an increase of the pressure, to achieve picometre level displacement measurements over several hundreds of millimetres.

The described approximation of the refractive index based on the extended Edlén formulae assumes the residual gas to be dry air. This assumption cannot be taken for granted for any vacuum chamber. Depending on the leakage rate of the chamber, the different pumping rates for particular molecules and outgassing rates of components inside the chamber, the composition of the residual gas can differ from the mole fractions of dry air. The portion of water and its decomposition products is higher inside the chamber without baking it out or using a cooling trap. A mass spectrometer can be used to determine the composition of the residual gas. An exemplary measurement is shown in figure 9.4(a). Using equation (9.5) and the refractivity of the gases at standard conditions, the refractive index of the residual gas inside the chamber can be calculated. The proportional constant of the refractive index is $\partial n/\partial p = 2.34 \times 10^{-9}/\text{Pa}$ for the example shown in figure 9.4(a).

In the case of contaminations caused by outgassing, a calculation of the refractive index tends to be more difficult. In particular, the presence of hydrocarbons limits the possibility of determining the specific molecules with a mass spectrometer and along with it the refractive index of the single fractions. Figure 9.4(b) shows an example, where a PVC hose, used to separate vibrations of the vacuum pump from the setup, caused contamination of the vacuum chamber with hydrocarbons. For this example, the proportional constant of the refractive index was experimentally determined to be $\partial n/\partial p = 2.99 \times 10^{-9}/\text{Pa}$. These two examples emphasise that determination of the pressure with a small uncertainty alone might not be sufficient for displacement measurements with an uncertainty at the picometre level, since the proportional constant of the refractive index can vary by up to 10% depending on

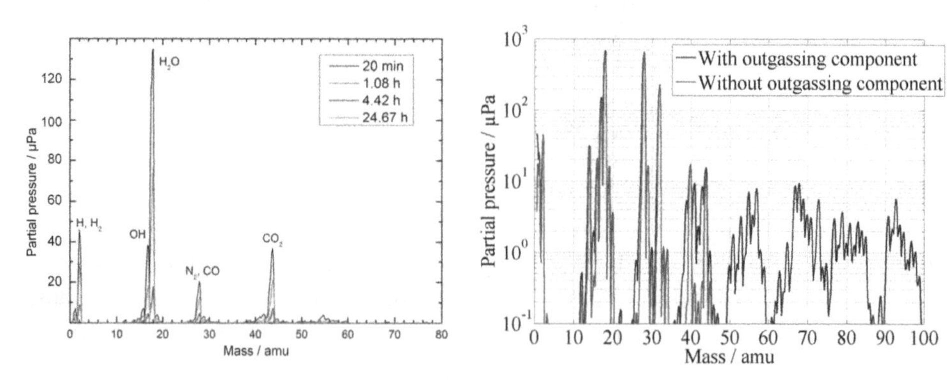


**Figure 9.4.** Examples of mass spectrometer measurements to determine the composition of the residual gas inside two vacuum chambers: (a) a chamber without cooling trap for different pumping times and (b) a chamber contaminated with hydrocarbons after a pumping time of 1 h.

the composition of the residual gas. Similary, the purity of a specific gas used to fill the chamber is ruined by leakages and outgassing components. If thermal balancing via convection is not essential for the experiment, the technically easiest solution would therefore be the integration of a turbo pump to achieve picometre level uncertainties.

### 9.2.1 Errors of the evaluated phase

Beside the above-mentioned influence of the laser source, namely its intensity noise, its frequency variations and its coherence length, other influences can limit the resolving capability of the interferometer system. Noise is, in principle, introduced by every component of the interferometer system, e.g. thermal noise of the mirrors, shot noise of the photodetectors, noise of the amplifiers, resolution and sampling noise of the analogue-to-digital converters (ADC) and jitter of the clock signal supplied to the ADCs. Typically, these influences can be reduced by averaging and do not limit picometre level displacement measurements nowadays—which should not imply, that a careful analysis of the different contributions is unnecessary. The phase meter can be characterised with current sources and function generators. The influence of the laser source, the photodetectors and the amplifiers can be determined by realising using a Mach–Zehnder interferometer setup. This Mach–Zehnder setup can successively be equipped with all the components of the displacement interferometer system and therefore be used to determine their respective influence, as exemplarily shown in [33]. Furthermore, the resolving capability of the displacement interferometer system can be analysed by fixing the mirrors or using a common mirror reflecting all beams [51]. As already stated in the introduction of this chapter, picometre level resolving capabilities have been demonstrated using many different setups independently of whether homodyne or heterodyne interferometers were investigated.

In contrast to the mentioned noise sources, there are systematic deviations of the evaluated phase, which cannot be reduced by averaging. These systematic effects must be addressed during the conceptional phase of the realisation of the displacement interferometer and are discussed in detail in the following section.

### 9.2.2 Periodic nonlinearities—an often discussed, (un)solved problem

Nonlinearities are errors in determining the fraction of a period. Depending on the source of the nonlinearities they can occur as periodic errors, drift of the interferometric signals or both. In any case they limit the precision of interferometric displacement measurements.

Every component of the interferometer system can cause periodic nonlinearities (pNLs). Assuming a constant moving speed of the mirror, these errors can be identified by means of the frequency spectrum of the signal [52]. Under the appearance of periodic nonlinearities, the spectra of the signal will show peaks apart the basic one. The basic peak is located at a frequency resulting from the sum of the beat frequency and the velocity dependent frequency shift calculated according to equation (9.2). The erroneous peaks are located at frequencies depending on the frequency shift multiplied with the order of the periodic nonlinearity. Since the moving speed of the mirror typically varies around an average value, a direct calculation of the frequency spectrum will lead to an underestimation of the amplitudes of the periodic nonlinearities. One solution to determine the correct amplitudes from the frequency spectrum is an interpolation of the interferometric signals on the basis of the average moving speed and a constant data acquisition rate [53]. A determination of the periodic nonlinearities and their origin is the first step for reducing their uncertainty contribution. One possibility to reduce their influence is to apply an ellipse fit for homodyne [54] as well as for heterodyne interferometers [55]. But an ellipse fit can only reduce the influence of some periodic nonlinearities of the first and second order. For higher and negative order periodic nonlinearities an ellipse fit can even increase the uncertainty contribution, for example for ghost reflections [56] or intermodulation caused by the amplifiers [57]. Therefore, it is essential to investigate the source of periodic nonlinearities to reduce their influence.

Using a heterodyne interferometer, the most significant source of periodic nonlinearities is optical or electrical cross talk, which is summarized in figure 9.5. In the case when interferometer beams with different frequencies are superposed but separated by their linear polarisation state, polarisation mixing effects cause periodic nonlinearities of first and second order. The polarisation mixing can be caused by

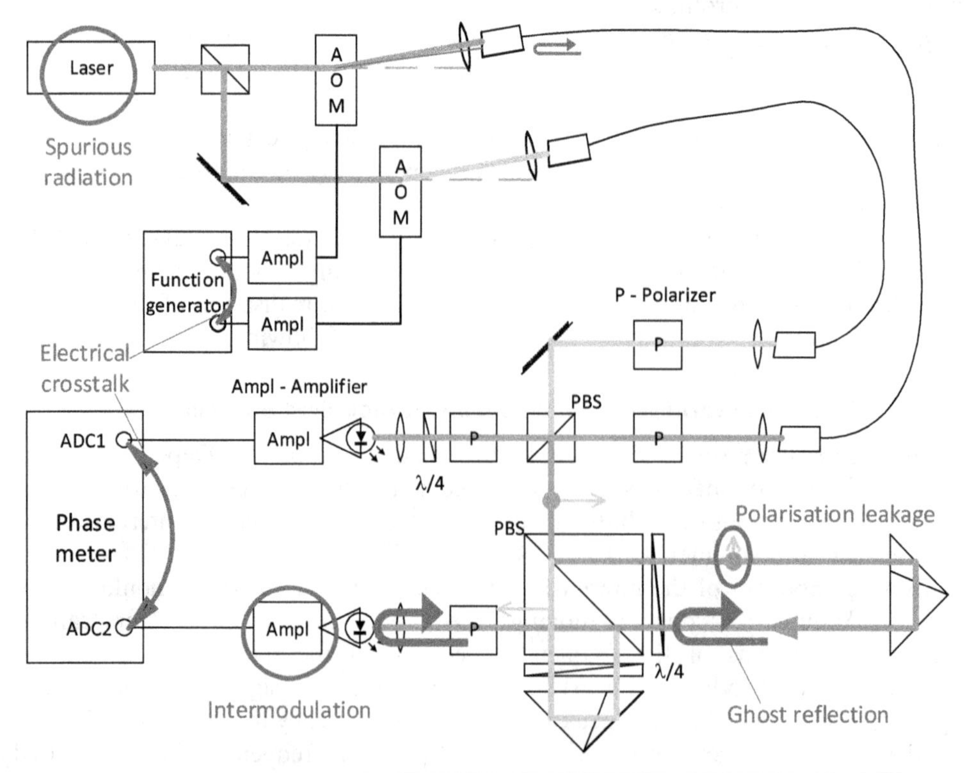


**Figure 9.5.** Overview of sources of periodic nonlinearities at heterodyne interferometers.

the imperfection of any optical component and increased by their misalignment. The influence of the laser head, polarisation beam splitters, polarisers, corner cubes and quarter-wave plates can be analysed by means of Jones matrices [58, 59, pp 24–34]. Since the beams of different frequencies are separated by their linear polarisation state, a polarisation leakage ends up in a frequency mixing. In this case, energy from different source frequencies is unintentionally reaching the detector on the same path. These kinds of periodic errors can be minimized to the sub-nanometre level with different techniques. The most successful one is the use of spatially separated input beams [60], since it also minimizes other resulting periodic nonlinearities of different order. The so-called dynamic periodic error and periodic nonlinearities resulting from intermodulation [61] are initiated by frequency mixing effects and cannot be minimized with other methods like the ellipse fit.

Other sources of periodic nonlinearities are ghost reflections [62]. Depending on the interface responsible for the ghost reflections, they can cause periodic nonlinearities of higher and negative order. In agreement with higher order pNLs caused by intermodulation the interferometric signal deviation cannot be described by an ellipse function. Consequently, the erroneous signal cannot be corrected using fitted ellipse parameters [56]. Multi-reflected beams inside the reference path of the interferometer result in a phase offset, which is constant in the first approximation since the reference mirror is typically not moving. But variations of the ambient conditions cannot be compensated for by multi-reflected beams. Therefore, ghost reflections also result in a system with higher sensitivity to variations of the ambient conditions. To achieve picometre level periodic nonlinearities and stability of the displacement interferometer, ghost reflections must be avoided. Since every interface reflects a certain portion of light various actions must be taken. In the case where the interferometer is fibre-coupled, the optical fibres must be angle polished. Nevertheless, for plane mirror displacement interferometers the influence of 'recycled light' can still cause deviations of the measured position of 10 pm [63]. This remaining error could be minimized under the acceptance of a cosine error for the experiment, which in principle can be compensated for when the tilt of the plane mirror is known. Another source of multi-reflected beams are the photodetectors, the window of their housing and the focusing lens in front of it. These ghost reflections can be suppressed for example by arranging a polariser and a quarter-wave plate in front of the photodetectors. All optical components should be arranged or designed in a way that avoids interfaces orthogonal to the beam paths, for example by using coated plates instead of cubes as beam splitters. Of course, a beam splitter cube can be rotated with respect to the beams to avoid orthogonal interfaces, but a rotation of the polarising beam splitter cube can also increase periodic nonlinearities caused by frequency mixing [59, p 33]. In an evacuated environment coated plates also have the advantage of a reduced sensitivity to variations of the ambient conditions. The thickness of the plate has to be chosen such that multi-reflected beams inside the plate do not reach the photodetectors. A tilt of the optical components with respect to the beams leads to multi-reflected beams with an angle relative to the general beam path. The influence of these ghost reflections can be suppressed by spatially filtering the beams, for example by means of a limited active area of the photodetector and the lens arranged in front of it.

Despite the avoidable effect of polarisation leakage, frequency mixing can also result from electrical cross talk. Electrical cross talk can, for example, occur if the beat frequency is generated by means of two acousto-optic modulators (AOM) [64]. AOMs shift the laser frequency by means of a moving phase grating of planes of compression and rarefaction, induced by ultrasonic waves [65]. Since the first order diffraction angle is different according to the frequency of the ultrasonic, one can spatially separate the beams of different frequencies. The diffraction angle difference can be comparably small between the desired radio-frequency (RF) signal applied to the AOMs and a contained spurious signal. Typically, the RF signals generated to drive the AOMs contain at least multiples of 10 MHz. When the desired frequency is, for example, close to 80 MHz the diffracted beam will not easily be separated from the unwanted leakage one of 80 MHz. This results in periodic nonlinearities and phase fluctuations in the setup. This error can be minimized by different approaches, like choosing a beat frequency in the MHz range to spatially separate unwanted diffractions or by using a multiple of 10 MHz as the frequency shift for the beam entering the measurement path of the interferometer. Since the displacement of the reference mirror is often comparably small during a measurement, a contamination of the reference beam 'only' causes instability of the system in the case of ambient condition variations.

Electrical cross talk can also occur for the detected interferometric signals [18], for example between the input channels of the phase meter board. The amplitude of the resulting periodic nonlinearities is proportional to the amplitude ratio of the electrical cross talk. For a heterodyne interferometer working at a wavelength of 532 nm, the amplitude ratio of the cross talk must be smaller than $184\,\mu V/1\,V$ (−74.7 dB) to have an amplitude of the pNL of less than 10 pm.

The necessity of the implementation of signal corrections via elliptical fitting depends on the design of heterodyne interferometer systems. At homodyne interferometer systems, the implementation of a correction is essential to reach picometre level displacement measurements [17]. The roughly approximated statistical error of the phase can be calculated from the signal-to-noise ratio of the interferometric signals, for example a ratio of 10 000 leads to a standard deviation of the phase of 5 pm at a single-pass interferometer with a period of 316.5 nm. The statistical error of the phase increases when elliptical fitting is used for the correction of the interferometric signals. The number of points in the fit-interval should be larger than 100 for a negligible influence on the statistical error, the phase lag ($\varphi_0$—difference from $\pi/2$) between the signals increases the error approximately by the factor $1/|\cos(\varphi_0)|$ and a correlation coefficient between the signals, for example caused by the laser intensity noise, can increase the statistical error up to 40% [66]. For a worst-case scenario of phase lag of $\varphi_0 = 45°$ and a correlation coefficient of 0.9, the roughly approximated error from the signal-to-noise ratio is doubled.

Following the concept of 'brute strength' avoiding 'gadgets' [67], periodic nonlinearities at heterodyne interferometers should be avoided instead of corrected. The concept of spatially separated input beams calls for an optical design consisting of more components, but is an effective method to minimise periodic nonlinearities caused by polarisation leakage and potentially resulting in periodic nonlinearities of higher order, for example caused by intermodulation. The sources of periodic nonlinearities are also often the roots of instabilities of the interferometer system, for example in the case of ghost reflections or in when the update rate of elliptical parameters is slower than laser intensity drifts at homodyne systems.

### 9.2.3 Stability—or interferometers are not just optics

The lack of stability of an interferometer system results in phase variations without an actual movement of the mirrors. Whether an effect is regarded as drift or noise depends on the acquisition rate and bandwidth of the phase measurement. While periodic nonlinearities are connected to a movement of the mirrors, drift or instability of the measured phase affects the reproducibility of displacement measurements mostly independent of the mirror position. However, drift effects can occur periodically when their stimulation is periodic. In principle, every component causing a temporal delay can affect the stability of the system for an uncompensated variation of this delay. Following, instabilities of the measured phase are discussed, which are caused by:

- unbalanced beam paths or a variation of the balance,
- optical fibres,
- amplifiers.

Unbalanced paths between the reference beam and measuring beam are caused by the intended displacement of the moving mirror. The effects of refractive index and laser frequency variations are discussed above. In the following, the effects on allegedly dead path free interferometer systems are discussed. Self-evidently the interferometer optics must be designed in a way, that the beam paths through glass are equal for the reference and measurement beam. Otherwise the thermal expansion and thermal variation of the refractive index of glass would directly lead to a phase drift proportional to the temperature variation. Therefore, any compensation plates should be arranged close to the other glass components, most suitable in thermal contact, to minimise the influence of temperature gradients. Nevertheless, different published interferometers (including commercial ones) with obviously balanced beam paths still exhibit thermal drift coefficients of several nanometres per Kelvin, e.g. 2.5 nm $K^{-1}$ [68] and 1.2 nm $K^{-1}$ [69]. These thermal drifts can be introduced by small misalignments [70], by the combination of the glass and holder material or an asymmetric fixing of the holders [71]. Depending on the homogeneity of the glass, balanced beam paths in the paper sketch might be unbalanced in the real setup. Additionally, the mounting of the glass components can introduce local stress. Along with the introduced stress the refractive index of glass varies locally on top of potential birefringence. Therefore, the thermal expansion coefficient of the glass should match with that of the holders, e.g. combining Invar steel with quartz glass. Beside the stress introduced by different thermal expansion coefficients, temperature gradients can be a major reason for instability of the interferometer system.

Therefore, it can be preferable to use a material with a high thermal conductivity like aluminium for the holders instead of one with a small thermal expansion coefficient [71]. In the ideal case, the holders are made from the same material as the optical functional parts and the optical functional parts do not have a direct connection to the machine structure. This is, for example, realized at the nanometer comparator [72], the LISA experiment [73] or the combined optical and x-ray interferometer setup of PTB [74]. All these setups are placed in an evacuated environment and their designs are adapted to the benefits of this ambient condition to improve the systems stability. Due to the evacuated environment, the influence of humidity variations on optical coatings and kit is avoided, which otherwise can also result in a phase shift. In addition, beam paths through glass are minimized. At interferometer setups designed for use in air, beam paths through glass are typically preferred to minimize the influences on refractive index gradients of air. In evacuated environments it is the other way around, since dependency of the refractive index of the residual gas on temperature variations is smaller than that of glass, e.g. at pressure of 2 Pa the dependency of the residual gas is $\Delta n/\Delta T = 20 \times 10^{-12}$/K for light with a wavelength of 532 nm, which is much smaller than that of Zerodur of $\Delta n/\Delta T = 13{\cdot}10^{-6}$/K [75]. This design choice also minimises the influence of temperature gradients.

An interferometer setup totally insensitive to variations of the ambient conditions cannot be realised. Therefore, a proper solution for the realisation of picometre level displacement measurements is to stabilise the ambient conditions. This can be achieved for example by spatially separating heat sources like the laser source and the photoreceivers from the optical setup using optical fibres. At heterodyne interferometers two fibres should be used, each to deliver the light of a specific frequency, to minimise the effect of frequency mixing and the resulting periodic nonlinearities [76]. Polarisation maintaining fibres should be used and a polarizer should be placed behind the collimators to suppress the influence of light scattered into the fast axis of the polarization-maintaining fibres. This scattered light can cause phase variations, because the splitting ratio of following non-polarizing beam splitters slightly depends on the polarisation state and a polarisation jitter implies variations of the phase difference between the linear (vertical and horizontal) polarization components of the input beams [77, 78]. To avoid frequency mixing errors two fibres are used. Each fibre induces different phase variations to the beams of different frequency along with variations of the ambient conditions like pressure and temperature variations [79] or vibrations [80]. Therefore, the reference phase of the interferometer must be picked behind the fibres to be able to compensate for these phase variations [68]. The phase perturbations introduced by the fibres are below 100 kHz [81, p 26]. Without harsh mechanical handling, the frequency variations introduced by the fibres are below 100 Hz [81, p 26]. The resulting phase variations approximated with equation (9.4) are negligible for picometre level interferometry. Therefore, heterodyne interferometers with a beat frequency in the MHz range the phase behind the fibres has not to be stabilised. Instead of picking the reference phase directly behind the fibres the reference beam paths can be used to compensate for thermal influences on the interferometer optics.

To further increase the stability of the interferometer, the superposed beams could be coupled into multi-mode fibres, separating the heat of the photoreceivers and working as spatial filter for multi-reflected beams with different angles due to the limited core diameter of the fibres. However, without careful adjustment the sensitivity of the interferometer to angle variations may increase. The interfering beams are usually not superposed perfectly. They exhibit an angle between their directions of propagation and an offset. This misalignment adds noise and drift when the superposed beams are coupled into multi-mode fibres. Due to their different angle and offset the two beams are coupled with different mode power distributions, which leads to different transmission characteristics despite the perpetual mixing process. This results in a partial separation of the superposed beams, different delays and consequently in noise and drift of the phase difference between the interferometric signals. Gradient-index fibres should be used instead of step index fibres, because they are inducing fewer disturbances and the phase variation remains smaller than ±20 pm [33]. The use of single-mode fibres would avoid this effect, but also has the drawback of a higher sensitivity to angle variations of the moving plane mirror.

The photoreceivers of the interferometer typically consist of photodiodes and transimpedance amplifiers. Both components are known noise sources, due to the associated shot noise, dark current noise and the specified current and voltage noise of the amplifiers. Nevertheless, both components are also responsible for a signal delay depending on their capacity (including the connecting cable) and the size of the feedback resistance. These parameters of the electrical components vary with temperature. Therefore, temperature variations change the delay introduced by the photoreceivers and along with this the phase offset between the reference and the measurement channel. To achieve picometre level stability the photoreceivers must be housed to damp temperature variations effectively or even actively temperature stabilised [82].

Another source of unintended phase variations is scattered light. Scattered light has similar influences like the described ghost reflections and can also cause frequency mixing. Additionally, scattered light can cause a significant relative error in case of a balanced beam path and the scattered light passes through the reference and measurement path [83].

## 9.3 Length-proportional error resulting for beam diffraction

In many experiments traceable measurements with uncertainties in the picometre range are required. Deviations of the wavefront from a plane wave lead to corrections of the observed period of the interferometer system, which has been a well-known fact for some decades [84, 85]. Assuming a Gaussian beam with a beam waist ($w_0$) and a perfectly aligned interferometer system the correction can be calculated

$$\frac{\Delta z}{z} = -\left(\frac{\lambda_0}{2\pi w_0}\right)^2. \tag{9.6}$$

Equation (9.6) states that the measured displacement is always too short when nonplanar waves are interfering and the periodicity of the interferometer is assumed to depend only on the vacuum wavelength of the light. In other words, the norm of the wavevector is smaller than it is expected to be from the vacuum wavelength. A smaller norm of the wavevector leads to smaller measured phase variation for a mirror moved over a certain distance. If now the displacement is calculated only on basis of the measured phase variation and the vacuum wavelength, the result would be too small. The derivation of this equation and its theoretical background is explained below.

The assumption of a perfectly aligned interferometer system with Gaussian beams leads to equation (9.6), which leaves the determination of the beam waist as the only source of uncertainty [86]. To give an example: with a beam waist of $w_0 = 2\,\text{mm}$, a wavelength of $\lambda = 532.245\,\text{nm}$ and a displacement of $z = 10\,\text{mm}$ the correction is $\Delta z = -18\,\text{pm}$. To achieve an uncertainty contribution of the correction below 1 pm the beam's waist would have to be determined with an uncertainty below 60 μm. Therefore, the determination of the beam parameters to realise a correction with a small uncertainty is an issue addressed below.

The discussion of other uncertainty contributions above emphasizes that interferometer systems violate the assumption of a perfectly aligned interferometer with Gaussian beams, in some cases even deliberately. When optical fibres are used the approximation of Gaussian beams is not valid on the full scale since the beam profile must be described by means of Bessel and modified Hankel functions [87]. For spatially separated input beams the beam profiles (waists) of the two input beams will deviate from each other. Additionally, the interfering beams will never be superposed perfectly. The presence of an offset ($x_0$) and an angle misalignment ($2\alpha$) between the interfering beams leads to additional corrections of the vacuum wavelength [88]. Therefore, the knowledge of the misalignment of the interferometer setup is an additional uncertainty contribution. If the Gaussian beams are non-coaxial, the correction has to be calculated according to equation (9.7) [88]:

$$\frac{\Delta\lambda}{\lambda} = \left(1 - \frac{x_0^2}{w_0^2}\right) \cdot \frac{\gamma_1^2 + \gamma_2^2}{2} + \frac{\alpha^2}{2}, \tag{9.7}$$

where the propagation distance is assumed to be much smaller than the Rayleigh length and $\gamma_{1,2}$ are the eigenvalues of the second central moment matrix of the beam angular power spectrum.

For traceable interferometric length measurements with relative uncertainties in the range of $10^{-9}$ the diffraction of the laser beams used is an important contributor to the uncertainty budget. Physical models of light propagation are assumed to estimate contributions to the measurement uncertainty. Depending on uncertainty demands the modelling effort increases to capture all relevant effects. The theoretical background of this systematic error is reviewed starting with a simple plane wave model. Then, the theoretical models are successively refined to capture the essence of this systematic error. Finally, the question of how to properly characterize interfering laser beams to yield an estimation for the relative length error and its uncertainty is discussed.

### 9.3.1 Theoretical background

*9.3.1.1 A simple plane wave model*

A common approximation for the description of two-beam interference is the assumption that the interfering beams are scalar plane waves with wavelength $\lambda$ inside a given medium which are defined as follows [89–91]:

$$E(x, y, z) = E_0 \exp[i(\omega t - k_x x - k_y y - k_z z)], \tag{9.8}$$

where $x, y, z$ are Cartesian coordinates, $\omega$ is the circular frequency, $t$ is time and $k_{x,y,z}$ are the components of the wavevector $\mathbf{k}$ of length $k = (k_x^2 + k_y^2 + k_z^2)^{1/2} = 2\pi/\lambda$. The amplitude $E_0$ is generally a complex number. Strictly speaking, plane waves are transversal vector waves and solutions to the wave equation also known as the Helmholtz equation, which in turn is derived from the Maxwell equations [89–91].

The time dependence of the exponent in equation (9.8) is not essential for the following treatment, therefore, it is omitted from now on. Furthermore, the wavevector is assumed to be parallel with the $z$-axis, i.e. $k_{x,y} = 0$ and $k = k_z$. It is straightforward to write $E_0$ in Euler form as $E_0 = |E_0|\exp(i\phi_0)$. The superposition of a plane wave with an identical wave shifted along the $z$-axis can be expressed as:

$$E(z_1, z_2) = |E_0|\{\exp[i(\phi_0 - kz_1)] + \exp[i(\phi_0 - kz_2)]\}, \tag{9.9}$$

where $z_{1,2}$ are distances as they appear, e.g. in a Michelson interferometer (figure 9.6).

The field $E(z_1, z_2)$ of the interfering light waves cannot be measured directly. Instead, the irradiance, i.e. light power on a surface per unit area, is measured. Here, for the sake of simplicity, it is written without proportional constants:

$$I(z_1, z_2) \propto E(z_1, z_2)E^*(z_1, z_2) = |E(z_1, z_2)|^2, \tag{9.10}$$

where the asterisk denotes the complex conjugate. Setting equation (9.9) into (9.10) one gets:

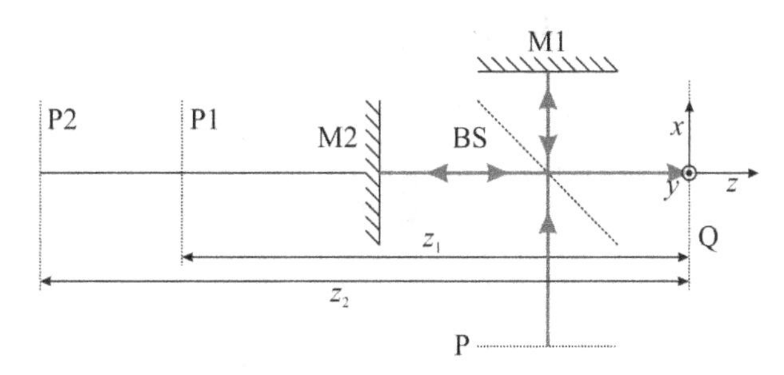


**Figure 9.6.** Geometry of a Michelson interferometer with mirrors M1,2 and beam splitter BS. The input field at plane P is observed at Q as two laterally shifted fields at the virtual planes P1,2 which are shifted from Q along $z$ by $-z_{1,2}$.

$$I(z_1, z_2) \propto |\, E_0 \,|^2 \{2 + \exp[ik(z_2 - z_1)] + \exp[ik(z_1 - z_2)]\}. \tag{9.11}$$

Using the exponential definition of the cosine function and defining $\Delta z = z_2 - z_1$ equation (9.11) reduces to:

$$I(\Delta z) \propto 2 \,|\, E_0 \,|^2[1+\cos(k\Delta z)]. \tag{9.12}$$

Therefore, one observes a sinusoidal variation of the irradiance with $\Delta z$ which is the well-known principle of interferometry. Several techniques exist to extract the phase $\varphi = k\Delta z$ of sinusoidal measurement signals. Usually the obtained phase is *wrapped* into an interval, e.g. $\varphi_w \in (-\pi, \pi)$, and must be *unwrapped.* The displacement can then be obtained from the unwrapped phase

$$\Delta z = \frac{\varphi_u}{k}. \tag{9.13}$$

Equation (9.13) is exact under the assumption of plane waves. However, an infinitely extended plane wavefront is simply not possible in the real world, e.g. laser beams always have a finite lateral size. This finiteness causes diffraction, i.e. the progressive bending of wavefronts, which can be understood intuitively by employing Huygens' principle [92] as it is depicted in figure 9.7.

The wavefield behind the aperture can be described as a superposition of plane waves propagating in different directions, described by an *angular spectrum* [89, pp 48–51]. This view is mathematically exact and completely equivalent to the Huygens' picture. While the latter is based on spherical waves in space, the angular spectrum is the counterpart in spatial frequency domain, i.e. the associated Fourier space. Following this view, each beam can be understood as a superposition of plane waves, which propagate in different directions [89, p 48]. One can imagine that there

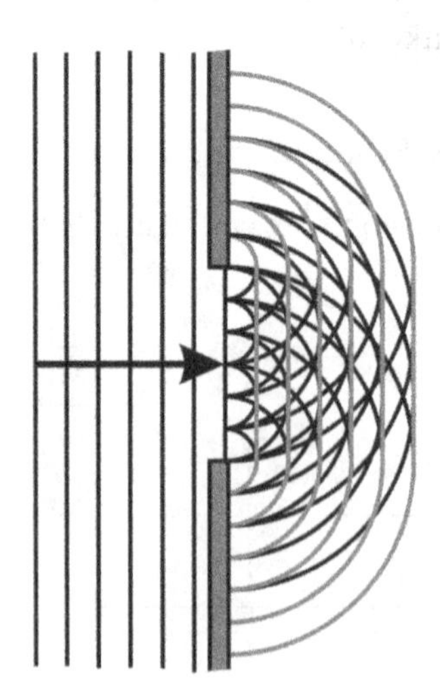

**Figure 9.7.** Intuitive explanation of diffraction by Huygens' principle [92]. A plane wave is incident on an aperture from the left. Spherical waves emanating from the plane wavefronts reconstruct the plane wave. Due to the aperture, only a finite part can contribute to the field on the right side causing the correspondingly reconstructed wavefronts (shown in red) to bend.

is a mean direction, which represents the direction of the whole beam. Then, an effective wavevector in beam direction can be assigned whose length is given by the average of the projection of all wavevectors onto this direction, weighted by the respective plane wave amplitudes. This effective wavevector is always shorter than the wavevectors of the involved plane waves (assuming a monochromatic wavefield), which is caused by the smaller lengths of the projected wavevectors. Using this picture, it is already understandable that more divergent beams, which consist of a broader angular spectrum, have a shorter effective wavevector. Indeed, the interference of two beams must be considered here, not just the propagation of one beam. It will be shown later how this effect influences the length measurement, and an expression is derived for the relative length error based on a rigorous evaluation within the scalar wave approximation.

*9.3.1.2 The relative length error of Gaussian beams*

Starting from the simple plane wave picture, equations (9.8) and (9.13), the next level of approximation to real laser beams are Gaussian beams which are solutions of the *paraxial* wave equation [93, pp 626–41]. The scalar field can now be described as follows:

$$E(r, z) = u(r, z)\exp(-ikz) = \frac{\exp\left[-ik\frac{r^2}{2q(z)}\right]}{q(z)}\exp(-ikz), \tag{9.14}$$

where $r = (x^2 + y^2)^{1/2}$ and the complex beam parameter $q(z)$ are used. The latter is defined by

$$q(z) = z + iz_R, \tag{9.15}$$

where

$$z_R = \frac{\pi w_0^2}{\lambda} \tag{9.16}$$

is the Rayleigh length of a Gaussian beam with waist radius $w_0$ (figure 9.8). Equation (9.14) contains no normalization or scaling to a certain amplitude as it is not relevant for the following considerations.

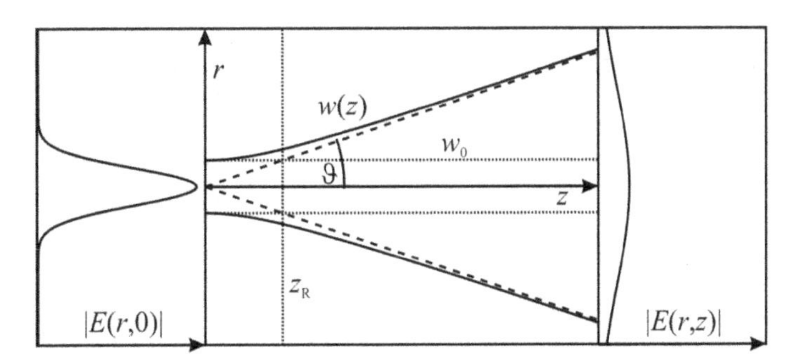


**Figure 9.8.** Amplitude $|E(r, z)|$ and beam width $w(z) = w_0[1+(z/z_R)^2]^{1/2}$ of a Gaussian beam. The Rayleigh length $z_R$ is defined via the position where the cylinder of radius $w_0$ intersects the circular cone with half apex angle $\vartheta = \lambda/\pi w_0$. The surface of revolution defined by $w(z)$ approaches this cone asymptotically for $z \to \infty$. The function $w(z)$ is symmetric with $z$ and $|E(r, z)|$ is symmetric with $r$ and $z$, while only the $+z$ part is shown.

The interference of a Gaussian beam with a longitudinally shifted copy of it (shift: $\Delta z = z_2 - z_1$) can be calculated in analogy to the plane wave model above. The irradiance is now a radial function, i.e. $I(r, z_1, z_2) = |E(r, z_1)+E(r, z_2)|^2$. The power $P(z_1, z_2)$ measured by a detector is thus proportional to the integral of $I(r, z_1, z_2)$ over all $r$. Here, it is worth noting that in a real setup it is required that the complete transverse beam profile is detected and the detector generates a signal linear to the incident light power. The limited efficiency of a real detector results only in a scale factor and does not introduce an additional length error. The power $P(z_1, z_2)$ can be expressed as a two-dimensional integral in cylindrical coordinates

$$P(z_1, z_2) = \int_0^{2\pi} d\theta \int_0^{\infty} drr[E(r, z_1) + E(r, z_2)][E^*(r, z_1) + E^*(r, z_2)], \tag{9.17}$$

which can be solved analytically by the steps shown below. First, equation (9.15) is set into (9.14) to obtain:

$$\begin{aligned} u(r, z) &= \frac{z - iz_R}{z^2 + z_R^2} \exp\left[-\frac{kr^2}{2(z^2 + z_R^2)}(z_R + iz)\right], \\ u^*(r, z) &= \frac{z + iz_R}{z^2 + z_R^2} \exp\left[-\frac{kr^2}{2(z^2 + z_R^2)}(z_R - iz)\right]. \end{aligned} \tag{9.18}$$

One can rewrite equation (9.17) as:

$$\begin{aligned} P(z_1, z_2) = 2\pi \int_0^{\infty} & drr\{u(r, z_1)u^*(r, z_1) + u(r, z_2)u^*(r, z_2) \\ & + u(r, z_1)u^*(r, z_2)\exp(ik\Delta z)+u(r, z_2)u^*(r, z_1) \\ & \exp(-ik\Delta z)\} \\ = 2\pi\Bigg\{ & \int_0^{\infty} drr[|u(r, z_1)|^2 + |u(r, z_2)|^2]+(12) \\ & \exp(ik\Delta z)\int_0^{\infty} drru(r, z_1)u^*(r, z_2)+(12) \\ & \exp(-ik\Delta z)\int_0^{\infty} drru(r, z_2)u^*(r, z_1)\Bigg\}. \end{aligned} \tag{9.19}$$

Setting equation (9.18) into (9.19) and solving the individual integrals analytically one obtains:

$$\begin{aligned} P(z_1, z_2) = P(\Delta z) &= 2\pi\left[\frac{1}{kz_R} + \frac{\exp(ik\Delta z)}{k(2z_R + i\Delta z)} + \frac{\exp(-ik\Delta z)}{k(2z_R - i\Delta z)}\right] \\ &= \frac{\lambda}{z_R}\left[1 + \frac{\exp(ik\Delta z)}{2 + i\dfrac{\Delta z}{z_R}} + \frac{\exp(-ik\Delta z)}{2 - i\dfrac{\Delta z}{z_R}}\right]. \end{aligned} \tag{9.20}$$

Writing the complex denominators in equation (9.20) in Euler notation, the exact and quite cumbersome result is obtained:

$$P(\Delta z) = \frac{\lambda}{z_R}\left\{1+\frac{1}{\sqrt{1+\frac{\Delta z^2}{4z_R^2}}}\cos[\varphi(\Delta z)]\right\}, \tag{9.21}$$

where

$$\varphi(\Delta z) = k\Delta z - \tan^{-1}(\Delta z/2z_R).$$

The inverse square root factor before the cosine function can be understood as a contrast term. The amplitude of the modulation is damped down when $\Delta z$ is comparable to $z_R$. This behaviour is intuitively accessible by noticing the appearance of circular fringes in the interference pattern, which reduces the contrast of total light power modulation [94]. Since one usually deals with displacements which are small compared to the Rayleigh length ($\Delta z \ll z_R$), equation (9.21) can be simplified:

$$P(\Delta z) \approx \frac{\lambda}{z_R}\{1+\cos[\varphi(\Delta z)]\}, \tag{9.22}$$

with $\varphi(\Delta z) \approx k\Delta z - \Delta z/2z_R$ when applying the small-angle approximation to the arc tangent term. It will now be shown that this approximation leads to a well-known result, which is a compact formula for Gaussian beams, first obtained by Dorenwendt and Bönsch [84]. To obtain the systematic length measurement error, a hypothetical interferometric measurement is considered, where from an unwrapped phase $\varphi_u(\Delta z)$ the erroneous displacement $\Delta z'$ is inferred. Hence, by applying equation (9.13) one obtains:

$$\Delta z' \approx \frac{k\Delta z - \Delta z/2z_R}{k} = \Delta z - \frac{\Delta z}{2kz_R}, \tag{9.23}$$

which deviates from the real distance $\Delta z$. Eventually, the relative length error $\varepsilon$ can be calculated by subtraction and normalization:

$$\varepsilon \equiv \frac{\Delta z' - \Delta z}{\Delta z} \approx -\frac{1}{2kz_R} = -\frac{1}{k^2 w_0^2}. \tag{9.24}$$

Equation (9.24) is identical to the result of Dorenwendt and Bönsch, here already presented in equation (9.6) where the notation of [84] for $z$ and $\Delta z$ is used which differs from the notation used in equation (9.24). Several authors acknowledged this result, investigated also misaligned and laterally displaced beams and expanded the model to higher order polynomial Gaussian beams or even non-Gaussian beams [85, 86, 88, 95–99]. The departure from the Gaussian beam model is already a relevant improvement, e.g., for the description of beams originating from single-mode fibres whose beam profiles cannot accurately be fitted by a single Gaussian function (a sum of two Gaussian functions, one narrow, one broad, fits much more closely). It should be noted that some authors interpret the systematic effect which leads to equation (9.24) by a change of wavelength, which differs from the plane wave wavelength $\lambda$ [88, 95–97, 99]. Then, the relative length error can be expressed using an *effective* wavelength $\lambda_e$ by

$$\varepsilon = -\frac{\lambda_e - \lambda}{\lambda}. \tag{9.25}$$

To describe arbitrary paraxial beams the framework of beam moments can be used [88, 97, 100–103]. The theoretical derivation is not reproduced here and only the key results are shown, which are needed to obtain the relative length error from the intensity of an input beam, which is fed into an interferometer. The light beam's intensity distribution is expressed in the Fourier plane, i.e. $\Psi(\alpha_x, \alpha_y)$ which is the power per solid angle and $\alpha_{x,\,y}$ are angles. One can calculate the distribution's centre of mass by [101]:

$$\alpha_{x,\,y;0} = \frac{\iint_{-\pi/2}^{\pi/2} \mathrm{d}\alpha_x \mathrm{d}\alpha_y \alpha_{x,\,y} \Psi(\alpha_x, \alpha_y)}{\iint_{-\pi/2}^{\pi/2} \mathrm{d}\alpha_x \mathrm{d}\alpha_y \Psi(\alpha_x, \alpha_y)}. \tag{9.26}$$

Now, the second moments are [101]:

$$\begin{aligned} \vartheta^2_{xx,\,yy} &= \frac{\iint_{-\pi/2}^{\pi/2} \mathrm{d}\alpha_x \mathrm{d}\alpha_y (\alpha_{x,\,y} - \alpha_{x,\,y;0})^2 \Psi(\alpha_x, \alpha_y)}{\iint_{-\pi/2}^{\pi/2} \mathrm{d}\alpha_x \mathrm{d}\alpha_y \Psi(\alpha_x, \alpha_y)}, \\ \vartheta_{xy} &= \frac{\iint_{-\pi/2}^{\pi/2} \mathrm{d}\alpha_x \mathrm{d}\alpha_y (\alpha_x - \alpha_{x;0})(\alpha_y - \alpha_{y;0}) \Psi(\alpha_x, \alpha_y)}{\iint_{-\pi/2}^{\pi/2} \mathrm{d}\alpha_x \mathrm{d}\alpha_y \Psi(\alpha_x, \alpha_y)}. \end{aligned} \tag{9.27}$$

For centrosymmetric (stigmatic) beams the mixed moments $\vartheta_{xy}$ are zero. Then, the second order moments are sufficient to describe the relative length error of the displacement measurement of an arbitrary stigmatic beam interfering with a copy of itself [96]:

$$\varepsilon \approx -\frac{\vartheta^2_{xx} + \vartheta^2_{yy}}{8}. \tag{9.28}$$

Equation (9.28) can be generalised to the case of misaligned and laterally displaced self-interference which has already been introduced above by equation (9.7) [88]. For a stigmatic zero-order Gaussian beam the second order moments $\vartheta^2_{xx} = \vartheta^2_{yy} = \vartheta^2$ are equal to the square of the beam divergence $\vartheta = \lambda/\pi w_0$. In this case, the relative length error $\varepsilon \approx -\vartheta^2/4$ is identical to equation (9.24).

### 9.3.2 Refinement of theory

Up to now, analytical methods could be used to tackle the problem, also because the paraxial approximation was assumed. Now, also interference of different arbitrary beams, albeit of same light frequency, must be investigated. Although it may be possible to do this analytically, the authors prefer a more direct numerical approach.

As a benefit, the paraxial approximation is no longer a requirement, which could potentially become important for future problems or quite unusual interferometer setups. Therefore, starting from the complex amplitude of two generally independent beams, which includes the case of identical beams, a numerical method to determine the relative length error is developed.

*9.3.2.1 The relative length error in scalar wave approximation*

Since independent beams are considered, their initial planes, i.e. the planes where the field complex amplitudes are defined, can be located at different positions $z_{0;1,2}$ along the $z$-axis (but are still perpendicular to the $z$-axis), with index 0 denoting initial positions:

$$E_{1,2}(x, y, z_{0;1,2}) = A_{1,2}(x, y, z_{0;1,2})\exp[i\phi_{1,2}(x, y, z_{0;1,2})], \tag{9.29}$$

where the real-valued functions $A_{1,2}(x, y, z_{0;1,2}) = |E_{1,2}(x, y, z_{0;1,2})|$ and $\phi_{1,2}(x, y, z_{0;1,2}) = \arg[E_{1,2}(x, y, z_{0;1,2})]$ are the local amplitude and the phase, respectively. The position of the observation plane is at $z$, i.e. where the beam interference is observed. The two propagation distances for the beams, $z_{1,2}$, are then defined by $z_{1,2} = z - z_{0;1,2}$. To propagate the fields, the angular spectrum [89, pp 48–51] is used, which is calculated by a Fourier transformation:

$$\tilde{E}_{1,2}(k_x, k_y, z_{0;1,2}) = \iint_{-\infty}^{+\infty} dx\, dy E_{1,2}(x, y, z_{0;1,2})\exp[i(k_x x + k_y y)]. \tag{9.30}$$

The inverse operation is defined by:

$$E_{1,2}(x, y, z_{0;1,2}) = \frac{1}{4\pi^2}\iint_{-\infty}^{+\infty} dk_x dk_y \tilde{E}_{1,2}(k_x, k_y, z_{0;1,2})\exp[-i(k_x x + k_y y)]. \tag{9.31}$$

The left side of equation (9.30) and the elements of the kernel in equation (9.31) are the plane waves of the angular spectrum which can be easily propagated to $z$ by:

$$\tilde{E}_{1,2}(k_x, k_y, z) = \tilde{E}_{1,2}(k_x, k_y, z_{0;1,2})\exp[-ik_z z_{1,2}]. \tag{9.32}$$

Now, the field at $z$ is directly obtained by [89, pp 50–51]:

$$\begin{aligned} E_{1,2}(x, y, z) = \frac{1}{4\pi^2}\iint_{-\infty}^{+\infty} & dk_x dk_y \tilde{E}_{1,2}(k_x, k_y, z_{0;1,2}) \\ & \exp[-i(k_x x + k_y y + k_z z_{1,2})]. \end{aligned} \tag{9.33}$$

Equations (9.30) and (9.33) are the basis of a practical propagation method [89, pp 50–51]. The numerical implementation involves two two-dimensional DFT (Discrete Fourier Transform) which can effectively be implemented by FFT (Fast Fourier Transform) [104]. The numerical implementation can be used to propagate fields over short distances, if sampling resolution and window size are chosen properly to guarantee that the field stays confined inside the window and does not 'leak' out over the propagated distance. The first FFT transforms the beam from the spatial representation into the angular space according to equation

(9.30). The second FFT transforms the propagated angular spectrum back to the spatial form as per equation (9.33). If apertures, which must be modelled in space, do not have to be considered, one can spare the second FFT to reduce the computational load. It is then sufficient to process the angular spectrum only which simplifies the further calculations. The power of the interference is then simply given by a complete integration of the sum of the angular spectra of beam 1 and 2 over $k_{x,y}$:

$$P \propto \iint_{-\infty}^{+\infty} \mathrm{d}k_x \mathrm{d}k_y \left|\tilde{E}_1(k_x, k_y, z_{0;1})\exp[-ik_z z_1] + \tilde{E}_2(k_x, k_y, z_{0;2})\exp[-ik_z z_2]\right|^2. \quad (9.34)$$

Using the Euler notation, equation (9.34) results in:

$$P \propto \iint_{-\infty}^{+\infty} \mathrm{d}k_x \mathrm{d}k_y \{|\tilde{E}_1|^2 + |\tilde{E}_2|^2 + 2|\tilde{E}_1||\tilde{E}_2|\cos[k_z(z_2 - z_1) - \tilde{\phi}_2 + \tilde{\phi}_1]\}, \quad (9.35)$$

where $\tilde{E}_{1,2} = \tilde{E}_{1,2}(k_x, k_y, z)$ and $\tilde{\phi}_{1,2} = \tilde{\phi}_{1,2}(k_x, k_y, z) = \arg(\tilde{E}_{1,2})$.

In general, the integral in equation (9.35) cannot be solved analytically. One way to retrieve the phase is numerical integration and making use of the well-known quadrature technique [105]. The constant term is dropped and only the sinusoidal variation, which remarkably depends only on $\Delta z = z_2 - z_1$, is considered. Therefore, it is indistinguishable which beam has propagated and which one has not, or if both beams have propagated. From now on, for the sake of brevity, we denote the relative phase as $\varphi_{\Delta z} = \varphi(\Delta z)$ which is obtained from the respective power modulation by:

$$\varphi_{\Delta z} = \tan^{-1}\left[\frac{\iint_{-\infty}^{+\infty} \mathrm{d}k_x \mathrm{d}k_y |\widetilde{E}_1||\widetilde{E}_2|\sin(k_z\Delta z + \widetilde{\phi}_1 - \widetilde{\phi}_2)}{\iint_{-\infty}^{+\infty} \mathrm{d}k_x \mathrm{d}k_y |\widetilde{E}_1||\widetilde{E}_2|\cos(k_z\Delta z + \widetilde{\phi}_1 - \widetilde{\phi}_2)}\right]. \quad (9.36)$$

Equation (9.36) is analogue to equation (9.7) in the paper of Dorenwendt and Bönsch [84], where the integration was performed in the spatial domain in the focal plane of a lens which in turn was approximated by the thin lens approximation. Others have also derived formulae like (9.36) but for self-interference only [95, 97]. The main difference is, therefore, the appearance of $\tilde{\phi}_{1,2}$ which infers the existence of a residual phase offset when the beam paths are equal:

$$\varphi_0 = \tan^{-1}\left[\frac{\iint_{-\infty}^{+\infty} \mathrm{d}k_x \mathrm{d}k_y |\widetilde{E}_1||\widetilde{E}_2|\sin(\widetilde{\phi}_1 - \widetilde{\phi}_2)}{\iint_{-\infty}^{+\infty} \mathrm{d}k_x \mathrm{d}k_y |\widetilde{E}_1||\widetilde{E}_2|\cos(\widetilde{\phi}_1 - \widetilde{\phi}_2)}\right], \quad (9.37)$$

which vanishes only if (for all components of the angular spectrum!) $\tilde{\phi}_1(k_x, k_y, z) = \tilde{\phi}_2(k_x, k_y, z)$, where $\tilde{\phi}_{1,2} \in (-\pi, \pi)$, i.e. for self-interference (or interference of identical beams).

From the equations (9.36) and (9.37) one can deduce the naively determined (erroneous) displacement:

$$\Delta z' = \frac{\varphi_{\Delta z} - \varphi_0}{k}. \tag{9.38}$$

With the definition in equation (9.24) one obtains for the relative length error:

$$\varepsilon_N = \frac{\varphi_{\Delta z} - \varphi_0}{k\Delta z} - 1, \tag{9.39}$$

where the label N indicates that it is obtained by the numerical procedure. If no wrapping occurs between $\varphi_{\Delta z}$ and $\varphi_0$ then equation (9.39) is exact. Wrapping issues can be avoided by, first, replacing the function $\tan^{-1}$ in equations (9.36) and (9.37) by *atan2*, which is the common name of a function implemented in many mathematical programming languages that employs a case differentiation to unambiguously map to the full interval $(-\pi, \pi)$ which replaces the ambiguous $(-\pi/2, \pi/2)$ mapping of $\tan^{-1}$. Secondly, by using this simple algorithm:

1. Since $\varepsilon$ is independent from $\Delta z$ (it is relative by definition) the latter can arbitrarily be chosen to satisfy $k\Delta z = 2\pi \Leftrightarrow \Delta z = \lambda$.
2. One must find the index $i$ of $\min(|\Phi_j|), j = 1,2,3$, where the array $\mathbf{\Phi}$ is defined by $\mathbf{\Phi} = (\varphi_\lambda - 2\pi - \varphi_0, \varphi_\lambda + 2\pi - \varphi_0, \varphi_\lambda - \varphi_0)$, where $\varphi_\lambda = \varphi(\Delta z = \lambda)$. The correct phase difference $\Phi_i$ replaces the numerator in equation (9.39).

The equations (9.36) and (9.37) must be solved numerically. To check how accurately this can be done a comparison of the numerical procedure to the analytical formula (9.24) for stigmatic zero-order Gaussian beams is instructive. From now on, the label A is used to denote the analytically determined relative length error of equation (9.24) by $\varepsilon_A$. Several Gaussian beams of wavelength $\lambda = 532$ nm and different waist radii $w_0$ are sampled at their waist positions with $99 \times 99$ points on $b^2 = 10w_0 \times 10w_0$ windows. From equation (9.14) one gets $E_2(x, y, 0) = E_1(x, y, 0) = \exp[-(x^2 + y^2)/w_0^2]$ and the interference is considered after propagation paths which differ by $\Delta z = \lambda$. Table 9.2 lists the corresponding relative discrepancies $|(\varepsilon_N - \varepsilon_A)/\varepsilon_A|$ for three examples. For beam sizes typically used for picometre level interferometry, the discrepancy is fully negligible.

**Table 9.2.** Comparison between numerical evaluation and analytical formula for the calculation of the relative length error of an interferometric measurement with Gaussian beams of wavelength $\lambda = 532$ nm, where $\varepsilon_N$ is the numerically and $\varepsilon_A$ the analytically determined relative length error. The rightmost column shows the corresponding relative discrepancies.

| $w_0$/mm | $\varepsilon_N$ | $\varepsilon_A$ | $\lvert(\varepsilon_N - \varepsilon_A)/\varepsilon_A\rvert$ |
|---|---|---|---|
| 0.5 | $-2.867\,632\,8 \times 10^{-8}$ | $-2.867\,632\,7 \times 10^{-8}$ | $3.5 \times 10^{-8}$ |
| 1 | $-7.169\,081\,8 \times 10^{-9}$ | $-7.169\,081\,7 \times 10^{-9}$ | $1.4 \times 10^{-8}$ |
| 2 | $-1.792\,270\,47 \times 10^{-9}$ | $-1.792\,270\,42 \times 10^{-9}$ | $2.8 \times 10^{-8}$ |
| 4 | $-4.480\,676\,6 \times 10^{-10}$ | $-4.480\,676\,0 \times 10^{-10}$ | $1.3 \times 10^{-7}$ |

To take also misaligned and sheared beams into consideration, it is not necessary to adapt the formulae (9.30), (9.36), (9.37) and (9.39). The only needed artifice is a transformation of the input field of one beam, given in the beam-fixed coordinate system, i.e. the system in which its mean propagation direction and centre of mass coincide with local $z$-axis and origin, respectively, into the beam-fixed coordinate system of the other beam and continue as usual. The necessary theory is treated by several authors for the problem of complex amplitude propagation between shifted and tilted planes [106–113]. If one implements the numerical solution of the integrals (9.30) and (9.33) as sums, the transformations are easily implemented. However, this implementation is very slow and it is desirable to use the FFT. These in turn, suffer from the fact that the discrete equidistant components of the angular spectrum, which must be transformed if the beam direction is changed, will no longer be equidistant after the transformation. Therefore, interpolation of the transformed spectrum on an equidistant grid is necessary [110]. Since the angular spectrum of well collimated large beams is usually concentrated on a few grid points, interpolation is even more problematic. This issue can be mitigated by use of a scaled convolution [114] where zooming into the relevant angular space enlarges the number of relevant grid points. Unfortunately, the details of the implementation of this technique go far beyond the scope of this chapter and the interested reader is referred to the literature [106–113].

The numerical procedure shown here is fast and flexible enough to be effectively used in Monte Carlo simulations, which is helpful to estimate also the uncertainty of the relative length error. For this task, also random perturbations, which can result from imperfect interferometer components, can easily be added to the complex amplitudes. Additionally, the coordinates can be scaled to mimic uncertain knowledge about the beam dimensions. Furthermore, a quadratic, i.e. defocus, or other functions can be added to the phase to introduce errors of wavefront measurements into the simulation.

#### *9.3.2.2 Consideration of the vectorial nature of light*

Ultimately, when the accuracy of sophisticated interferometric experiments must be increased even more, the question arises whether even the scalar approximation is still justified. Light waves are transversally polarized [90, pp 23–32]. When laser beams are refracted at an interface of an interferometer setup the Fresnel coefficients determine the amount of reflection and transmission depending on incidence angle and polarization [90, pp 36–51]. While most surfaces are plane and traversed by the large and well collimated beams at almost normal incidence there is one case where curved surfaces may lead to a significant violation of the scalar approximation. This case is investigated in the following text.

For the numerical procedure described above it was assumed that the total power of the interference signal is captured. How is this achieved in experiment? One could use a detector large enough to avoid truncation of the beams. In heterodyne interferometers with beat frequencies of several MHz such a large detector could be too slow because its impedance is coupled to its size [115]. A lens can be used to focus a large beam to a small and fast detector. One can imagine that such a lens can modulate the angular spectrum of the interfering beams due to the angular dependence of the reflection coefficients on the lens interface. This modulation can even be different for both beams. The theory to describe such effects must therefore necessarily include the vectorial nature of light.

The vectorial ray-based diffraction integral (VRBDI) [116, 117] enables a fully vectorial description of monochromatic light propagation in optical systems consisting of linear, isotropic and dielectric media such as, e.g. lens systems. This method can be described as follows: an input field is decomposed optionally either into plane waves (by FFT) or simply into spherical waves, i.e. point sources which are defined by the given sampling points. Then each of these components is represented by many sequentially traced finite-differential ray tubes which are each aimed at a point of interest, usually a point on a two-dimensional Cartesian grid on the output plane. There, the field contributions of all components are calculated by a matrix-optics evaluation of the traced ray tubes and summed up. The VRBDI can be used along sequences of interfaces whose apertures do not cut off significant beam parts and not in focal regions [116, 117]. Furthermore, if the output of a simulated optical system is a collimated beam further propagation would be dominated by diffraction which is not well represented by the outgoing ray tubes because the full region which is reached by diffraction cannot be reached by the constrained ray tubes. Then, it is advised to put the output plane directly behind the last lens interface. From there, any free-space propagation can be used.

One can check if a focusing lens affects the resulting relative length error if the situation is simulated with the VRBDI. The names of the noteworthy MATLAB functions used for field-propagation in the following simulation are given in round brackets. These functions are freely available and can be downloaded [118].

In figure 9.9 the simulated setup is sketched. In table 9.3 the parameters of optical system and simulation are listed.

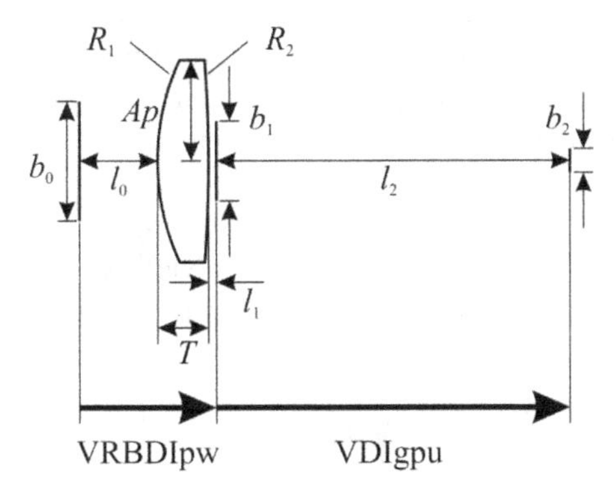


**Figure 9.9.** Simulation of beam focusing onto a detector in laser interferometry. The parameters are listed in table 9.3. The vectorial ray-based diffraction integral with plane wave decomposition of the input field (MATLAB function: VRBDIpw) is used for the propagation through the lens to an intermediate plane [116, 117]. From there a vectorial diffraction integral (MATLAB function: VDIgpu) is used to propagate the field to the detector plane in the focal region. The propagation methods are freely available MATLAB functions and can be downloaded [118].

**Table 9.3.** Parameters of the simulated setup illustrated in figure 9.9. The parameter $f_B$ denotes the back focal length, $R_{1,2}$ are radii of spherical curvature. The other parameters are self-explanatory by looking at figure 9.9.

| $Ap$ | 12.7 mm |
|---|---|
| $R_1$ | 30.06 mm |
| $R_2$ | −172 mm |
| $T$ | 6.5 mm |
| Lens material | N-BK7 |
| Immersion medium | Vacuum |
| $f_B$ | 46.4 mm |
| $l_0$ | 10 mm |
| $l_1$ | 0.1 mm |
| $l_2$ | $f_B - l_1$ |
| $b_0$ | $10 \times \max(w_{0;s}, w_{0;R})$ |
| $b_1$ | 10 mm |
| $b_2$ | 0.2 mm |
| Sampling points | 99 × 99 |

As an example, the lens LBF254-050-A from THORLABS[1] is chosen. The aperture $Ap$ is large enough to safely neglect beam truncation for the chosen input beams. Different Gaussian beams defined by equation (9.14) are chosen as input beams. In addition, optional misalignment and shear is allowed individually for both beams, which is realized by equation (9.30), subsequent coordinate transformation and final numerical integration of the transformed spectral components on the input plane (PWIgpu). By using equations (9.30) and (9.33) beam 2 is propagated by $\Delta z$ and additionally four $\lambda/4$ steps at zero and at $\Delta z$ displacement. For each step, the angular spectra of beam 1 and 2 are propagated through the lens to an intermediate plane by using the vectorial ray-based diffraction integral with plane wave decomposition of the input field (VRBDIpw). From there a vectorial diffraction integral (VDIgpu) is used to propagate the field to the detector plane in the focal region. The input fields are assumed to be linearly polarized along the $x$-axis. The output fields generally contain also $y$- and $z$-components. To calculate the power on the detector plane the magnetic field components are also needed, which can be obtained from the vectorial angular spectrum of the electric field via:

$$\tilde{\mathbf{H}}(k_x, k_y) = \sqrt{\frac{\epsilon_0}{\mu_0}} \hat{\mathbf{k}} \times \tilde{\mathbf{E}}(k_x, k_y), \tag{9.40}$$

[1] The authors do not recommend certain products from certain companies. This design has been chosen because it was freely available on the company's web shop. The authors' intention is to enhance transparency and to allow reproducibility, not advertising. Comparable designs may be found in web shops of other companies.

where $\hat{\mathbf{k}} = \mathbf{k}/k$, $\epsilon_0$ and $\mu_0$ are vacuum permittivity and permeability, respectively [91, p 297]. It is worth noting that equations (9.30) and (9.31) can be used for each component of the electromagnetic field which allows the transformation of the field between spatial and angular domain.

The Poynting vector [91, pp 259–65]:

$$\mathbf{S}(x, y) = \frac{1}{2}\mathrm{Re}[\mathbf{E}(x, y) \times \mathbf{H}^{*}(x, y)] \tag{9.41}$$

is dot-multiplied to the normal $\hat{\mathbf{N}}_{\mathrm{D}}$ of the detector plane to obtain the sought irradiance which must be integrated over the full detector plane to get the power of the simulated interference signal:

$$P_{0,\,\Delta z} = \iint_{-b_2/2}^{b_2/2} \mathrm{d}x\mathrm{d}y\mathbf{S}_{0,\,\Delta z}(x, y) \cdot \hat{\mathbf{N}}_{\mathrm{D}}. \tag{9.42}$$

The phase of the interference signal at zero and $\Delta z$ displacement is retrieved by using the well-known four-step-formula [119]:

$$\varphi_{0,\,\Delta z} = \tan^{-1}\left[\frac{P_{0,\,\Delta z}(\lambda) - P_{0,\,\Delta z}(\lambda/2)}{P_{0,\,\Delta z}(\lambda/4) - P_{0,\,\Delta z}(3\lambda/4)}\right]. \tag{9.43}$$

To avoid wrapping problems $\Delta z = N\lambda$ is chosen, where $N$ is an integer. Then, the relative length error of the VRBDI simulation can be directly obtained by

$$\varepsilon_V = \frac{\varphi_{\Delta z} - \varphi_0}{k\Delta z} = \frac{\varphi_{\Delta z} - \varphi_0}{2\pi N}. \tag{9.44}$$

Normally, $\varepsilon_V$ is so small that $N$ would need to be extremely large for wrapping to occur. Here, $N = 2 \times 10^4$ is deliberately chosen, which is still small enough.

In table 9.4 the input beam parameters and corresponding relative length errors with their relative discrepancies are listed. Interestingly, in simulation 4 positive relative length errors appear, which is caused by the large shear in that case. In simulation 3 the error is more negative than in simulation 2 (ideally aligned) because the misalignment dominates over shear. One can also see that there is practically no relevant difference between fully vectorial simulation and scalar approximation realized by numerical implementation of equations (9.30), (9.36), (9.37) and (9.39). Thus, at present, provided that reasonably large and well collimated beams are used, the challenges to reach even smaller uncertainties in traceable interferometric length measurements will be elsewhere. For instance, the experimental characterization of laser beams has orders of magnitude larger issues to consider. Therefore, attention is now turned to experimental techniques, which gather relevant information from real laser beams.

### 9.3.3 Characterization of laser beams

In this section the different methods to characterize laser beams as well as their pros and cons are presented. A considerable fraction of them intrinsically assumes that the investigated laser beams are Gaussian beams while other techniques are only

**Table 9.4.** Input beam parameters for the simulation according to figure 9.9 and table 9.3, where $w_{0;1,2}$ are the waist radii of the simulated Gaussian beams, $z_{0;1,2}$ their initial propagation distances (before shifting beam 2), $\delta x_{1,2}$ and $\delta y_{1,2}$ are lateral shifts with respect to the ideal axis of the optical system, $\xi_{1,2}$ and $\eta_{1,2}$ rotate the propagation directions of beam 1,2 (initially parallel to $z$) consecutively around the $x$- and $y$-axis, respectively, $\varepsilon_V$ and $\varepsilon_N$ denote the differently obtained relative length errors defined by equations (9.42)–(9.44) and (9.30), (9.36), (9.37) and (9.39), respectively.

| Simulation | 1 | 2 | 3 | 4 |
|---|---|---|---|---|
| $w_{0;1}$/mm | 1 | 1.1 | 1.1 | 1.1 |
| $w_{0;2}$/mm | 1 | 0.9 | 0.9 | 0.9 |
| $z_{0;1}$/mm | 100 | 200 | 200 | 200 |
| $z_{0;2}$/mm | 100 | 130 | 130 | 130 |
| $\delta x_1$/mm | 0 | 0 | 0.13 | 0.7 |
| $\delta x_2$ /mm | 0 | 0 | −0. 09 | −0. 6 |
| $\delta y_1$ /mm | 0 | 0 | −0. 11 | −0. 4 |
| $\delta y_2$ /mm | 0 | 0 | 0.05 | 0.5 |
| $\xi_1$/arcsec | 0 | 0 | 7 | 15 |
| $\xi_2$/arcsec | 0 | 0 | −1 | −10 |
| $\eta_1$/arcsec | 0 | 0 | 12 | 12 |
| $\eta_2$ /arcsec | 0 | 0 | −2 | −19 |
| $\varepsilon_V \times 10^9$ | −7.169 080 | −7.097 89 | −7.5053 | +1.066 |
| $\varepsilon_N \times 10^9$ | −7.169 082 | −7.097 86 | −7.5045 | +1.067 |
| $\lvert(\varepsilon_V - \varepsilon_N)/\varepsilon_N\rvert$ | $3 \times 10^{-7}$ | $4 \times 10^{-6}$ | $1 \times 10^{-4}$ | $1 \times 10^{-3}$ |

restricted by the paraxial and the scalar wave approximation or only by the latter one. It is also shown below why and how the assumption of Gaussian beams limits the attainable accuracy.

*9.3.3.1 Methods assuming Gaussian beams*

Soon after the realization of lasers, the need to characterize their beams has led to the description by Gaussian beams emerging as the fundamental mode of their optical resonators [120, 121]. Inside this model, only the parameters waist radius $w_0$ and wavelength must be known to describe the propagation of laser beams. Assuming prior knowledge of the wavelength, only the waist radius of a beam must be determined. In principle, one could measure beam profiles at different locations along the beam path and locally fit $w(z) = w_0[1+(z/z_R)^2]^{1/2}$ to obtain a least square mean $w_0$. One can imagine that the uniqueness of the fit would be far better if the whole Rayleigh range, i.e. where $z\in[-z_R, z_R]$, would be accessible (cf figure 9.8). However, this is not always possible or practical. For example, the waist of a laser beam, which originates directly from a laser, lies inside the resonator! In addition, if a beam is large and well collimated, the Rayleigh range can easily exceed the size of a laboratory.

Among these considerations, it is important to clearly define the meaning of 'well collimated'. In the picture of Gaussian beams the following definition is instructive:

a well collimated beam originates from a collimator, which is adjusted such that the beam has the maximally possible Rayleigh length $z_R$. This means that the waist $w_0$ is maximized, too. How is a collimator adjusted to achieve this and where is the location of the waist? If one considers a fibre collimator, the distance of the fibre end with respect to the collimator lens(es) must be optimized. Moving the fibre away from the optimal position introduces defocus into the wavefront of the collimated beam. The defocus always shortens $z_R$, which means a different Gaussian beam with shorter Rayleigh length and smaller waist is formed, which has a larger relative length error. Therefore, optimizing collimation means minimizing defocus. Then, the waist is in the lens plane of the collimator, which is again not directly accessible.

To avoid the need to search and measure the waist region directly, the Fourier-transforming properties of lenses are exploited by several techniques. As pointed out by Kogelnik [120], one finds that the beam radius in the rear focal plane of an ideal lens is independent of the distance of this lens to the location of the beam waist in front of the lens. This finding also follows from the fact that a lens does a Fourier transform of the incident beam whereby the result is found in the Fourier plane, i.e. the focal plane [89, pp 77–100, 102, 103, 122]. The Fourier transform yields the angular spectrum, which is, apart from a phase factor, not dependent on the propagation in free-space. If the focal length $f$ of the lens is known and the beam profile is measured in the focal plane, one can determine $w_0$. Several techniques are reported to measure beam profiles, e.g. translating a knife-edge, a slit or another aperture perpendicular to the beam [123] or illuminating areal detectors such as charge-coupled devices (CCD) or complementary metal-oxide semiconductor (CMOS) sensors [102, 103]. When the full intensity distribution in the Fourier plane is captured, the second moments $\vartheta^2_{xx,\,yy}$ and $\vartheta_{xy}$ can be obtained for arbitrary, e.g. also non-Gaussian, beams which makes the method quite general. Strictly speaking this method is then only limited by the paraxial approximation and does not necessarily assume Gaussian beams. The crucial points are: measuring as close as necessary to the position of the focal plane, knowing the numerical value of $f$ well enough and suppressing the contribution of noise to the second moments [101–103, 122].

A very similar technique involves scanning through the Rayleigh range of the beam transformed by a well-known lens. The waist behind the lens is the image of the waist of the input beam. It is worth noting that, in general, the image is not at the focal plane. When the transformed beam is found by a fit to the scanned beam profiles, the incident beam can be calculated by knowing the lens transformation and applying the inverse operation. This is usually done in paraxial approximation by use of ABCD matrices or by assuming a thin lens with known $f$ [120]. Several commercial suppliers offer systems, which use this procedure. However, the beam transformation by lenses adds systematic effects to the procedure, which increases the uncertainty. Especially for relatively well collimated large beams that produce relative length errors in the range of several parts in $10^9$, the detection limit prevents resolving the beam parameters of the input beam.

A rough estimation can be given by comparing the respective change in divergence with the well-known Rayleigh criterion. For example, the distinction between a Gaussian beam with $w_0 = 2$ mm and a Gaussian beam with $w_0 = 2.1$ mm at $\lambda = 532$ nm corresponds to a change of divergence by 4 μrad and different relative length errors of $-1.8 \times 10^{-9}$ and $-1.6 \times 10^{-9}$. The resolution limit $\delta\theta$ of a lens with a diameter $D = 50.8$ mm can be estimated by $\delta\theta \approx 1.22\lambda/D \approx 13$ μrad. From the first beam a Gaussian beam with $w_0 = 1.7$ mm would just start to be distinguishable with this lens, while the relative length error differs by $0.7 \times 10^{-9}$. This error contribution would already be the dominant one at an interferometer system addressing picometre level displacement measurements. However, the real limitation is the size of the input beam. Therefore, even increasing the lens diameter does not help and the real resolution limit is likely much worse (about one order of magnitude) than in this rough estimation.

Furthermore, in this simple configuration, the technique is limited to stigmatic (rotationally symmetric) or simple astigmatic beams (different waist radii in $x$- and $y$-direction but with waist at same $z$-position). Generally astigmatic beams [121] must be handled by more complicated systems containing sequences of cylindrical lenses [124] making the method quite cumbersome.

Some authors report the knife-edge scanning technique to determine an approximate waist radius by waiving the lens and assuming to measure close to the waist [86]. Depending on the individual case, this procedure can be a satisfying approximation. However, one should be aware of the relevant uncertainty according to $u_\varepsilon = 2|\varepsilon|u_{w_0}/w_0$ which can be obtained from equation (9.24). For example, if the discrepancy of the measured waist radius of a Gaussian beam from the correct value is +5% the relative length error will deviate from the true value by −10%. However, already the assumption that a beam is Gaussian can be wrong or does not represent the full picture [103]. The following example will illustrate the impact of falsely assuming a Gaussian beam.

In figure 9.10 the field amplitude of a step index glass fibre is shown. The blue continuous curve shows the exact solution [125, pp 9–29], which has been determined from the manufacturer values for numerical aperture $NA = 0.09$ and cut-off wavelength $\lambda_c = 420$ nm. The guided wavelength is $\lambda = 532$ nm. The orange dashed curve represents a Gaussian beam at its waist where the manufacturer value mode-field diameter (MFD) is used to define the waist radius according to $w_0 = \mathrm{MFD}/2$. One can see that the exact curve has a broader tail than the Gaussian curve. The exact mode field is used as input for a VRBDI simulation of an ideally adjusted and aligned fibre collimator. The irradiance of the collimated beam and its wavefront is shown in figure 9.11. Using the equations (9.30), (9.36), (9.37) and (9.39) a relative length error of $\varepsilon_N = -1.93 \times 10^{-9}$ is obtained. Thus, a realistic beam has been rigorously simulated, which incidentally is not a Gaussian beam.

Following the method presented in the appendix of reference [86], a simulation of the capping of an ideally aligned perfectly linear knife-edge scan parallel to $x$ on the irradiance of the simulated beam (figure 9.11, left) is performed and the results presented in figure 9.12. The left plot shows the total power in the progressively clipped aperture. Using the inverse normal (norm. inv.) one obtains the right plot. Then, from the slope $s = -0.8788$ of the linear fit to the norm. inv. One gets for the waist radius $w_0 = -2/s = 2.28$ mm [86]. Using equation (9.24) a relative length error

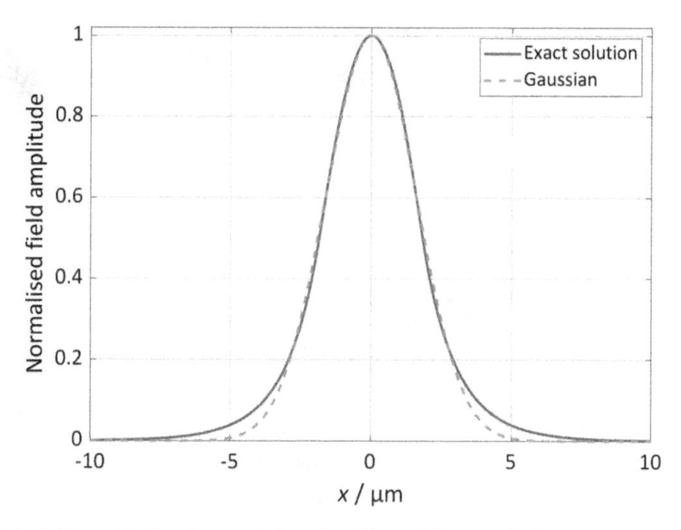


**Figure 9.10.** Mode-field amplitude of a step index glass fibre with numerical aperture $NA = 0.09$ and cut-off wavelength $\lambda_c = 420$ nm. The exact solution [125] has a broader tail than the approximative Gaussian profile which assumes a Gaussian beam at its waist having the same diameter as the mode-field. The guided wavelength is $\lambda = 532$ nm.

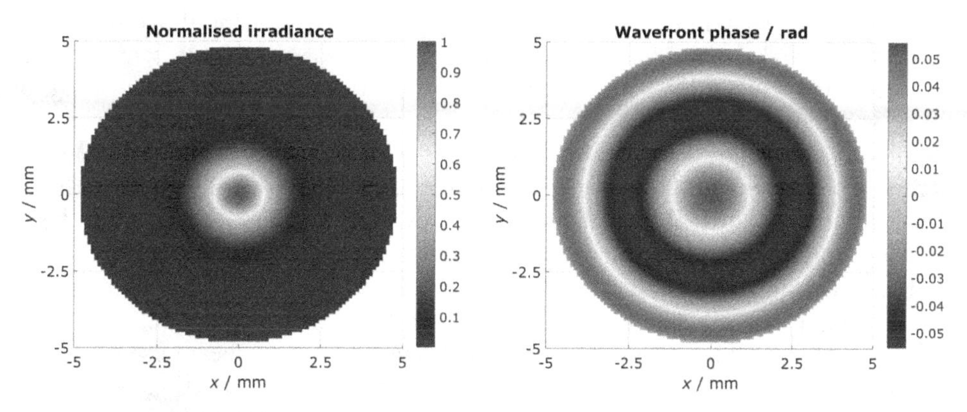


**Figure 9.11.** Simulated irradiance (left) and wavefront phase (right) of an ideally adjusted and aligned fibre collimator. The wavelength is $\lambda = 532$ nm.

of $\varepsilon_A = -1.38 \times 10^{-9}$ is obtained, which deviates from the correct value by almost 30% despite of ideal conditions without any noise influences! Looking carefully at the right plot in figure 9.12 one can see that the linear fit does not represent the norm. inv. well which oscillates around the fitted line. It is instructive to assume that this oscillation stems from the non-Gaussian beam profile. However, in real measurements such subtle deviations can easily be obfuscated by noise. Therefore, one would not even recognize that the linear fit is improper and that the investigated beam is not really a Gaussian beam. Nowadays, beam profilers (BP) are widely available, e.g. based on scanning slit mechanisms. Therefore, a quick check if the measured

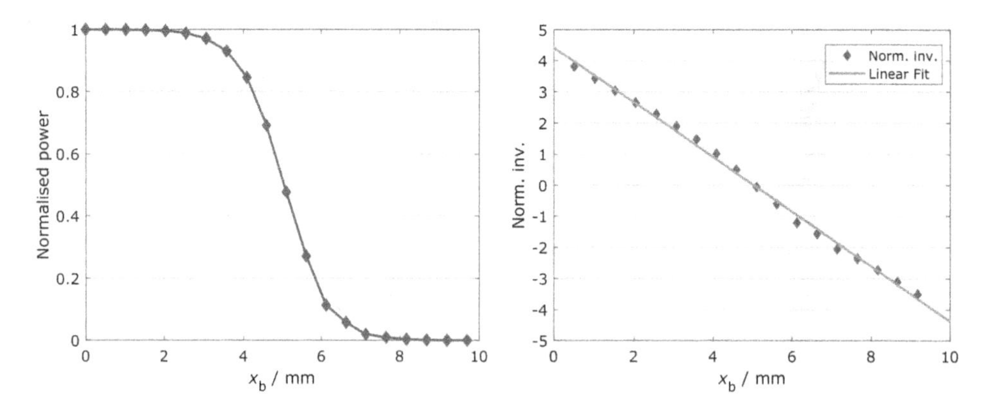


**Figure 9.12.** Simulation of the knife-edge scanning method to determine the waist radius of a Gaussian beam. Here it is applied to the irradiance of a non-Gaussian beam. The normalised power (left) and inverse normal (right) is plotted over the knife-edge block position $x_b$. From the slope of the linear fit to the inverse normal the supposed waist radius can be obtained [86].

beam profile can be fitted well by a single Gaussian function should help to avoid falling into this kind of trap.

### *9.3.3.2 A method assuming arbitrary scalar waves*

An alternative technique to the classical methods described above has been developed. The key element is an additional device: a wavefront sensor. Some authors describe a combination of wavefront and irradiance measurements, e.g. by using a Shack–Hartmann sensor (SHS) [126, 127], to determine the complex amplitude of a beam and finally derive the beam moments [128, 129]. Here, a method, which is also based on the combination of wavefront and irradiance measurements, is described. However, the evaluation does not involve beam moments (a paraxial concept) but uses equations (9.30), (9.36), (9.37) and (9.39) to directly calculate the relative length error from analytic fit functions to irradiance distribution(s) and wavefront(s) of the measured beam(s).

Under the conditions of a single source of a collimated laser beam used in a two-arm interferometer consisting only of plane interfaces, neglecting any beam truncation and parasitic reflections, it is sufficient to characterize the ingoing laser beam to yield the relative length error of the interferometer.

First, the Cartesian beam-fixed coordinates $(x_0, y_0, z_0)$ are defined, where $z_0$ lies on an axis parallel to the mean beam propagation direction and $(x_0 = 0, y_0 = 0)$ coincides with the centre of mass of the irradiance distribution. Furthermore, the third dimension is omitted and the two-dimensional coordinates of the measurement are defined as $(x_m, y_m) = (x_0 + x_c, y_0 + y_c)$.

Now, the constituents of the complex amplitude (9.29) can be related to a measured wavefront $W(x_m, y_m)$ and a measured irradiance distribution $I(x_m, y_m)$ by:

$$A(x_0, y_0) \propto \sqrt{I(x_m - x_c, y_m - y_c)}, \tag{9.45}$$

$$\phi(x_0, y_0) = -\frac{2\pi}{\lambda} W(x_m - x_c, y_m - x_c) + \phi_{\text{const.}}, \tag{9.46}$$

where $\phi_{\text{const.}}$ is unknown but of no practical relevance and can be omitted. It is worth noting that $(x_m, y_m)$ and $(x_c, y_c)$ can be different or identical between equations (9.45) and (9.46) if the corresponding measurements are done with different devices or the same device, which delivers both measurements in one run, respectively. The latter case can be realized, e.g. with a SHS calibrated for both $W(x_m, y_m)$ and $I(x_m, y_m)$.

Although the relations (9.45) and (9.46) exist, one cannot directly use them for the numerical implementation of equations (9.30), (9.36), (9.37) and (9.39). An important condition must be met by the input complex amplitude $E(x_0, y_0)$. Apart from being properly sampled $E(x_0, y_0)$ should also be *bandlimited*: its angular spectrum must be nonzero only in a finite region of angular space [89, p 25]. In practice, the following conditions should be met:

1. $A(x_0, y_0)$ should be smooth (no discontinuities!) while all significant parts should be confined in a finite region and approach zero inside this region.
2. The region in which $\phi(x_0, y_0)$ is given should at least include significant parts of $A(x_0, y_0)$.

If these conditions are met, the associated (discrete!) angular spectrum $\tilde{E}(k_x, k_y)$, i.e. the 2D-FFT of sampled $E(x_0, y_0)$, follows a square-integrable function. However, measured data cannot directly fulfil these conditions. Firstly, the measured irradiance is superposed by detection noise. Secondly, the dynamical range of the irradiance detector is limited and only a fraction of the beam is measured, i.e. low-irradiance cut-off which violates the first condition. Such a low-irradiance cut-off is typical also for SHS, where the second condition is then violated. The sharp edges in space, regardless of whether it is for amplitude or phase, result in an infinite spatial frequency spectrum. Indeed, this is not true for a discrete spectrum obtained by FFT. However, the choice of the size of the calculation windows for the angular spectrum (9.30) and for the numerical integrations of equations (9.36) and (9.37) would always influence the resulting $\varepsilon$. Therefore, analytical functions, which satisfy the above conditions, are fitted to the measured data. These functions are then properly sampled and serve as input for the numerical implementation of equations (9.30), (9.36), (9.37) and (9.39).

The improvement of this concept is still work in progress. Optimal measurement conditions to devise a best practice guide must be investigated. Furthermore, proper fit functions must be found and the measurement uncertainty is to be estimated. The relevant dependences are the residual calibration errors as well as the extrapolation of fit functions. First approaches using a sum of two concentric Gaussian functions for $A(x_0, y_0)$ and the first nine Zernike polynomials for $\phi(x_0, y_0)$ have been quite successful for beams of high quality as can be seen in figure 9.13. For the shown fit residua for wavefront and irradiance the discrepancies are 0.7% RMS (6.2% PV) and 0.8% RMS (7.6% PV), respectively. While the irradiance can also be measured with a BP which is not severely affected by low-irradiance cut-off, the wavefront data is limited to the shown region. While the obtained beam profiles of BP and SHS agree

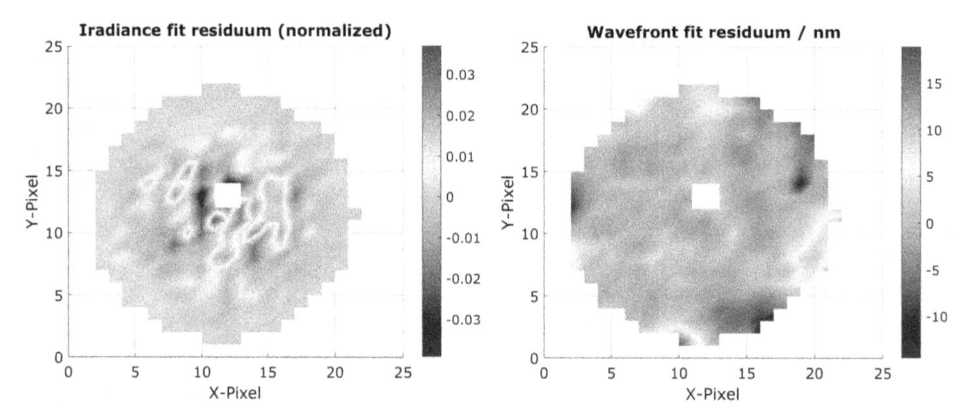


**Figure 9.13.** Fit residua for irradiance (left) and wavefront (right). The measurement data has been obtained from a calibrated Shack–Hartmann sensor (SHS) and a well collimated laser beam at 532 nm wavelength. The size of a pixel is 256 μm in both dimensions. The white square indicates an intentional defect in the microlens array which serves as reference position. The irradiance fit function is a sum of two centred 2D Gaussian functions which are applied to the square root of the measured irradiance. For the wavefront fit the first nine Zernike terms have been used namely: piston, tilt $x$, $y$, defocus, astigmatism $x$, $y$, coma $x$, $y$ and primary spherical.

better than 10%, which is the uncertainty given by the manufacturer of the BP, the wavefront can only be extrapolated blindly to the unknown region. Extrapolation of Zernike polynomials is a dangerous endeavour because of the increasingly higher powers of the higher terms. If the investigated beams can be sufficiently well fitted in the accessible region with the first nine Zernike terms, namely: piston, tilt $x$, $y$, defocus, astigmatism $x$, $y$, coma $x$, $y$ and primary spherical, as shown in figure 9.13, then the extrapolation is still manageable and overshoots are mostly harmless. This has been checked by use of the VRBDI for *ab initio* Monte Carlo simulations of a fibre collimator (including tilts and displacements of the fibre end) yielding simulated SHS data for different cut-off-levels. A contribution to the relative uncertainty of the relative length error smaller than 6% for a low-irradiance cut-off-level of 5% is obtained. It is worth noting, that the unit disk radius of the Zernike fit must be matched to the available data region. Then the extrapolated region lies outside the unit disk. One finds that the accuracy of extrapolation suffers greatly if this rule is not followed.

## 9.4 Conclusion

The realisation of displacement sensors, either encoder systems or interferometers, with a precision at the picometre level is state-of-the art. Nevertheless, many effects have to be considered to build a displacement interferometer with a resolution, a stability and linearity in the picometre range. Beside the amplifier noise and the shot noise limitation of heterodyne interferometers, the linewidth and the intensity noise of the laser source can limit the resolving capability of a displacement interferometer. Using a phase evaluation electronics with a bandwidth permitting mirror movements of several millimetres per second and above, a laser source with a linewidth below 1 MHz and a relative intensity noise below −135 dB $Hz^{-1}$ has to be picked to offer picometre level resolution. The stability of the interferometer system is one of the limiting uncertainty contributions at long-term measurements. The stability of the measured mirror displacement can be increased with the degree of constancy of environmental parameters like temperature, pressure fluctuations (turbulences) and humidity. This requirement is not limited to the beam paths, which are situated in an evacuated environment with a pressure below 0.001 Pa in the best case. The interferometer optic, the amplifiers and the laser head should be situated in a temperature stabilized environment, too. Periodic nonlinearities thwart a linear relation between the evaluated phase and the mirror displacement. At heterodyne interferometers, they should be minimized using an optical design based on spatially separated input beams. In other cases, a fitted ellipse can be used to minimize them and enable displacement measurements with picometre resolution. But a correction of the interferometric signals with an ellipse can only minimize periodic nonlinearities of the first and second order. Other orders, for example caused by ghost reflections or intermodulation, have to be avoid by the optic design, alignment and adequate chosen components.

For the realisation of traceable displacement measurements with a relative uncertainty better than $10^{-9}$, the period of the displacement interferometer has to be known on top of the influence of the systems' precision. The period of the interferometer is defined by the laser frequency, the refractive index of the ambient medium, potential relativistic corrections in case of movement speeds above 0.1 m $s^{-1}$ and the wavefront of the interfering beams. Therefore, the laser has to be stabilized to a recommended frequency standard, the interferometer should be placed in an evacuated environment and a proper diffraction correction, which is based on thorough beam characterization, is necessary. The relevant methods for the beam characterization are already an integral part of sophisticated interferometry experiments [102, 103, 130] and subject to ongoing research and development.

Another method based on Shack–Hartmann sensor measurements is limited to beams of high quality. To assess also more complicated beams improved inter- and extrapolation techniques or the implementation of HDR (high dynamic range) into the data retrieval would be rewarding. The latter could mitigate or even avoid low-irradiance cut-off and the necessity of extrapolation. Then, also higher order Zernike terms for $\phi(x_0, y_0)$ could be included without risking overshoots or splines could be used instead.

The *ab initio* simulation of a fibre collimator and the measurement procedure can produce simulated data, which can be compared to measured data. A fit procedure in the simulation process could be used to find a matching configuration, i.e. a deviation inside tolerances from the known collimator design. This configuration can be used to calculate the full beam, as it is required for the numerical evaluation. Still, the uniqueness of a found configuration would be in question and might never be guaranteed.

The realization of traceable measurements with a relative uncertainty better than $10^{-9}$ is of particular importance for the comparability of fundamental scientific results, like the Avogadro experiment and Watt-balance experiments for the redefinition of the kilogram, as well as for the comparability of different national standards disseminating the unit of length. An industrial application of displacement interferometers with picometre level precision is photomask metrology tools. At these machines self-calibration methods are used to guarantee the precision of the tools [131] and a 'golden mask' is used to match the scale of different tools [132]. This solution without the need of traceability is sufficient, as long as the measurement results of the machines do not have to match with those of other suppliers. But these machines are also impressive examples, that displacement interferometers have been understood for decades, conceptually. Their implementation—at the sub-nanometre level—is entirely a different story. The mechanical issues to deliver displacement strokes of sufficient straightness have not been considered here. At least, it is worth noting that these issues bring about cosine and Abbe errors, which represent relevant limitations to achieve picometre level uncertainty for millimetre scan ranges.

## Acknowledgements

The authors would like to thank Jens Flügge, Rainer Köning, Paul Köchert, Karl Meiners-Hagen, Ulrich Kuetgens, Susanne Quabis and Axel Wiegmann for years of scientific collaboration, for educating discussions and for their support on cordial terms.

## References

[1] Holzapfel W 2008 Advancements in displacement metrology based on encoder systems *Proc. 23rd Annual ASPE Meeting* 71–4

[2] Köchert P, Köning R, Weichert C, Flügge J and Manske E 2015 An upgraded data acquisition and drive system at the Nanometer Comparator *Proc. of ASPE 5th Topical Meeting: Precision Interferometric Metrology* 61–6

[3] Yacoot A and Cross N 2003 Measurement of picometre nonlinearity in an optical grating encoder using x-ray interferometry *Meas. Sci. Technol.* **14** 148–52

[4] Halbgewachs C, *et al* 2016 Qualification of HEIDENHAIN linear encoders for picometer resolution metrology in VTF Etalons *Ground-based and Airborne Instrumentation for Astronomy VI* **9908H**

[5] Kunzmann H, Pfeifer T and Flügge J 1993 Scales vs. laser interferometers performance and comparison of two measuring systems *CIRP Ann. Manuf. Technol.* **42** 753–67

[6] Teimel A 1992 Technology and applications of grating interferometers in high-precision measurement *Precis. Eng.* **14** 147–54

[7] Gao W, *et al* 2015 Measurement technologies for precision positioning *CIRP Ann.* **64** 773–96

[8] Bryan J B 1979 The Abbé principle revisited: An updated interpretation *Precis. Eng.* **1** 129–32

[9] Bosman N, Qian J and Reynaerts D 2015 Self-calibration of the non-linear length deviation of a linear encoder by using two reading heads *Proc. of the 2015 ASPE Annual Meeting* 469–73

[10] Tiemann I *et al* 2008 An international length comparison using vacuum comparators and a photoelectric incremental encoder as transfer standard *Precis. Eng.* **32** 1–6

[11] Steinmetz C R 1990 Sub-micron position measurement and control on precision machine tools with laser interferometry *Precis. Eng.* **12** 12–24

[12] Bobroff N 1993 Recent advances in displacement measuring interferometry *Meas. Sci. Technol.* **4** 907–26

[13] Badami V and De Groot P 2013 Displacement measuring interferometry *Handbook of Optical Dimensional Metrology* ed K Harding (London: Taylor & Francis) ch 4, pp 157–238

[14] Mielke S L and Demarest F C 2008 Displacement measurement interferometer error correction techniques *Proc. of the ASPE Topical Meeting on Precision Mechanical Design and Mechatronics for Sub-50 nm Semiconductor Equipment* **vol 43** 113–16

[15] Köchert P, Flügge J, Köning R, Weichert C and Manske E 2013 An ultra-precision positioning device using interferometric feedback signals and a moving coil actuator *Proc. of Precision Control for Advanced Manufacturing Systems, ASPE Spring Topical Meeting 2013* 39–44

[16] Chen W and Ellis J D 2016 Data age error compensation for nonconstant velocity metrology *IEEE Trans. Instrum. Meas.* **65** 2601–11

[17] Hori Y, Gonda S, Bitou Y, Watanabe A and Nakamura K 2018 Periodic error evaluation system for linear encoders using a homodyne laser interferometer with 10 picometer uncertainty *Precis. Eng.* **51** 388–92

[18] Wu C, Lawall J and Deslattes R D 1999 Heterodyne interferometer with subatomic periodic nonlinearity *Appl. Opt.* **38** 4089–94

[19] Heilmann R K, Konkola P T, Chen C G and Schattenburg M L 2000 Relativistic corrections in displacement measuring interferometry *J. Vac. Sci. Technol.* B **18** 3277–81

[20] Heydemann P L M 1981 Determination and correction of quadrature fringe measurement errors in interferometers *Appl. Opt.* **20** 3382–84

[21] Editor's Note 1984 Documents concerning the new definition of the metre *Metrologia* **19** 163–77

[22] Quinn T J 1994 Mise en Pratique of the definition of the metre (1992) *Metrologia* **30** 523–41

[23] Quinn T J 1999 Practical realization of the definition of the metre (1997) *Metrologia* **36** 211–44

[24] Quinn T J 2003 Practical realization of the definition of the metre, including recommended radiations of other optical frequency standards (2001) *Metrologia* **40** 103–33

[25] Felder R 2005 Practical realization of the definition of the metre, including recommended radiations of other optical frequency standards (2003) *Metrologia* **42** 323–5

[26] Riehle F, Gill P, Arias F and Robertsson L 2018 The CIPM list of recommended frequency standard values: guidelines and procedures *Metrologia* **55** 188–200

[27] Stone J A, Decker J E, Gill P, Juncar P, Lewis A, Rovera G D and Viliesid M 2008 Advice from the CCL on the use of unstabilized lasers as standards of wavelength: the heliumneon laser at 633 nm *Metrologia* **46** 11–8

[28] Koning J, Schellekens P H J and McKeown P A 1979 Wavelength stability of He-Ne lasers used in interferometry: limitations and traceability *Technische Hogeschool Eindhoven*

[29] Lea S N, Rowley W R C, Margolis H S, Barwood G P, Huang G, Gill P, Chartier J-M and Windeler R S 2003 Absolute frequency measurements of 633 nm iodine-stabilized helium–neon lasers *Metrologia* **40** 84–8

[30] Lee W-K, Suh H S and Kang C S 2011 Vacuum wavelength calibration of frequency-stabilized He-Ne lasers used in commercial laser interferometers *Opt. Eng.* **50** 054301–4

[31] Salvadé Y and Dändliker R 2000 Limitations of interferometry due to the flicker noise of laser diodes *J. Opt. Soc. Am.* A **17** 927–32

[32] Salvadé Y *et al* 2016 Interferometric measurements beyond the coherence length of the laser source *Opt. Express* **24** 21729–43

[33] Weichert C, Köchert P, Köning R and Flügge J 2014 Stability of a fully fibre-coupled interferometer *Proc. of the 58th Ilmenau Scientific Colloquium*

[34] Yokoyama T, Araki T, Yokoyama S and Suzuki N 2001 A subnanometre heterodyne interferometric system with improved phase sensitivity using a three-longitudinal-mode He-Ne laser *Meas. Sci. Technol.* **12** 157–62

[35] Meiners-Hagen K and Abou-Zeid A 2008 Refractive index determination in length measurement by two-colour interferometry *Meas. Sci. Technol.* **19** 084004

[36] Bönsch G and Potulski E 1998 Measurement of the refractive index of air and comparison with modified Edlén's formulae *Metrologia* **35** 133–9

[37] Schödel R, Walkov A and Abou-Zeid A 2006 High-accuracy determination of water vapor refractivity by length interferometry *Opt. Lett.* **31** 1979–81

[38] Sommargren G E 1987 A new laser measurement system for precision metrology *Precis. Eng.* **9** 179–84

[39] Egan P and Stone J A 2011 Absolute refractometry of dry gas to ±3 parts in $10^9$ *Appl. Opt.* **50** 3076–86

[40] Estler W T 1985 High-accuracy displacement interferometry refin air *Appl. Opt.* **24** 808–15

[41] Weichert C, Köchert P, Schötka E, Flügge J and Manske E 2018 Investigation into the limitations of straightness interferometers using a multisensor-based error separation method *Meas. Sci. Technol.* **29** 064001

[42] Bobroff N 1987 Residual errors in laser interferometry from air turbulence and nonlinearity *Appl. Opt.* **26** 2676–82

[43] Bryan J B 1979 Design and construction of an ultraprecision 84inch diamond turning machine *Precis. Eng.* **1** 13–7

[44] Schödel R, Walkov A, Voigt M and Bartl G 2018 Measurement of the refractive index of air in a low-pressure regime and the applicability of traditional empirical formulae *Meas. Sci. Technol.* **29** 064002

[45] Ciddor P E 1996 Refractive index of air: new equations for the visible and near infrared *Appl. Opt.* **35** 1566–73

[46] Jäger G, Manske E, Hausotte T, Müller A and Balzer F 2016 Nanopositioning and nanomeasuring machine NPMM-200—a new powerful tool for large-range micro-and nanotechnology *Surf. Topogr.: Metrol. Prop.* **4** 034004

[47] Schödel R, *et al* 2012 A new Ultra Precision Interferometer for absolute length measurements down to cryogenic temperatures *Meas. Sci. Technol.* **23** 094004

[48] ISO 3529-1 1981 Vacuum technology — Vocabulary — Part 1: General terms

[49] Tilford C R 1992 *Physical Methods of Chemistry* ed B W Rossiter and R C Baetzold (New York: Wiley) ch 2 (Pressure and vacuum measurements)

[50] Cusco L *et al* 1998 *Guide to the Measurement of Pressure and Vacuum* (London: The Institute of Measurement and Control)

[51] Weichert C, Köchert P, Köning R, Flügge J, Andreas B, Kuetgens U and Yacoot A 2012 A heterodyne interferometer with periodic nonlinearities smaller than ±10pm *Meas. Sci. Technol.* **23** 094005

[52] Badami V G and Patterson S R 2000 A frequency domain method for the measurement of nonlinearity in heterodyne interferometry *Precis. Eng.* **24** 41–9

[53] Yang H, Weichert C, Köchert P and Flügge J 2016 Nonlinearity of a double-path interferometer qualified with a non-constant moving speed *Opt. Lett.* **41** 5478–81

[54] Wu C-M, Su C-S and Peng G-S 1996 Correction of nonlinearity in one-frequency optical interferometry *Meas. Sci. Technol.* **7** 520–4

[55] Eom T B, Choi T Y, Lee K H, Choi H S and Lee S K 2002 A simple method for the compensation of the nonlinearity in the heterodyne interferometer *Meas. Sci. Technol.* **13** 222–5

[56] Fu H, Wang Y, Hu P, Tan J and Fan Z 2018 Nonlinear errors resulting from ghost reflection and its coupling with optical mixing in heterodyne laser interferometers *Sensors* **18** 758

[57] Fu H, Hu P, Tan J and Fan Z 2017 Nonlinear errors induced by intermodulation in heterodyne laser interferometers *Opt. Lett.* **42** 427–30

[58] de Groot P 1999 Jones matrix analysis of high-precision displacement measuring interferometers *Proc. 2nd Topical Meeting on Optoelectronic Distance Measurement and Applications (9-14, Pavia, Italy)*

[59] Cosijns S J A G 2004 Displacement laser interferometry with sub-nanometer uncertainty *Doctoral Thesis* (Technische Universiteit Eindhoven)

[60] Tanaka M, Yamagami T and Nakayama K 1989 Linear interpolation of periodic error in a heterodyne laser interferometer at subnanometer levels *IEEE Trans. Instrum. Meas.* **38** 552–4

[61] Schmitz T L and Beckwith J F 2003 An investigation of two unexplored periodic error sources in differential-path interferometry *Precis. Eng.* **27** 311–22

[62] Wu C-M 2003 Periodic nonlinearities resulting from ghost reflections in heterodyne interferometry *Opt. Commun.* **215** 17–23

[63] Cavagnero G, Mana G and Massa E 2005 Effect of recycled light in two-beam interferometry *Rev. Sci. Instrum.* **76** 053106

[64] Lawall J and Kessler E 2000 Michelson interferometry with 10 pm accuracy *Rev. Sci. Instrum.* **71** 2669–76

[65] Korpel A 1981 Acousto-optics—a review of fundamentals *Proc. IEEE* **69** 48–53

[66] Köning R, Wimmer G and Witkovský V 2015 The statistical uncertainty of the Heydemann correction: a practical limit of optical quadrature homodyne interferometry *Meas. Sci. Technol.* **26** 084004

[67] Bryan J B 1980 The benefits of brute strength *Precis. Eng.* **2** 173

[68] Weichert C, Flügge J, Köchert P, Köning R and Tutsch R 2011 Stability of a fiber-fed heterodyne interferometer *10th IMEKO Symp. Laser Metrology for Precision Measurement and Inspection in Industry* (Düsseldorf: VDI) pp 243–50

[69] Fu H, Wu G and Hu P 2018 Thermal drift of optics in separated-beam heterodyne interferometers *IEEE Trans. Instrum. Meas.* **67** 1446–50

[70] Voigt D, Ellis J D, Verlaan A L, Bergmans R H, Spronck J W and Munnig Schmidt R H 2011 Toward interferometry for dimensional drift measurements with nanometer uncertainty *Meas. Sci. Technol.* **22** 094029

[71] Hosoe S 1995 Highly precise and stable displacement-measuring laser interferometer with differential optical paths *Precis. Eng.* **17** 258–65

[72] Flügge J, Weichert C, Hu H, Köning R, Bosse H, Wiegmann A, Schulz M, Elster C and Geckeler R D 2008 Interferometry at the PTB nanometer comparator: design, status and development *Proc. SPIE* **7133** 713346

[73] Gohlke M, Schuldt T, Döringshoff K, Peters A, Johann U, Weise D and Braxmaier C 2015 Adhesive bonding for optical metrology systems in space applications *J. Phys.: Conf. Ser.* **610** 012039

[74] Weichert C, Köchert P, Quabis S and Flügge J 2017 A displacement interferometer for the calibration of the silicon lattice parameter *Proc. of the 17th Int. Conf. of the European Society for Precision Engineering and Nanotechnology* pp 327–8

[75] Jedamzik R 2007 TIE-43: Optical Properties of ZERODUR *Technical report* Schott AG

[76] Knarren B A W H 2003 Application of optical fibers in precision heterodyne laser interferometry *PhD Thesis* Technische Universiteit Eindhoven

[77] Weichert C, Flügge J, Köning R, Bosse H and Tutsch R 2009 Aspects on design and characterization of a high resolution heterodyne interferometer *Fringe 2009 Proc. 6th Int. Workshop on Advanced Optical Metrology* pp 263–8

[78] Massa E, Mana G, Krempel J and Jentschel M 2013 Polarization delivery in heterodyne interferometry *Opt. Express* **21** 27119–26

[79] Hocker G B 1979 Fiber-optic sensing of pressure and temperature *Appl. Opt.* **18** 1445–8

[80] Stone J 1988 Stress-optic effects, birefringence, and reduction of birefringence by annealing in fiber Fabry–Perot interferometers *J. Lightwave Technol.* **6** 1245–8

[81] Smith R C G 2013 Physical optics analysis of a fiber-delivered displacement interferometer *Dissertation* University of Rochester, Institute of Optics

[82] Yang H, Zhu P, Tan J, Hu P and Fan Z 2017 High stable remote photoelectric receiver for interferometry *Rev. Sci. Instrum.* **88** 033105

[83] Mana G, Massa E and Sasso C P 2018 Forward scattering in two-beam laser interferometry *Metrologia* **55** 222–8

[84] Dorenwendt K and Bönsch G 1976 Über den Einfluß der Beugung auf die interferentielle Längenmessung *Metrologia* **12** 57-60

[85] Monchalin J P, Kelly M J, Thomas J E, Kurnit N A, Szoke A, Zernike F, Lee P H and Javan A 1981 Accurate laser wavelength measurement with a precision two-beam scanning Michelson interferometer *Appl. Opt.* **20** 736–57

[86] van Westrum D and Niebauer T M 2003 The diffraction correction for absolute gravimeters *Metrologia* **40** 258–63

[87] Gloge D 1971 Weakly guiding fibers *Appl. Opt.* **10** 2252–8

[88] Cavagnero G, Mana G and Massa E 2006 Aberration effects in two-beam laser interferometers *J. Opt. Soc. Am.* A **23** 1951–9

[89] Goodman J W 1968 *Introduction to Fourier Optics* (New York: McGraw-Hill)

[90] Born M and Wolf E 1964 *Principles of Optics* (Oxford: Pergamon)

[91] Jackson J D 1999 *Classical Electrodynamics* (New York: Wiley)

[92] Huygens C 1690 *Traité de la lumière* (Leiden: Pieter van der Aa)

[93] Siegman A E 1986 *Lasers* (Mill Valley, CA: University Science Books)

[94] Tango W J and Twiss R Q 1974 Diffraction effects in long path interferometers *Appl. Opt.* **13** 1814–9

[95] Mana G 1989 Diffraction effects in optical interferometers illuminated by laser sources *Metrologia* **26** 87–93

[96] Bergamin A, Cavagnero G, Cordiali L and Mana G 1997 Beam astigmatism in laser interferometry *IEEE Trans. Instrum. Meas.* **46** 196-200

[97] Bergamin A, Cavagnero G, Cordiali L and Mana G 1999 A Fourier optics model of two-beam scanning laser interferometers *Eur. Phys. J.* D **5** 433–40

[98] Robertsson L 2006 On the diffraction correction in absolute gravimetry *Metrologia* **44** 35–9
[99] D'Agostino G and Robertsson L 2011 Relative beam misalignment errors in high accuracy displacement interferometers: calculation and detection *Appl. Phys.* B **103** 357–61
[100] Hodgson N and Weber H 2005 *Laser Resonators and Beam Propagation* (New York: Springer)
[101] Mana G, Massa E and Rovera A 2001 *Appl. Opt.* **40** 1378–85
[102] Sasso C P, Massa E and Mana G 2016 Diffraction effects in length measurements by laser interferometry *Opt. Express* **24** 6522–31
[103] Mana G, Massa E, Sasso C P, Andreas B and Kuetgens U 2017 A new analysis for diffraction correction in optical interferometry *Metrologia* **54** 559–65
[104] Cooley J W and Tukey J W 1965 An algorithm for the machine calculation of complex Fourier series *Math. Comput.* **19** 297–301
[105] Downs M J and Raine K W 1979 An unmodulated bi-directional fringe-counting interferometer system for measuring displacement *Precis. Eng.* **1** 85–8
[106] Tommasi T and Bianco B 1993 Computer-generated holograms of tilted planes by a spatial frequency approach *J. Opt. Soc. Am.* A **10** 299–305
[107] Delen N and Hooker B 1998 Free-space beam propagation between arbitrarily oriented planes based on full diffraction theory: a fast Fourier transform approach *J. Opt. Soc. Am.* A **15** 857–67
[108] Delen N and Hooker B 2001 Verification and comparison of a fast Fourier transform-based full diffraction method for tilted and offset planes *Appl. Opt.* **40** 3525–31
[109] Matsushima K, Schimmel H and Wyrowski F 2003 Fast calculation method for optical diffraction on tilted planes by use of the angular spectrum of plane waves *J. Opt. Soc. Am.* A **20** 1755–62
[110] Matsushima K 2008 Formulation of the rotational transformation of wave fields and their application to digital holography *Appl. Opt.* **47** D110–6
[111] Yamamoto K, Ichihashi Y, Senoh T, Oi R and Kurita T 2012 Calculating the Fresnel diffraction of light from a shifted and tilted plane *Opt. Express* **20** 12949–58
[112] Zhang S, Asoubar D, Hellmann C and Wyrowski F 2016 Propagation of electromagnetic fields between non-parallel planes: a fully vectorial formulation and an efficient implementation *Appl. Opt.* **55** 529–38
[113] Stock J, Worku N G and Gross H 2017 Coherent field propagation between tilted planes *J. Opt. Soc. Am.* A **34** 1849–55
[114] Nascov V and Logofătu P C 2009 Fast computation algorithm for the Rayleigh-Sommerfeld diffraction formula using a type of scaled convolution *Appl. Opt.* **48** 4310–9
[115] Yariv A 1997 *Optical Electronics in Modern Communications* (New York: Oxford University Press)
[116] Andreas B, Mana G and Palmisano C 2015 Vectorial ray-based diffraction integral *J. Opt. Soc. Am.* A **32** 1403–24
[117] Andreas B, Mana G and Palmisano C 2016 Vectorial ray-based diffraction integral: erratum *J. Opt. Soc. Am.* A **33** 559–60
[118] Andreas B 2015 *Vectorial ray-based diffraction integral (VRBDI)* http://www.mathworks.com/matlabcentral/fileexchange/52210
[119] Wyant J C 1982 Interferometric optical metrology: basic principles and new systems *Laser Focus* **18** 65–71

[120] Kogelnik H 1965 Imaging of Optical Modes—Resonators with Internal Lenses *Bell Syst. Tech. J.* **44** 455–94

[121] Kogelnik H and Li T 1966 Laser beams and resonators *Appl. Opt.* **5** 1550–67

[122] Day G W and Stubenrauch C F 1978 Laser far-field beam-profile measurements by the focal plane technique *National Bureau of Standards (U.S.)* Techn. Note 1001

[123] Arnaud J A, Hubbard W M, Mandeville G D, de la Clavière B, Franke E A and Franke J M 1971 Technique for fast measurement of Gaussian laser beam parameters *Appl. Opt.* **10** 2775–6

[124] Nemes G and Siegman A E 1994 Measurement of all ten second-order moments of an astigmatic beam by the use of rotating simple astigmatic (anamorphic) optics *J. Opt. Soc. Am.* A **11** 2257–64

[125] Jeunhomme L B 1990 *Single-Mode Fiber Optics* (New York: Marcel Dekker)

[126] Hartmann J 1900 Bemerkungen über den Bau und die Justierung von Spektrographen *Z. Instrumk.* **20** 47–58

[127] Shack P V and Platt B C 1971 Production and use of a lenticular Hartmann screen *J. Opt. Soc. Am.* **61** 656

[128] Neal D R, Alford W J, Gruetzner J K and Warren M E 1996 Amplitude and phase beam characterization using a two-dimensional wavefront sensor *Proc. SPIE* 2870 72–82

[129] Schäfer B and Mann K 2000 Investigation of the propagation characteristics of excimer lasers using a Hartmann-Shack sensor *Rev. Sci. Instrum.* **71** 2663–8

[130] Kuetgens U, Andreas B, Friedrich K, Weichert C, Köchert P and Flügge J 2018 Measurement of the silicon lattice parameter by scanning single photon x-ray interferometry *Proc. of CPEM 2018: Conf. on Precision Electromagnetic Measurements*

[131] Huebel A, Schellhorn U, Arnz M, Klose G and Beyer D 2009 Calibration strategies for precision stages in state-of-the-art registration metrology *Proc. of SPIE Photomask and Next-Generation Lithography Mask Technology XVI* **vol 7379** p 737914

[132] Beyer D, Bläsing C, Boehm K, Heisig S and Seidel D 2013 Fleet matching performance for multiple registration measurement tools *Proc. of SPIE*vol 8880 (Photomask Technology 2013) p 88801W

CPSIA information can be obtained
at www.ICGtesting.com
Printed in the USA
BVHW010906150422
634412BV00010B/127